RESEARCH ETHICS FORUM

Volume 3

This Series, Research Ethics Forum, aims to encourage discussion in the field of research ethics and the ethics of research. Volumes included can range from foundational issues to practical issues in research ethics. No disciplinary lines or borders are drawn and submissions are welcome from all disciplines as well as scholars from around the world. We are particularly interested in texts addressing neglected topics in research ethics, as well as those which challenge common practices and beliefs about research ethics. By means of this Series we aim to contribute to the ever important dialogue concerning the ethics of how research is conducted nationally and internationally. Possible topics include: Research Ethics Committees, Clinical trials, International research ethics regulations, Informed consent, Risk-benefit calculations, Conflicts of interest, Industry-funded research, Exploitation, Qualitative research ethics, Social science research ethics, Ghostwriting, Bias, Animal research and Research participants

More information about this series at http://www.springer.com/series/10602

Françoise Baylis • Angela Ballantyne
Editors

Clinical Research Involving Pregnant Women

Springer

Editors
Françoise Baylis
Novel Tech Ethics, Faculty of Medicine
and Department of Philosophy
Faculty of Arts and Social Sciences
Dalhousie University
Halifax, Nova Scotia, Canada

Angela Ballantyne
Department of Primary Health Care
and General Practice
University of Otago
Wellington, New Zealand

ISSN 2212-9529 ISSN 2212-9537 (electronic)
Research Ethics Forum
ISBN 978-3-319-79965-0 ISBN 978-3-319-26512-4 (eBook)
DOI 10.1007/978-3-319-26512-4

Printed on acid-free paper

This Springer imprint is published by Springer Nature
The registered company is Springer International Publishing AG Switzerland

Acknowledgments

The idea for this book grew out of two panels held in June of 2012 at the World Congress on Feminist Approaches to Bioethics and the World Congress of Bioethics, in Rotterdam, the Netherlands. The first panel included Françoise Baylis, Angela Ballantyne, and Ruth Macklin and addressed the question of what would be "A Just Research Agenda for Pregnant Women Supporting Appropriate Health Care Now and in the Future." The second panel included Françoise Baylis, Angela Ballantyne, Ruth Macklin, and Ruth Faden and tackled the issue of "Fair Inclusion of Pregnant Women in Research." We thank the colleagues who participated in these panels as well as audience members who provided helpful comments.

More specifically, we thank our peer reviewers who provided timely, thorough, and constructive reviews for each of the chapters in this book. Your time, effort, and diligence have contributed to making this book a worthy scholarly resource for others.

Last but not least, we thank Samantha Copeland and Timothy Krahn who completed much of the final copyediting for the book and whose attention to detail in managing the review and contributing to the editing processes helped bring the many parts of this volume together and ready for publication.

In closing, we salute the *Second Wave Initiative* (secondwaveinitiative.org), cofounded by Margaret Olivia Little, Anne Drapkin Lyerly, and Ruth Faden in 2009. The initiative seeks to advance the responsible inclusion of pregnant women in medical research. We hope this book, which might well be characterised as *Surfing the Second Wave*, proves to be an important contribution to this most worthy goal.

Contents

Part III

Part IV

About the Authors

Richard Ashcroft is Professor of bioethics at Queen Mary University of London, where he teaches in the School of Law. He has published widely on research ethics and public health ethics. He is a past member of both the ethics committee of the Royal College of Obstetricians and Gynaecologists and of the UK Gene Therapy Advisory Committee. He is currently an investigator on the European Commission-funded EVERREST programme, which is investigating an innovative therapy in foetal growth restriction, a serious disorder of the uterine blood supply during pregnancy leading in many cases to stillbirth or premature delivery of a baby with very low birthweight. He leads the workstream on ethics, law, and social aspects of the experimental treatment of this disorder.

Angela Ballantyne is a Senior Lecturer in bioethics in the Department of Primary Health Care and General Practice at the University of Otago in Wellington, New Zealand. She has worked in schools of medicine, primary health care, and philosophy in New Zealand, Australia, England, and the United States and in Human Genetics at the World Health Organization in Geneva. Her work focuses on research ethics, especially issues of fairness, exploitation, and vulnerability. She is president of the International Association of Bioethics.

Christine Barthow is a registered nurse and Research Fellow in the Wellington Asthma Research Group, Department of Medicine, at the University of Otago in Wellington, New Zealand. Her research focuses on primary prevention studies. She is a co-investigator for Probiotics in Pregnancy: Improving Health During Pregnancy and Preventing Infant Eczema and Allergy. This is a randomised controlled trial, investigating the use of probiotics to prevent the development of allergic disease in infants and to reduce gestational diabetes and bacterial vaginal infections in pregnant women.

Françoise Baylis is Professor and Canada Research Chair in bioethics and philosophy at Dalhousie University in Halifax, Canada, and founder of the Novel Tech Ethics research team. She is an elected fellow of the Royal Society of Canada and the Canadian Academy of Health Sciences. Her research focuses on ethical issues in women's health, and she has a particular interest in assisted human reproduction and research participation. She has long advocated for the just inclusion of both women and pregnant women in clinical trials. In addition to her academic research and writing, she contributes to national policymaking via government research contracts, membership on national committees, and public education. Her website is www.noveltechethics.ca. Her blog is impactethics.ca.

Nikola Biller-Andorno is Director of the Institute of Biomedical Ethics and History of Medicine at the University of Zurich in Switzerland, a WHO collaborating centre for bioethics. Before her appointment in Switzerland, she held positions at the Charité Berlin and the University of Gottingen in Germany. Most recently, she has been a Commonwealth Fund Harkness Fellow (2012–2013), a Safra Network fellow (2013–2014), and a visiting professor of biomedical ethics (2012–2014) at Harvard. Together with Verina Wild, she has pioneered the discussion on the ethics of including pregnant women in clinical trials in Germany.

Joyce L. Browne is a PhD candidate at the University Medical Center Utrecht, the Netherlands. She is writing her dissertation on hypertensive disorders in pregnancy in Ghana. Additionally, she is one of the teachers in the online course Introduction to Global Health. She has previously received her MSc in social epidemiology from University College London and received her MD/MSc in medicine/clinical research from Utrecht University. She is currently working as a resident at the Department of Obstetrics and Gynecology.

Carolyn Ells is an Associate Professor of medicine at the Biomedical Ethics Unit and associate member of the Division of Experimental Medicine, at McGill University, in Montreal, Canada, and research associate at the Lady Davis Institute for Medical Research. She has many years of experience chairing university and hospital research ethics boards and has contributed to several publications and policy analyses on research ethics review. She is a coauthor of a widely used online research ethics tutorial for research ethics board members and support staff. She currently chairs the Interagency Advisory Panel on Research Ethics, which advises Canada's federal research funding agencies on research ethics policy.

Ruth R. Faden is the Philip Franklin Wagley Professor of biomedical ethics and founding director of the Johns Hopkins Berman Institute in Baltimore, the United States. She is the author and editor of many books and articles on biomedical ethics and health policy. A member of the Institute of Medicine and a fellow of the Hastings Center, she has served on numerous national advisory committees and commissions, including President Clinton's Advisory Committee on Human Radiation Experiments, which she chaired. She is a cofounder, along with Anne Drapkin Lyerly and Margaret Olivia Little, of the Second Wave Initiative.

Ruth M. Farrell is a staff member in the Departments of Obstetrics and Gynecology and Bioethics at the Cleveland Clinic. She is an assistant professor of surgery at the Cleveland Clinic Lerner College of Medicine and a faculty affiliate of the Center for Genetic Research Ethics and Law (CGREAL) at Case Western Reserve University, in Cleveland, the United States. She is a practicing obstetrician-gynaecologist and clinical researcher who examines ethical and clinical issues associated with the integration of new reproductive and genetic technologies into women's health. She has served on the Committee on Ethics of the American Congress of Obstetricians and Gynecologists and the Ethics Committee for the American Society of Reproductive Medicine in addition to ethics committees at the Cleveland Clinic and University MacDonald Women's Hospital.

Rebecca Flyckt is Associate Staff at the Cleveland Clinic in the Departments of Obstetrics and Gynecology, Section of Reproductive Endocrinology and Infertility, in the United States. Her interests include reproductive surgery, endometriosis, fibroids, female infertility, and in vitro fertilisation. She is director of the Fertility Preservation and Cancer Program in the Women's Health Institute of the Cleveland Clinic and is actively involved in the ovarian tissue freezing programme for women and young girls diagnosed with cancer.

Lisa H. Harris is Associate Professor of obstetrics and gynaecology and of women's studies at the University of Michigan in the United States. She is a practicing obstetrician-gynaecologist and a cultural historian. She teaches students in many schools across the University of Michigan's campus. In her research, she examines issues at the intersection of clinical care and law, policy, politics, ethics, history, and sociology. She focuses in particular on abortion stigma and its consequences, and on the social construction of assisted reproductive technologies.

David Healy is a Professor of psychiatry at Cardiff University in Wales. He is a former secretary of the British Association for Psychopharmacology and author of over 200 peer-reviewed articles, 200 other pieces, and 22 books, including *The Antidepressant Era* and *The Creation of Psychopharmacology* from Harvard University Press, *The Psychopharmacologists* volumes 1–3, *Let Them Eat Prozac* from New York University Press, *Mania* from Johns Hopkins University Press, and *Pharmageddon*. His main areas of research are clinical trials in psychopharmacology, the history of psychopharmacology, and the impact of both trials and psychotropic drugs on our culture. He is involved in raising awareness of how pharmaceutical companies sell drugs by marketing diseases and co-opting academic opinion leaders, ghostwriting their articles.

L. Syd M. Johnson is an Assistant Professor of philosophy and bioethics at Michigan Technological University in the United States. Her work in research ethics considers the persistent underrepresentation of women in biomedical research and particularly the harms of excluding pregnant women. Her research on the ethical implications of using reproductive technologies for offspring selection focuses on sex selection and selection for and against disability. She also works on the ethical implications of neuroscientific developments related to disorders of consciousness and sport-related concussion.

Chris Kaposy is an Assistant Professor of health-care ethics at Memorial University in St. John's, Newfoundland, Canada. He is also a member of the Novel Tech Ethics research team and is a practicing clinical ethicist with the Provincial Health Ethics Network of Newfoundland and Labrador. His research interests include ethical issues in pregnancy particularly the inclusion of pregnant women in research, issues in abortion access, and the ethics of prenatal testing.

Rebecca Kukla is Professor of philosophy and senior research scholar at the Kennedy Institute of Ethics at Georgetown University in Washington, the United States. She received her PhD in philosophy from the University of Pittsburgh and completed a Greenwall postdoctoral fellowship in bioethics and health policy at the Johns Hopkins School of Public Health. Within bioethics, she has published on reproductive health ethics and the culture of motherhood and pregnancy, methodological issues in medical research, public health ethics, risk communication, and patient agency. Her broader research interests include social epistemology, philosophy of the special sciences, philosophy of language, bioethics, and gender theory. She is the author of *Mass Hysteria: Medicine, Culture, and Mothers' Bodies*, among other publications. She is the editor of the *Kennedy Institute of Ethics Journal* and a former co-coordinator of the Feminist Approaches to Bioethics Network.

Lucy Langston is a PhD student in philosophy and bioethics at McMaster University in Hamilton, Canada, and is also completing a graduate diploma in gender studies and feminist research. She is writing her dissertation on how the rise of novel health-related technologies – such as the clinical trial – interacts, changes, and challenges the moral concepts and categories associated with pregnancy. Her research interests beyond bioethics include philosophy of technology, social epistemology, feminist theory, and philosophy of biology.

Margaret Olivia Little is Director of the Kennedy Institute of Ethics and a member of the Philosophy Department at Georgetown University in Washington, the United States. Her research interests include issues in reproduction, clinical research ethics, and the structure of moral theory. A fellow of the Hastings Center, she has twice served as visiting scholar in residence at the National Institutes of Health Department of Bioethics. Together with Ruth Faden and Anne Drapkin Lyerly, she cofounded the Second Wave Initiative, which works to promote responsible research into the health needs of pregnant women.

Anne Drapkin Lyerly is Associate Professor of social medicine and associate director of the Center for Bioethics at the University of North Carolina at Chapel Hill, the United States. An obstetrician-gynaecologist and bioethicist, she has published widely on morally complex issues in reproduction including her recent book,

A Good Birth: Finding the Positive and Profound in Your Childbirth Experience (Penguin Group). She cofounded, with Ruth Faden and Margaret Olivia Little, the Second Wave Initiative, an effort to ensure that the health interests of pregnant women are fairly represented in biomedical research and drug device policies.

Caroline Lyster is a recent graduate of McGill University, where she completed her MA in philosophy with a specialisation in biomedical ethics. While at McGill, she was a research fellow at the Institute for Health and Social Policy, a position that pushed her graduate research toward a critical engagement with health policy. She has written on the eugenic implications associated with unrestricted access to abortion and has also engaged with the problematic implications of both provincial and federal end of life legislation. She currently resides in Alberta, Canada, where she is involved with the SSHRC-CURA-funded Living Archives on Eugenics in Western Canada.

Robyn MacQuarrie is a PhD candidate whose research interests focus on the exclusion of pregnant women from clinical trials and specifically vaccine trials. She is completing her PhD programme through the Interdisciplinary PhD Program at Dalhousie under the supervision of Françoise Baylis and Scott Halperin from the Canadian Center for Vaccinology. She has previously completed her MASc in biomedical engineering (Dalhousie), as well as an undergraduate degree in microbiology and immunology (Dalhousie). Robyn also works part-time as a practicing obstetrician-gynaecologist in Bridgewater, Nova Scotia, Canada.

Derelie Mangin is the David Braley and Nancy Gordon Chair in family medicine at McMaster University in Hamilton, Canada, and a practicing family doctor and researcher. She was a member of the Pharmaceutical and Therapeutic Products Advisory Committee to PHARMAC and served on the New Zealand Southern Region Ethics Committee. Her research centres on rational prescribing, drug effectiveness and safety, innovative models of primary care delivery, and the influences of science, policy, and commerce on the nature of care. She is chief medical officer for RxISK.org, which provides and collects information on drug adverse events from direct patient reports.

Anna C. Mastroianni is Professor of Law at the University of Washington School of Law, in Seattle, the United States. Prior to her academic career, she held a number of legal and federal policy positions in Washington, DC, including staff leadership positions with a presidential commission and the US Institute of Medicine, and as a practicing attorney with law firms specialising in health law. She has been nationally recognised for her contributions to health policy, law, and bioethics as a Fellow of the American Association for the Advancement of Science. Her publications include six books and numerous peer-reviewed articles on law, medicine, and bioethics, with a special emphasis on the legal and ethical challenges in public health, research with human subjects, and assisted reproductive technologies.

Wendy Rogers is ARC Future Fellow, Professor of clinical ethics, and Deputy Director of the Research Centre for Agency, Values, and Ethics at Macquarie University, Sydney, Australia. Her research draws on feminist bioethical theory to focus on practical issues in health care. She has a long-standing interest in various topics in research ethics, including treatment of vulnerable populations, agenda setting and justice in research, surgical innovations, evidence-based medicine, and research ethics guidelines. As well as her academic writing and teaching, she contributes to policymaking through membership on local and national committees including the Australian Health Ethics Committee.

Johannes J.M. van Delden is Professor of medical ethics at the Julius Center Department of Medical Humanities, at the University Medical Center Utrecht, the Netherlands. Until recently, he has worked as a nursing home physician. His special fields of interest are research ethics, moral problems at the end of life, and moral problems in the care for the elderly. He was secretary of the International Association of Bioethics. He is a member of the Dutch Central Committee on Research Involving Human Subjects (CCMO) and has over 15 years of experience in RECs. Since November 2010, he is the chair of the Council for International Organizations of Medical Sciences (CIOMS).

Rieke van der Graaf is an Assistant Professor of research ethics at the Julius Center Department of Medical Humanities, at the University Medical Center Utrecht, the Netherlands. She has written a PhD thesis on equipoise and human dignity. Her current research interests are global justice in international research, the integration of care and research, and the ethics of innovative research designs. She is teaching medical ethics at the UMC Utrecht and is a member of the UMC Utrecht's Research Ethics Committee (METC). She is also the secretary of the Working Group on the Revision of the CIOMS Guidelines for biomedical research.

Indira S.E. van der Zande is a PhD candidate in research ethics at the Julius Center Department of Medical Humanities, at the University Medical Center Utrecht, the Netherlands. She is writing her dissertation on ethically sound inclusion of pregnant women in medical research. She has previously received her MA in applied ethics and her MSc in international development studies, both from the University of Utrecht.

Kristin Wickens is a Senior Research Fellow at University of Otago in Wellington, New Zealand. She has expertise in the epidemiology of asthma, eczema, and other allergic diseases, including study design, preparing funding applications, study management, quality control of protocol adherence, data analysis, and manuscript preparation. Over the years, her research interest has evolved from case-control, observational and cohort studies looking at risk factors for asthma to randomised controlled trials of probiotic interventions for the primary and secondary prevention of eczema and allergy in infants and children through to age six years.

Verina Wild is Senior Teaching and Research Associate at the Institute of Biomedical Ethics, University of Zurich, Switzerland. Formerly, she worked as a physician of internal medicine in Berlin, Germany. She has been a visiting researcher at the Mailman School of Public Health, Columbia University, New York; an intern at the Ethics and Health Unit of the World Health Organization, Geneva; and a medical trainee in various hospitals in Africa, South America, and Europe. For her work on drug research on pregnant women, she received the German Young Scholars Award in Medical Ethics.

Chapter 1
Missed Trials, Future Opportunities

Françoise Baylis and Angela Ballantyne

Pregnant women deserve more from clinical research. Justice requires a research agenda that adequately addresses the health needs of pregnant women, and fair inclusion criteria that support the safe and responsible participation of pregnant women in relevant research. In recent years, there have been successful global efforts to expand paediatric clinical research[1] and to achieve appropriate gender balance in clinical trials. Significant challenges remain, however, with respect to the fair inclusion of pregnant women in clinical research. Indeed, pregnant women continue to be routinely excluded from such research *without justification* beyond the generic belief that vulnerable foetuses must be protected from research-related harms and that one effective way to meet this obligation is to exclude pregnant women from clinical research.

At the present time, pregnancy care and advice are driven by the precautionary principle (Kukla 2005). This principle advocates action to reduce threats of potentially serious, irreversible harm, before there is strong evidence of such harm (Harremoës et al. 2002). With the precautionary principle there is a reversal of the standard burden of proof – advocates need to demonstrate safety, rather than critics needing to demonstrate predictable harm. Precaution is usually applied in cases where unintended harms (or accidents) would be potentially catastrophic, for exam-

[1] Following the US National Institutes of Health, clinical research includes: 1. Patient-oriented research (which in turn includes mechanisms of human disease, therapeutic interventions, clinical trials, or development of new technologies). 2. Epidemiological and behavioural studies. 3. Outcomes research and health services research. National Institutes of Health. Glossary. http://grants.nih.gov/grants/policy. Accessed 17 May 2016.

F. Baylis, CM, ONS, PhD, FRSC, FCAHS (✉)
Novel Tech Ethics, Faculty of Medicine and Department of Philosophy, Faculty of Arts and Social Sciences, Dalhousie University, Halifax, Nova Scotia, Canada
e-mail: francoise.baylis@dal.ca

A. Ballantyne, BSc, PhD
Department of Primary Health Care and General Practice, University of Otago, Wellington, New Zealand

F. Baylis, A. Ballantyne (eds.), *Clinical Research Involving Pregnant Women*, Research Ethics Forum 3, DOI 10.1007/978-3-319-26512-4_1

ple nuclear power, genetic engineering, and pregnancy. The underlying philosophy is perhaps best summed up in the proverb 'better safe, than sorry'. For example, when large epidemiological studies showed no evidence of caffeine-related harm at low doses, but showed increased rates of miscarriage at moderate to high doses, the message communicated to all pregnant women was 'avoid all caffeine'. The absence of evidence confirming potential harm at low doses was not taken as evidence of safety. Using a precautionary approach, evidence of potential harm with moderate to high doses of caffeine suggested that pregnant women should avoid all caffeine (Lyerly et al. 2009). The precautionary principle is key to understanding the reluctance to include pregnant women in clinical research.

Two now classic cases changed the way we perceive risk during pregnancy. Indeed, the histories of thalidomide and diethylstilboestrol (DES) are among the more significant barriers to the routine inclusion of pregnant women in clinical research (see Langston 2016). Both of these tragedies, which highlight foetal vulnerability, continue to influence research today, despite the fact that neither of these cases were the result of research-related harm. In the 1950s, thalidomide was prescribed to pregnant women to treat nausea, without prior safety studies having been completed. Tragically, this resulted in severe birth defects in over 10,000 children (Macklin 2010). From the 1940s through to the 1960s, DES was prescribed to millions of women to prevent miscarriage. In 1971, evidence emerged linking DES to several adverse effects, including vaginal and cervical cancer in young women exposed to DES during foetal development (Swan 2000).

These examples, and subsequent research, have clearly demonstrated that the foetus is not 'a bun in the oven' that floats suspended in a bubble until it is born. The foetus grows out of the pregnant woman. Even before it implants, the blastocyst is receiving and responding to environmental cues (Armant 2005). Once the embryo implants, it begins to function as part of the pregnant woman. There is no clear boundary or distinction between the pregnant woman and the foetus. Understandably, this physiological inter-connectedness leads to a focus on the behaviours of pregnant women and the conditions they experience, as these may have profound and lasting effects on the subsequent child (or children).[2]

The precautionary principle as an over-riding principle governing clinical research involving pregnant women gained ground after the thalidomide and DES cases demonstrated foetal vulnerability. Precaution is now deeply embedded in the ethos of pregnancy and clinical research during pregnancy. Indeed, efforts to protect the foetus from potential, rather than demonstrated, harm include increasing prohibitions on acceptable behaviour during pregnancy that go well beyond clinical research participation (Kukla 2005). In this age of 'intensive motherhood', with the

[2] We explicitly avoid the language of 'lifestyle choices' here because many behaviours that affect foetal health are the result of external factors (for example, employment stress, financial insecurity, ill health, domestic abuse) or habits (for example, diet, exercise, sleep) that have little to do with conscious, intentional deliberation and choice. Pregnancy is certainly a time when women become more conscious of their behaviours and have higher motivation for changing behaviours (for example smoking, see WHO 2013). But, pregnancy also entails barriers to health-related behaviour change (Sui et al. 2013). Despite increased consciousness regarding the importance of behaviour during pregnancy, many behaviours are still driven primarily by habit, environmental stimuli, and unconscious motivations.

burgeoning growth in pregnancy and infant-related health advice (Lee et al. 2014), there are instructions on virtually all aspects of a pregnant woman's life. For example, pregnant women are routinely given advice on diet (e.g., eat plenty of green, leafy vegetables, avoid eating hummus), exercise (e.g., do this in moderation, don't go horseback riding), work, including unpaid housework (e.g., avoid exposure to dangerous chemicals, reduce work hours), sleep (e.g., not on your back during the third trimester), prescription and over-the-counter drugs (e.g., avoid most medications, take care with others), tobacco (e.g., stop smoking, avoid second-hand smoke), alcohol (stop drinking), recreational drugs (stop taking them), and sex (continue as comfortable) (see, for example, Baylis and Sherwin 2002, 287–288).

While some behaviours during pregnancy may pose immediate physiological harm to the developing foetus (for example, eating certain foods increases the risk of listeria, and sleeping on one's back during the third trimester restricts blood flow to the foetus), other potential harms operate via epigenetic programming during foetal development. Epigenetic programming can have significant and long-lasting effects on mental and physical health through the course of the future child's life (Gluckman et al. 2008). Sleep, stress, diet, drug use, and exercise can all affect the growing foetus. For example, it has been shown that stress during pregnancy, triggered by domestic violence, changes the cortisol receptors of offspring as observed during adolescence (Radtke et al. 2011). As well, the diet of pregnant women has been shown to correlate to epigenetic changes in DNA programming at birth that predict the child's vulnerability to later obesity and metabolic disease (Godfrey et al. 2011). Evidence of these sorts of correlations between the experiences of pregnant women and the future child's (or children's) health drive a distorted and erroneous view of the ethics of pregnancy according to which 'good' pregnant women are those who avoid all risks. The reality is much more complex, however. For example, for some pregnant women, many risky behaviours are unavoidable (e.g., driving, experiencing domestic violence), or difficult to define (e.g., healthy eating), or hard to change (e.g., weight management). More generally, few pregnant women could manage to follow the entire range of health advice they might be given (Baylis and Sherwin 2002).

Consider, for example, advice regarding diet. An overwhelming majority of pregnant women do not meet current pregnancy diet guidelines (Callaway et al. 2009; Blumfield et al. 2011). For instance, in New Zealand, only 3 % of pregnant women meet national dietary targets for all four food groups (Morton et al. 2014). In Australia, 2 % of pregnant women meet national guidelines for vegetable consumption and 10 % meet guidelines for meat consumption (Mishra et al. 2015). Achieving the designated behaviour is challenging to say the least. Further, recent clinical research shows that following pregnancy diet guidelines is sometimes unwise. For example, while pregnant women are susceptible to listeria food poisoning (and miscarriage) and are advised to avoid high risk food, clinical research has shown that following this advice results in pregnant women consuming fewer essential nutrients (Pezdirc et al. 2012). A similar story has emerged in relation to fish consumption in the United States. Pregnant women are instructed to limit fish consumption (Hibbelin et al. 2007) and to avoid specific species of fish during pregnancy in order to reduce the threat of mercury-related adverse effects to the foetus. But avoidance in this context has proven to be misguided. Overall, dietary intake of omega-3 fatty acids by pregnant and postpartum women in the United States falls short of recommended 'safe' levels (Benisek et al. 2000). These examples demonstrate the influence of the precautionary principle in pregnancy. Pregnant women are told to avoid multiple behaviours, often based on theoretical risks or preliminary evidence. However, avoidance is often impractical, and in some cases counterproductive. The same problems can occur when the precautionary principle is applied to clinical research with pregnant women.

If pregnant women should avoid eating certain foods on the grounds of foetal risk, it may seem obvious that they should avoid participating in clinical research. Moreover, *prima facie* this might seem much easier than avoiding stress, unhealthy food, or other potentially harmful exposures during pregnancy. From an individual perspective, participation in clinical research is 'unnecessary' insofar as research is designed primarily to benefit future generations, rather than the research participants themselves. As well, in almost all cases, research participation during pregnancy is simple to avoid.[3] Protecting foetuses from research-related risks, by excluding pregnant women from clinical research, therefore appears like an easy win for all who are rightly concerned with foetal and maternal wellbeing, including pregnant women, their families, their clinicians, and the community more broadly.

But not so fast; there are at least two problems here. First, all clinical research in humans involves a trade-off between risk borne by current research participants and potential benefits to future generations who may gain access to safe and effective treatments stemming from research. It follows that we can protect foetuses – as a population – by accepting some risk to current foetuses in order to generate knowledge that improves foetal safety in the future. We routinely accept the need for these sorts of trade-offs when it comes to doing clinical research involving other research populations.

Second, pregnant women, clinicians, and the community often are unclear about the potential benefits and risks of offered or recommended treatments. Studies show, for example, that in some cases pregnant women over-estimate the risks of drugs and other treatments used in clinical practice (Nordeng et al. 2010). But other studies suggest that pregnant women may have undue confidence in interventions seemingly offered as part of clinical care. Pregnant women use on average 2.6 medications (prescription and non-prescription) during pregnancy (Mitchell et al. 2011). This is despite the fact that greater than 98 % of medicines have no, or insufficient, safety data or pharmacokinetic data to guide dosing during pregnancy (McCormack and Best 2014). Consider the ongoing and controversial clinical use of the drug dexa-

[3]There are some exceptions to this general rule, however. Consider, for example, a pregnant woman diagnosed with a life-threatening condition where receiving an experimental intervention in a trial may be in her and her foetus' best interests. Medications to prevent perinatal transmission of HIV are some of the best-studied drugs in pregnant women. Consensus around the high risk of untreated HIV was enough to overcome the standard aversion to clinical research during pregnancy. But much clinical research is optional and therefore framed as an unnecessary risk.

methasone (DEX) to prevent virilisation of female foetuses affected by congenital adrenal hyperplasia and to prevent miscarriage for IVF patients. There is significant ethical debate in the literature, not only about the objective of preventing virilisation, but also about whether there is sufficient data regarding safety and efficacy to offer DEX as clinical treatment. A number of influential medical societies have concluded that DEX should only be offered in the context of approved research protocols (Witchel and Miller 2012). Yet many patients who are offered or recommended DEX will be unaware of this controversy and assume that DEX is safe and well established (Dreger 2015).

Reluctance to enrol pregnant women in clinical research is understandable, and the underlying concerns about potential foetal harm are valid. The widespread exclusion of pregnant women from clinical research results in its own harms, however, as when clinical care is compromised due to a lack of evidence about how to safely and effectively treat conditions affecting women during their pregnancies. This can result in a variety of problems, including the prescription of unsafe drugs because the health care provider is unaware of the risks, dangerous delays in the provision of medical treatment, and refusal to prescribe clinically indicated drugs. The resulting sub-optimal clinical care affects both the pregnant women and their developing foetuses. As a matter of justice, pregnant women are entitled to high-quality evidenced-informed care (see Baylis and MacQuarrie 2016). Clinical research involving pregnant women is an effective means to this end.

In 2009, the *Second Wave Initiative* at Georgetown University started to develop an ethical framework to support the increased inclusion of pregnant women in clinical research (Lyerly et al. 2008, 2009, 2012; Little 2011). In the United States, the Office of Research on Women's Health supported work focused on overcoming barriers to the inclusion of pregnant women in clinical research (ORWH 2011). More generally, for some time now, a number of academics have been advocating for the fair inclusion of pregnant women in clinical research (Chambers et al. 2008; Lyerly et al. 2008; Baylis 2010; Macklin 2010).

To date, much of this literature has focused on the *why* of including pregnant women in clinical research. As recently summarised by Lyerly and colleagues (2008), the benefits of this research include: developing effective treatments for women during pregnancy; promoting foetal safety; reducing harm to women and foetuses resulting from suboptimal care; and allowing access to the benefits of research participation. Notably, while there is still much resistance to the idea of including pregnant women in clinical research, increasingly there are some who are convinced of the need for such research. They understand and endorse the *why*; they are committed to the development of safe and effective treatments for pre-existing conditions in women who become pregnant, for medical conditions of pregnancy, and for conditions that threaten the successful outcome of pregnancy. To make meaningful progress on this front, however, they need to know more about the *how* (Baylis and Halperin 2012).

This book interrogates both the *why* and the *how* of clinical research involving pregnant women. In this way, the book contrasts markedly from much of the existing literature in support of clinical research involving pregnant women, which focuses predominantly on *why* the inclusion of pregnant women in clinical research is necessary. Particularly important with respect to the *how* are practical issues such as priority setting, research design, and research recruitment. Equally important, however, is research ethics oversight. This includes guidelines, and regulations, as well as their implementation through the work of research ethics review committees.

Research ethics oversight arose in response to unethical research over the last 80 years. For example, the Nazi medical research war crimes led to the *Nuremberg Code* (Annas and Grodin 1992). The Tuskegee syphilis study in the United States led to the *Belmont Report* (United States 1979). And, in New Zealand, the cervical cancer research at National Women's Hospital led to the *Cartwright Inquiry* (Cartwright 1988) and the *Code of Health and Disability Services Consumers' Rights* (New Zealand 1996, 2004). Public anger and dismay over the breach of trust by clinicians in these studies drove both the regulation separating clinical practice from research (United States 1979) and the insistence that vulnerable groups be protected from research related harms. This explains, in part, why contemporary research ethics guidelines (and legislation) continue to overemphasise the potential harms of research and underemphasise the social value of research. As a result, most guidelines (and legislation) have a distorted view of the dominant ethics of pregnancy focusing myopically on risk avoidance. This view informs the misguided belief that clinical research during pregnancy is either unnecessary or dangerous, rather than a social good. In combination, these perspectives effectively prohibit most clinical research involving pregnant women.

For example, the *Common Rule* in the United States lists pregnant women as vulnerable. But the concept of vulnerability is under-theorised in the literature and it is not clear what this vulnerability derives from or amounts to. For example, pregnant women and their foetuses are more physiologically vulnerable than non-pregnant adults, but are pregnant women also more morally vulnerable due to reduced capacity to consent, and if so, why? Many of the chapters in this book offer rich and diverse accounts of the concept of vulnerability. For example, the relationships between vulnerability and exploitation (see Ballantyne and Rogers 2016), vulnerability and informed consent (see Wild and Biller-Andorno 2016; Johnson 2016), and vulnerability and empowerment are explored in this book (see Ballantyne and Rogers 2016; Little et al. (2016).

While the pregnant woman is the so-called vulnerable research participant, the primary concern for many is the vulnerable foetus. Indeed, it is widely assumed that concerns about foetal vulnerability explain why research ethics review committees do not approve studies in pregnancy, why clinicians do not assist in recruiting their pregnant patients for such research, and why pregnant women do not volunteer to participate in such research. While concerns for foetal vulnerability are understandable, this book systematically challenges the continued routine exclusion of pregnant women from clinical research by arguing that routine exclusion is harmful, unfair, and illogical. The ethical alternative is fair, respectful, and responsible inclusion in appropriate clinical research.

1.1 Routine Exclusion Is Harmful

The use of medication during pregnancy (and lactation) is one of the least-developed areas of clinical pharmacology and drug research (Buhimschi and Weiner 2009). Changes in pharmacokinetics during pregnancy, correct therapeutic dosage, and compliance during pregnancy are not well understood. Due to a lack of robust evidence, many pregnant women are refused medically important drugs, are subject to dangerous delays in getting drugs, or are prescribed drugs that are thought 'safe' despite evidence of possible teratogenicity (see Baylis and MacQuarrie 2016; Ballantyne and Rogers 2016).

1.2 Routine Exclusion Is Unfair

Ethical research must meet the demands of justice. Justice requires a research agenda that fairly addresses the needs of diverse populations, and fair inclusion criteria that adequately reflect the intervention's intended (i.e., targeted) or likely patient population. The widespread exclusion of most populations from clinical research except for young or middle-aged white males over the last 60 years has resulted in a disproportionate body of evidence regarding the health of young or middle-aged white men (Dresser 1992). Indeed, as a direct consequence of entrenched exclusionary practices, in some areas, current clinical guidelines continue to be based on clinical research that under-represents women and excludes pregnant women (Baylis 2010; Ballantyne and Rogers 2011; Baylis and Halperin 2012). Efforts to rebalance clinical research include policies advocating for, or requiring, more clinical research involving women (NIH 1994). As yet, however, pregnant women remain unfairly excluded from clinical research. Protective research ethics guidelines and regulations are motivated by concerns for the well-being of pregnant women and their foetuses. The net effect of these guidelines and regulations, however, is unjust – unjust because pregnant women thereby lack safe and effective treatment options, or lack information about the ways in which treatment options developed for non-pregnant persons might be appropriately modified for, and made available to, pregnant women.

1.3 Routine Exclusion Is Illogical

In some circumstances – for example pregnant women with an underlying health condition that requires ongoing treatment – the manner in which the precautionary principle is applied to clinical research involving pregnant women is illogical. Not only does exclusion from clinical research increase the risks to pregnant women as already argued, it may also increase the risks to developing foetuses. Here it is worth repeating that the foetus is not 'a bun in the oven'. The foetus is a physiological, functional part of the pregnant woman. The foetus' presence significantly affects the pregnant woman's bodily processes and her health and wellbeing significantly affect the foetus in myriad and complex ways that we are only just beginning to understand. The physiological inter-connectedness of the foetus and the pregnant woman cannot be set aside. Excluding pregnant women with underlying health conditions that require ongoing treatment from clinical research does not protect developing foetuses from potential harm. When these pregnant women are excluded from clinical research, the risk of untested interventions is shifted from the context of a carefully controlled and monitored study, to potentially inconsistent off-label use in the context of clinical treatment (Baylis 2010; and Baylis and MacQuarrie 2016). In other words, research exclusion is precautionary about one sort of risk, and entirely ignores a parallel (and arguably greater) risk simply because the latter obtains outside the official realm of research.

More generally, it can be argued that the risk to pregnant women and their foetuses arises primarily from the lack of evidence about medical treatment during pregnancy, not necessarily from clinical research itself. Untreated or under-treated diseases, suboptimal care, and off-label prescription of untested drugs, can all pose harm to the foetus. A philosophy of extreme risk aversion may appear lofty, but it is unattainable and often counterproductive. Pregnant women need to make decisions involving complex trade-offs throughout their pregnancies, and these trade-offs often involve the use of medication (Lyerly et al. 2009). If precaution were really the guiding principle, then a thorough assessment of the risks and potential benefits of clinical research versus whatever intervention might be offered or recommended – which is sometimes nothing – would be required to determine which approach would be overall most precautionary.

1.4 The Book

Having discussed some of the background reasons for excluding pregnant women from clinical research, as well as some of the motivating reasons for advancing a discussion of both the *why* and the *how* of including pregnant women in clinical research, we now turn our attention to the ways in which this book contributes to the laudatory goal of promoting just research in this patient population. The book is original in three key ways. First, it provides bioethicists, clinicians, researchers, research ethics review committees, and health policy experts with an unparalleled depth of analysis regarding the ethics of clinical research involving pregnant women. To do so, it brings together many of the key authors in this field as well as experts in research ethics and vulnerability who have not previously applied their work to clinical research involving pregnant women. Second, the book incorporates innovative theoretical work in ethics and detailed disease-specific case studies that together highlight the complexity of clinical research involving pregnant women. The results of this integration include identifying conceptual priorities for future ethics research and practical priorities for future clinical research. Third, the book includes a nuanced assessment of arguments both for and against including pregnant women in various kinds of clinical research. Analysis of the complex trade-offs associated with how, where, and when to safely include pregnant women in research are addressed across and within chapters, thus allowing readers to fairly consider arguments from multiple perspectives.

The book is divided into four parts. The first part advocates for fair, respectful, and responsible inclusion of pregnant women in appropriate clinical research. Here the authors describe the *status quo*, drawing on critical historical analysis of the thalidomide and DES scandals to help explain current exclusion practices. Françoise Baylis and Robyn MacQuarrie (2016) briefly describe problems arising from routine exclusion and then explain why clinicians and women should support clinical research in pregnancy. Lucy Langston (2016) argues that stigma around pharmaceutical use during pregnancy does not empower pregnant women or their clinicians to make good decisions about research participation or medical treatment during pregnancy. Chris Kaposy (2016) describes a new model of presumptive inclusion. These chapters paint a vision of a better model of pregnancy research and care that provides pregnant women with evidence-informed clinical care.

The second part reviews current justifications for the exclusion of pregnant women from clinical research and thereby exposes contemporary barriers to such research. Indira van der Zande and colleagues (2016) provide a systematic review of reported reasons for exclusion and suggest practical solutions to some of these barriers. Next, Carolyn Ells and Caroline Lyster (2016) explore the role of research ethics review committees as barriers to clinical research. They highlight problems with current research ethics guidelines and then offer guidance for improved ethics oversight as an alternative to the routine exclusion of pregnant women from clinical research. A crucial piece of the puzzle is pregnant women's own views about evidence, risk, and research. Verina Wild and Nikola Biller-Andorno (2016) present empirical results from a qualitative research study involving pregnant women in Germany regarding their thoughts and experiences with decision-making during pregnancy. They confirm that pregnant women are initially averse to the vague idea of research, but are more willing to participate in clinical research when the burdens and potential benefits of specific trials are explained to them.

Part three describes ways forward in how to undertake fair, respectful, and responsible inclusion of pregnant women in clinical research. These chapters probe important theoretical problems at issue in research involving pregnant women and how these can be overcome. Here the authors push the boundaries of our understanding of key concepts of vulnerability, risk, and equipoise and describe the normative nature of the maternal-foetal relationship in terms of moral status, autonomy, and guardianship of foetal interests. These chapters also scrutinise different research methods in order to better understand the goals, parameters, and limitations of competing processes of evidence generation. Angela Ballantyne and Wendy Rogers (2016) argue that while pregnant women may experience inherent, situational, or pathogenic vulnerability, in general they are not at risk of exploitation during clinical research. L. Syd Johnson (2016) also explores the notion of vulnerability, but from a different tack. She views the classification of pregnant women as vulnerable research participants as a direct threat to pregnant women's autonomy. Rebecca Kukla (2016) focuses on equipoise and uncertainty in clinical research, underlining the importance of empowering pregnant women to make informed, autonomous decisions about research participation by including them in the early phases of research design. Finally, David Healy and Derelie Mangin (2016) highlight the shortcomings of a specific research design, namely the randomised controlled trial. In their view, when randomised controlled trials are used indiscriminately, their adverse effects may outweigh their benefits. Together these chapters suggest elements of an ethical framework for the future of clinical research involving pregnant women.

Part four moves the discussion from a careful review of theoretical and conceptual issues to a discussion of practical issues embedded in specific case studies that span the range of low to high risk research interventions. For example, Angela Ballantyne and colleagues (2016) write about clinical research on the use of probiotic supplements, which can be thought of as a lifestyle intervention. Ruth Farrell and Rebecca Flyckt (2016) write about clinical research involving reproductive medicine with a focus on uterine transplantation, the newest assisted reproductive technology which involves a complex combination of new and established fertility procedures and surgeries. In between these chapters, there is a chapter on clinical research involving women with, or at risk of contracting, HIV by Margaret Little and colleagues (2016), a chapter by Richard Ashcroft (2016) on clinical research involving maternal gene transfer with a view to improving foetal growth, and a chapter by Lisa Harris (2016) on clinical research involving women seeking abortion services. Together, these chapters show that clinical research can sometimes be effectively carried out under the existing oversight mechanisms, but they also highlight where guidelines and regulations unnecessarily hinder clinical research in pregnant women. Drilling down into the detail of specific cases brings to life the complexity and nuance of the ethical challenges facing clinical research involving pregnant women and showcases some inventive solutions to some of these challenges.

Taken together, these chapters represent a rich and diverse investigation of the ethical challenges associated with integrating pregnancy into the global clinical research agenda. Many chapters tell stories of the work of ethicists and researchers addressing questions of clinical importance for pregnant women. Their successes and innovative solutions to the restrictive regulatory environment should give us hope. The scholarship here challenges us to keep dismantling the harmful, unfair, and illogical barriers to the inclusion of pregnant women in clinical research and to build a framework for fair, respectful, and responsible clinical research during pregnancy.

References

Annas, G.J., and M.A. Grodin. 1992. *The Nazi doctors and the Nuremberg Code*. New York: Oxford University Press.

Armant, D.R. 2005. Blastocysts don't go it alone. Extrinsic signals fine-tune the intrinsic developmental program of trophoblast cells. *Developmental Biology* 280(2): 260–280.

Ashcroft, R. 2016. Ethical issues in a trial of maternal gene transfer to improve foetal growth. In *Clinical research involving pregnant women*, eds. F. Baylis and A. Ballantyne, 247–263. Cham: Springer.

Ballantyne, A.J., and W.A. Rogers. 2011. Sex bias in studies selected for clinical guidelines. *Journal of Women's Health* 20(9): 1297–1306.

Ballantyne, A., C. Barthow, and K. Wickens. 2016. Research into lifestyle changes in pregnancy. In *Clinical research involving pregnant women*, eds. F. Baylis and A. Ballantyne, 211–225. Cham: Springer.

Baylis, F. 2010. Pregnant women deserve better. *Nature* 465(7299): 689–690.

Baylis, F., and S. Halperin. 2012. Research involving pregnant women: Trials and tribulations. *Clinical Investigation* 2(2): 139–146.

Baylis, F., and R. MacQuarrie. 2016. Why physicians and women should want pregnant women included in clinical trials. In *Clinical research involving pregnant women*, eds. F. Baylis and A. Ballantyne, 17–31. Cham: Springer.

Ballantyne, A., and W. Rogers. 2016. Pregnancy, vulnerability, and the risk of exploitation in clinical research. In *Clinical research involving pregnant women*, eds. F. Baylis and A. Ballantyne, 139–159. Cham: Springer.

Baylis, F., and S. Sherwin. 2002. Judgements of non-compliance in pregnancy. In *Ethical issues in maternal-foetal medicine*, eds. D. Dickenson, 285–301. Cambridge: Cambridge University Press.

Benisek, D., J. Shabert, and R. Skornik. 2000. Dietary intake of polyunsaturated fatty acids by pregnant or lactating women in the United States. *Obstetrics and Gynecology* 95(4, Suppl 1): S77–S78.

Blumfield, M.L., A.J. Hure, L.K. Macdonald-Wicks, A.J. Patterson, R. Smith, and C.E. Collins. 2011. Disparities exist between national food group recommendations and the dietary intakes of women. *BMC Womens Health* 11(37): 1–9.

Buhimschi, C.S., and C.P. Weiner. 2009. Medications in pregnancy and lactation: Part 1, teratology. *Obstetrics and Gynecology* 113(1): 166–188.

Callaway, L., P.B. Colditz, and N.M. Fisk. 2009. Folic acid supplementation and spontaneous preterm birth: Adding grist to the mill? *PLoS Medicine* 6(5): e1000077.

Cartwright, S. 1988. *The report of the committee of inquiry into allegations concerning the treatment of cervical cancer at National Women's Hospital and into other related matters 1988*. Auckland: GPO.

Chambers, C.D., J.E. Polifka, and J.M. Friedman. 2008. Drug safety in pregnant women and their babies: Ignorance is not bliss. *Clinical Pharmacology & Therapeutics* 83(1): 181–183.

Dreger, A. 2015. *Galileo's middle finger: Heretics, activists, and the search for justice in science*. New York: Penguin Press.

Dresser, R. 1992. Wanted. Single, white male for medical research. *The Hastings Center Report* 22(1): 24–29.

Ells, C., and C. Lyster. 2016. Research ethics review of drug trials targeting medical conditions of pregnant women. In *Clinical research involving pregnant women*, eds. F. Baylis and A. Ballantyne, 95–118. Cham: Springer.

Farrell, R.M., and R. Flyckt. 2016. Research on uterine transplantation: Ethical considerations. In *Clinical research involving pregnant women*, eds. F. Baylis and A. Ballantyne, 285–297. Cham: Springer.

Gluckman, P.D., M.A. Hanson, C. Cooper, and K.L. Thornburg. 2008. Effect of in utero and early-life conditions on adult health and disease. *The New England Journal of Medicine* 359(1): 61–73.

Godfrey, K.M., A. Sheppard, P.D. Gluckman, K.A. Lillycrop, G.C. Burdge, C. McLean, et al. 2011. Epigenetic gene promoter methylation at birth is associated with child's later adiposity. *Diabetes* 60(5): 1528–1534.

Harremoës, P., D. Gee, M. MacGarvin, A. Stirling, J. Keys, B. Wynne, and C.M.P. Vaz (eds.). 2002. *The precautionary principle in the 20th century. Late lessons from early warnings*. London: Earthscan Publications Ltd.

Harris, L.H. 2016. Clinical research involving pregnant women seeking abortion services: United States perspectives. In *Clinical research involving pregnant women*, eds. F. Baylis and A. Ballantyne, 265–284. Cham: Springer.

Healy, D., and D. Mangin. 2016. Does my bias look big in this? In *Clinical research involving pregnant women*, eds. F. Baylis and A. Ballantyne, 197–208. Cham: Springer.

Hibbelin, J.R., J.M. Davis, C. Steer, P. Emmett, I. Rogers, C. Williams, and J. Golding. 2007. Maternal seafood consumption in pregnancy and neurodevelopmental outcomes in childhood (ALSPAC study): An observational cohort study. *Lancet* 369(9561): 578–585.

Johnson, L.S.M. 2016. When hypothetical vulnerability becomes actual: Research participation and the autonomy of pregnant women. In *Clinical research involving pregnant women*, eds. F. Baylis and A. Ballantyne, 161–178. Cham: Springer.

Kaposy, C. 2016. Presumptive inclusion and legitimate exclusion criteria. In *Clinical research involving pregnant women*, eds. F. Baylis and A. Ballantyne, 51–62. Cham: Springer.

Kukla, R. 2005. *Mass hysteria: Medicine, culture, and mothers' bodies*. Oxford: Rowman & Littlefield Publishers.

Kukla, R. 2016. Equipoise, uncertainty, and inductive risk in research involving pregnant women. In *Clinical research involving pregnant women*, eds. F. Baylis and A. Ballantyne, 179–196. Cham: Springer.

Langston, L. 2016. Better safe than sorry: Risk, stigma, and research during pregnancy. In *Clinical research involving pregnant women*, eds. F. Baylis and A. Ballantyne, 33–50. Cham: Springer.

Lee, E., J. Bristow, C. Faircloth, and J. Macvarish. 2014. *Parenting culture studies*. Basingstoke: Palgrave.

Little, M.O. 2011. Treating important medical conditions during pregnancy. In *Enrolling pregnant women: Issues in clinical research*. U.S. Department of Health and Human Services, Public Health Service, National Institutes of Health, Office of Research on Women's Health, 23–26. Bethesda: National Institutes of Health.

Little, M.O., A.D. Lyerly, A.C. Mastroianni, and R.R. Faden. 2016. Ethics and research with pregnant women: Lessons from HIV/AIDS. In *Clinical research involving pregnant women*, eds. F. Baylis and A. Ballantyne, 227–246. Cham: Springer.

Lyerly, A.D., M.O. Little, and R. Faden. 2008. The second wave: Toward responsible inclusion of pregnant women in research. *International Journal of Feminist Approaches to Bioethics* 1(2): 5–22.

Lyerly, A.D., L.M. Mitchell, E.M. Armstrong, L.H. Harris, R. Kukla, M. Kuppermann, and M.O. Little. 2009. Risk and the pregnant body. *Hastings Center Report* 39(6): 34–42.

Lyerly, A.D., E.E. Namey, B. Gray, G. Swamy, and R.R. Faden. 2012. Women's views about participating in research while pregnant. *IRB* 34(4): 1–8.

Macklin, R. 2010. Enrolling pregnant women in clinical research. *Lancet* 375(9715): 632–633.

McCormack, S.A., and B.M. Best. 2014. Obstetric pharmacokinetic dosing studies are urgently needed. *Frontiers in Pediatrics* 2(9): 1–9.

Mishra, G.M., D. Schoenaker, S. Mihrshahi, and A.J. Dobson. 2015. How do women's diets compare with the new Australian dietary guidelines? *Public Health Nutrition* 18(2): 218–225.

Mitchell, A.A., S.M. Gilboa, M.M. Werler, K.E. Kelley, C. Louik, S. Hernández-Diaz, and National Birth Defects Prevention Study. 2011. Medication use during pregnancy, with particular focus on prescription drugs: 1976–2008. *American Journal of Obstetrics and Gynecology* 205(1): 51.e1–51.e8.

Morton, S.M., C.C. Grant, C.R. Wall, P.Z. Carr, D.K. Bandara, J.M. Schmidt, et al. 2014. Adherence to nutritional guidelines in pregnancy: Evidence from the Growing Up in New Zealand birth cohort study. *Public Health Nutrition* 17(9): 1919–1929.

New Zealand. 1996, 2004. *Health and disability commissioner (Code of health and disability services consumers' rights) regulations 1996*. http://www.legislation.govt.nz/regulation/public/1996/0078/latest/whole.html. Accessed 7 Apr 2016.

NIH (National Institutes of Health). 1994. *NIH guidelines for the inclusion of women and minorities as subjects in clinical research*. http://grants.nih.gov/grants/guide/notice-files/not94-100.html. Accessed 11 Apr 2016.

Nordeng, H., E. Ystrøm, and A. Einarson. 2010. Perception of risk regarding the use of medications and other exposures during pregnancy. *European Journal of Clinical Pharmacology* 66(2): 207–214.

ORWH (United States. Department of Health and Human Services, Public Health Service, National Institutes of Health, Office of Research on Women's Health). 2011. *Enrolling pregnant women: Issues in clinical research*. Bethesda: National Institutes of Health.

Pezdirc, K.B., A.J. Hure, M.L. Blumfield, and C.E. Collins. 2012. Listeria monocytogenes and diet during pregnancy; balancing nutrient intake adequacy *v.* adverse pregnancy outcomes. *Public Health Nutrition* 15(12): 2202–2209.

Radtke, K.M., M. Ruf, H.M. Gunter, K. Dohrmann, M. Schauer, A. Meyer, and T. Elbert. 2011. Transgenerational impact of intimate partner violence on methylation in the promoter of the glucocorticoid receptor. *Translational Psychiatry* 1(e21): 1–6.

Sui, Z., D. Turnbull, and J. Dodd. 2013. Enablers of and barriers to making healthy change during pregnancy in overweight and obese women. *The Australasian Medical Journal* 6(11): 565–577.

Swan, S.H. 2000. Intrauterine exposure to diethylstilbestrol: Long-term effects in humans. *APMIS* 108(12): 793–804.

United States (National Commission for the Protection of Human Subjects of Biomedical and Behavioral Research). 1979. *The Belmont Report: Ethical principles and guidelines for the protection of human subjects of research*. http://www.hhs.gov/ohrp/humansubjects/guidance/belmont.html. Accessed 4 Apr 2016.

van der Zande, I.S.E., R. van der Graaf, J.L. Browne, and J.J.M. van Delden. 2016. Fair inclusion of pregnant women in clinical research: A systematic review of reported reasons for exclusion. In *Clinical research involving pregnant women*, eds. F. Baylis and A. Ballantyne, 65–94. Cham: Springer.

Wild, V., and N. Biller-Andorno. 2016. Pregnant women's views about participation in clinical research. In *Clinical research involving pregnant women*, eds. F. Baylis and A. Ballantyne, 119–136. Cham: Springer.

Witchel, S.F., and W.L. Miller. 2012. Prenatal treatment of congenital adrenal hyperplasia – Not standard of care. *Journal of Genetic Counseling* 21(5): 615–624.

World Health Organization. 2013. *WHO recommendations for the prevention and management of tobacco use and second-hand smoke exposure in pregnancy*. Geneva: WHO. http://www.ncbi.nlm.nih.gov/books/NBK190304/pdf/TOC.pdf. Accessed 4 Apr 2016.

Part I

Chapter 2
Why Physicians and Women Should Want Pregnant Women Included in Clinical Trials

Françoise Baylis and Robyn MacQuarrie

Abstract *A direct consequence of the routine exclusion of pregnant women from clinical trials is pregnant women using over-the-counter and prescription medications in the absence of population-specific clinical trial data about the potential benefits and harms of these medications for themselves, their foetuses, and their future children. In our view, pregnant women are as entitled as other patient populations to robust clinical trial data about safety and dosing on the basis of which to make evidence-informed decisions. To this end, we maintain that pregnant women should be presumed eligible to participate in clinical trials. This chapter asks and answers the following questions: Why are clinical trials in pregnancy important from a physician's perspective? And, why are clinical trials in pregnancy important from a pregnant woman's perspective? Having addressed these questions, we next consider why pregnant women might choose not to participate in clinical trials, and what can be done to encourage their participation.*

There are many reasons why pregnant women are routinely excluded from clinical trials of medications and vaccines including the fact that manufacturers, regulators, sponsors, researchers, and research ethics review committees would prefer to avoid the scientific, legal, and ethical complexities and the costs associated with research in pregnancy. In this chapter, we do not review these reasons which are well documented by others (Lyerly et al. 2008; Shields and Lyerly 2013; see also van der Zande et al. 2016). Instead, we critically examine why access to robust clinical trial data detailing the safety and effectiveness of drugs used during pregnancy

F. Baylis, CM, ONS, PhD, FRSC, FCAHS (✉)
Novel Tech Ethics, Faculty of Medicine and Department of Philosophy, Faculty of Arts and Social Sciences, Dalhousie University, Halifax, Nova Scotia, Canada
e-mail: francoise.baylis@dal.ca

R. MacQuarrie, MASc, MD, FRCS(S)
Nova Scotia Health Authority, Bridgewater, Nova Scotia, Canada

PhD(c) Dalhousie University, Halifax, Nova Scotia, Canada

F. Baylis, A. Ballantyne (eds.), *Clinical Research Involving Pregnant Women*, Research Ethics Forum 3, DOI 10.1007/978-3-319-26512-4_2

should be a high priority for physicians (and not just obstetricians)[1] and for pregnant women.[2] Physicians should want to provide their pregnant patients with the same standard of care they provide their non-pregnant patients, and pregnant women should demand no less. From this perspective, it is important to exert pressure on manufacturers, regulators, sponsors, researchers, and research ethics review committees to change the *status quo* so that pregnant women are presumed eligible to participate in all Phases of clinical research (Blehar et al. 2013).

In our view, pregnant women should not only be included in clinical trials specifically targeting pregnant women, they should also be included in clinical trials targeting the general population. In both instances, careful attention should be given to issues of trial design and to the timing of participation in research by pregnant women in order to build on knowledge gained from prior research in the general population (Baylis and Halperin 2012). We recognise that pregnant women may legitimately be excluded from specific clinical trials on scientific and ethical grounds, such as trials involving the use of drugs for which there is evidence of teratogenicity or evidence of foetal risk. However, the inclusion of pregnant women in clinical trials should be the rule, rather than the exception (see Kaposy 2016). This view is consistent with that of the World Health Organization (WHO) in the *International Ethical Guidelines for Biomedical Research Involving Human Subjects* of the Council for International Organizations of Medical Sciences (CIOMS). These guidelines state unequivocally that "[p]regnant women should be presumed to be eligible for participation in biomedical research"[3] (CIOMS 2002). Consistent with this directive, exclusion from a specific clinical trial would require an explicit rationale that references scientifically and ethically defensible exclusion criteria.

At the present time, a majority of pregnant women use over-the-counter and prescription medications in the absence of relevant clinical trial data confirming the potential benefits and harms of these medications for themselves, their foetuses, and their future children. Indeed, recent data confirm that more than 90 % of pregnant women in the United States take one or more over-the-counter or prescription medications for both obstetrical and non-obstetrical illnesses (Mitchell et al. 2011). As well, in recent years, the average number of medications taken during pregnancy has increased. For example, first-trimester use of prescription medications has increased by more than 60 % in the last 30 years and, during this same timeframe, the number of pregnant women taking four or more medications

[1] While many health care providers can be involved in the care of pregnant women, we focus on physicians as these are the health care providers authorised to prescribe medications. As well, though many of the claims advanced in this chapter apply equally to vaccines, our focus is on medications.

[2] Arguably, this claim applies to women and transgender men, some of whom have experienced pregnancy and some of whom may experience pregnancy. This is beyond the scope of this chapter.

[3] See Macklin (2010) for a brief discussion of ambiguity in the CIOMS guidelines. Also, at the time of writing the CIOMS guidelines are under review. Changes to the guidelines for research involving pregnant women are anticipated.

has tripled (Mitchell et al. 2011). We assume a similar practice pattern in many high- and middle-income countries.

The significant use of over-the-counter medications during pregnancy should not be surprising. Pregnancy occurs over nine months, and it would be unusual for anyone (including pregnant women) not to take any over-the-counter medications for the greater part of a year. Among the most commonly used over-the-counter medications are acetaminophen, ibuprofen, pseudoephedrine, and aspirin. Data from the Slone Epidemiology Center Birth Defects Study and the National Birth Defects Prevention Study show "that approximately two-thirds of women take acetaminophen and that approximately 1 in 6 women takes a decongestant or ibuprofen during pregnancy" (Werler et al. 2005).

As well, many pregnant women take prescription medications for acute or chronic obstetrical and non-obstetrical medical conditions. Among these women are those who are unable or unwilling to tolerate the side-effects of pregnancy. A serious example of this is women with extreme nausea and vomiting that results in weight loss and dehydration, and often requires hospitalisation. Moreover, there are women with underlying health conditions who require continued medical treatment during pregnancy. These women often will continue the use of their pre-pregnancy prescription medications (with or without changes in dosing). This includes women with diabetes, hypertension, epilepsy, asthma, depression, and anxiety. Among the most commonly prescribed medications in the first trimester of pregnancy are amoxicillin and other antibiotics used to treat a variety of infections (including bladder infections). As well, progesterone is commonly used throughout pregnancy to prevent preterm labour and in the first trimester to provide placental support in pregnancies resulting from assisted reproduction (Mitchell et al. 2011).

Three phenomena explain the increasing number of women taking medications during pregnancy, and the increasing number of medications being taken by them. First, women in high- and middle-income countries are delaying childbearing and, typically, older pregnant women have more health challenges than their younger counterparts (for example, hypertension, pre-existing diabetes, hypothyroidism) (Martin et al. 2012, 2015). Second, women with chronic health conditions for which physicians would have actively discouraged pregnancy are now choosing to become pregnant owing to improved management options for their underlying disease (for example, women with Crohn's disease, Factor IV Leiden, and congenital heart disease). Third, women with poorly controlled (i.e., difficult to manage) health conditions that previously precluded pregnancy are now able to become pregnant using fertility drugs (for example, women with obesity, polycystic ovarian syndrome, and uterine fibroids). Taken together, these discrete phenomena have resulted in an increase in the number of women with underlying health challenges that need to be managed during pregnancy. As one of us has noted previously, "pregnant women get sick, and sick women get pregnant" (Baylis 2010), and this is now happening in increasing numbers.

The problem for pregnant women who use over-the-counter and prescription medications is that, for the most part, robust clinical trial data confirming the safety and effectiveness of the drugs used during pregnancy do not exist. Data from retrospective observational studies and adverse event registries are sometimes available to guide physicians and patients in making treatment decisions during pregnancy. Sometimes, however, there are good reasons to question the quality of some of these data and to demand additional research. Not only is there the problem of recall bias with retrospective studies but, in many cases, data are inconsistent among drug safety databases that pool all known studies, including animal studies and retrospective studies. As well, data may be inconclusive within any one registry. In such cases, available data may be of limited value (see Ballantyne and Rogers 2016; Healy and Mangin 2016). This makes it difficult for physicians to offer sound clinical recommendations based on a clear understanding and evaluation of the potential benefits and harms. This situation contrasts markedly with medications available to the general population, as these medications are approved for use following the completion of a series of clinical trials that typically move through four Phases.

As described by the US Food and Drug Administration, the aim of Phase I clinical trials is to establish the safety of a new drug (FDA 2014). Typically, 20–80 healthy volunteers are recruited to Phase I trials to study the pharmacokinetic and pharmacodynamic properties of a new drug in order to identify acute side-effects. Pharmacokinetics refers to the process by means of which the body absorbs, distributes, metabolises, and eliminates a drug. Pharmacodynamics refers to the biochemical and physiological effects that a drug may have on the body. If a new drug passes this Phase, a Phase II trial is the next step. With Phase II clinical trials, the number of research participants is greater than those involved in Phase I trials – between 100 and 300 participants. In this research Phase, patients who suffer from the disease or condition for which the drug is being developed, rather than healthy volunteers, are recruited. The goal is to determine whether the drug under study is efficacious in treating the condition and whether, in addition to the desired therapeutic effect, there are undesirable side-effects. Next, there are Phase III clinical trials that involve a larger number of research participants – somewhere between 1,000 and 3,000 people. This Phase allows for a more robust assessment of the efficacy and dosing of the drug. As well, more information can be gathered on less common side-effects because the drug is being studied in more people over more time. Phase III is also when a drug may be compared to an available competitor drug to assess relative value – that is, comparative efficacy and effectiveness. Once a drug has been approved, the research that follows is generally described as a Phase IV trial. In this fourth Phase, a new drug is assessed for long-term safety and effectiveness, while considering the different ways in which the drug may be administered. To be clear, post-marketing 'research' doesn't resemble or recruit like the other research Phases described above.

This phased approach to the research and post-approval marketing of drugs for diverse patient populations typically does not occur for pregnant women. A direct consequence of not including pregnant women in clinical trials is that most drugs are used in this patient population not only without the benefit of robust evidence about safety and effectiveness (for pregnant women, their developing foetuses, or their future children), but also without population-specific information about the pharmacodynamics and pharmacokinetics of the drugs to know how they are processed in a pregnant woman's body. This is a serious lacuna considering the major physiological changes that occur during pregnancy (Carlin and Alfirivic 2008).

During pregnancy, women experience increased plasma volume, body weight, body fat, metabolism, and hormone levels. For example, during pregnancy a woman's blood plasma levels increase by 50 % and her cardiac output increases by 30–50 %. Her blood pressure dips in the second trimester and potentially increases in the third. As well, her lung volumes are diminished. Her glomerular filtration rate increases by 40–50 % and her renal plasma flow increases by up to 65 %. As well, her gastrointestinal system has decreased motility. Arguably these changes "make it impossible to calculate dose and safety information by extrapolating from data on men and non-pregnant women" (Baylis 2010), and yet extrapolation is exactly what physicians do when they recommend or prescribe medications off-label as would be the case with any medication not labelled (i.e., specifically approved) for use in pregnancy (which is the vast majority of medications). Off-label prescribing occurs when a physician prescribes an approved medication for (i) an unapproved condition, or (ii) an approved condition, but in an unapproved patient population, or at an unapproved dose, or in an unapproved way (i.e., form of administration).

A current example of off-label prescribing in pregnancy is the drug ondansetron. This is an anti-nausea drug labelled to treat nausea and vomiting in chemotherapy and surgery patients. It is being prescribed off-label to pregnant women with extreme nausea and vomiting. It has been suggested that this drug may be responsible for birth defects (Anderka et al. 2012). A recent retrospective study, however, suggests that there are no significant side-effects with use of this medication during pregnancy (Pasternak et al. 2013). What is particularly challenging for physicians and pregnant women in this scenario is that the only evidence available is contradictory and substandard. Had pregnant women been included in meaningful ways in clinical trials of ondansetron, prior to the drug coming to market, there would be reliable, prospective safety data to guide decision-making. Instead, the only data available is contextual, retrospective, and limited.

Off-label use of medications *de facto* results in unmonitored and unregulated experimentation in an unsuspecting population – patients who do not appreciate that they are individually participating in a 'trial of one'. Here we use the term 'trial of one' to refer to a practice where a person receives an intervention outside of a formal clinical trial in a context where knowledge regarding the potential benefits and harms of the intervention do not satisfy standards for therapeutic use in the patient population of which the person is a member. A 'trial of one' is not the same as an n-of-1 study, which is far more rigorous and systematic. A common feature of a 'trial of one' is that the patient mistakenly believes that she is receiving a therapeutic intervention. This is an instance of therapeutic misperception[4] – believing that a

[4]The term therapeutic misconception refers to the mistaken belief that a research intervention in a clinical trial is a therapeutic intervention. The term therapeutic misperception introduced here, refers to the mistaken belief that off-label use of a drug, biologic, or device in a patient population for which data about safety and efficacy is lacking (experimentation rather than research proper) is a *bona fide* therapeutic intervention.

physician's willingness to recommend a medication off-label while relying on data from sources other than randomised controlled trials is as good as clinical trial data confirming safety and effectiveness. This perception is deeply problematic and clinical trials are specifically intended to supplant this kind of decision-making. Another common feature of a 'trial of one' is that data cannot be efficiently collated and analysed to produce generalizable knowledge to validate the intervention as therapeutic.

In support of our claim that pregnant women should be presumed eligible for proper and full participation in all Phases of clinical research (and should not find themselves routinely participating in 'trials of one'), we address the following questions: (1) Why are clinical trials in pregnancy important from a physician's perspective? And, (2) Why are clinical trials in pregnancy important from a pregnant woman's perspective? Having addressed these questions, we next engage the moral imagination to ask and answer two further questions that we believe will become relevant at some future time when pregnant women are routinely invited to participate in all Phases of clinical research: (3) Why might pregnant women choose not to participate in clinical trials? And, (4) What can be done to encourage pregnant women to participate in clinical trials?

2.1 Why Are Clinical Trials in Pregnancy Important from a Physician's Perspective?

The short answer to this question is that physicians who treat pregnant women, like all physicians, should practice and promote evidence-informed decision-making. For them to do so, they need robust evidence regarding the safety and effectiveness of therapeutic interventions for pregnant women and their developing foetuses. Only in this way, can physicians offer their pregnant patients sound professional recommendations regarding the use of over-the-counter and prescription medications.

Evidence-informed decision-making is a term recently introduced in response to the backlash against the concept of evidence-based medicine (Miles and Loughlin 2011). As defined by Sackett and colleagues "[e]vidence based medicine is the conscientious, explicit, and judicious use of current best evidence in making decisions about the care of individual patients. The practice of evidence-based medicine means integrating individual clinical expertise with the best available external clinical evidence from systematic research" (Sackett et al. 1996). For years, many (mis) interpreted the reference to "best available external clinical evidence" in this definition of evidence-based medicine as requiring evidence from randomised controlled trials or meta-analyses. In an effort to underscore the fact that in some instances other types of evidence could satisfy the standard of 'best available external clinical evidence' (and thus appropriately guide clinical decision making) a new term was coined – evidence-informed decision-making (see Healy and Mangin 2016).

Currently, physicians who treat pregnant women rely on (often uncollated) data from animal studies, case reports, retrospective observational studies, adverse event registries, and some poorly supported meta-analyses. In some instances, this evidence suffices, as when a drug is clearly identified as causing harm through adverse events registries. In other instances, however, available evidence simply isn't good enough for the treatment of medical conditions complicating pregnancy. In some cases, without the benefit of evidence from randomised clinical trials, physicians are unable to offer their pregnant patients sound professional recommendations regarding the use of available drugs. Indeed, they are *de facto* precluded from practicing and promoting evidence-informed decision-making (Shields and Lyerly 2013). In many (if not most) instances, physicians treating pregnant women have few courses of action available to them. For pregnant women who are not taking medications, they can recommend over-the-counter or prescription medications based on 'no' or 'limited' evidence of potential benefits or harms. Alternatively, they can promote a philosophy of 'less is best' (Thall Bastow and Holmes 2016) and discourage pregnant women from taking any over-the-counter or prescription medications (consistent with the view that 'the safest pregnancy-related pharmacy is as little pharmacy as possible'). And, for pregnant women on a prior drug regimen, they can recommend the *status quo*, they can recommend a change in medication(s) or continuation of the same medication(s) but at a different dosage, or they can recommend discontinuation of the medication(s). Whatever physicians decide, however, they are invariably doing so in the absence of solid evidence.

As noted above, there are significant changes to women's physiology during pregnancy. In the absence of clinical trial data, in many instances, physicians won't know how these physiological changes affect the pharmacokinetics and pharmacodynamics of a drug in a pregnant body. For example, the known significant increase in plasma volume during pregnancy may dilute the concentration of a drug in the plasma. In addition, the amount of drug in the pregnant woman could be reduced further by the significant increase in glomerular filtration, and the subsequent renal processing of the drug which is then eliminated from the body in urine. The way in which a drug is metabolised is an important component of a clinical trial in which a drug's half-life is determined. The half-life of a drug – a measure of how long it takes for half of the drug to be cleared from active circulation – may be altered by these significant physiological variations. Understanding how a drug circulates in a pregnant body, and whether it crosses the placenta, is critical to evidence-informed decision-making.

In sum, physicians who care for pregnant women are regularly required to provide advice on the use of over-the-counter and prescription medications. Too frequently they do so in the absence of high-quality clinical trial data. This state of affairs is a direct result of intentional decision making on the part of manufacturers, regulators, sponsors, researchers and research ethics review committees to routinely exclude pregnant women from trial participation. Physicians should not accept this *status quo*, which effectively forces them to rely on lower standards of clinical evidence for the treatment of pregnant patients than would be the case for any other patient population. They should demand better for their patients and for themselves.

2.2 Why Are Clinical Trials in Pregnancy Important from a Pregnant Woman's Perspective?

Most pregnant women try to stay well during pregnancy, for their own sake, as well as the sake of their developing foetuses and future children. This can be a serious challenge, however, as when pregnant women take (or stop taking) medications without the benefit of good clinical trial data on toxicity and dosing.

Many pregnant women using prescribed medication(s) for underlying health conditions will modify the standard dosage of their prescription medication(s) when they become pregnant, or they will discontinue their prescription medication(s). Some pregnant women will make these decisions without seeking professional advice. Perhaps they rely on information available on the internet, or they simply act on their intuitions. Other pregnant women will ask their physicians to help them weigh the potential harms of untreated illness against the potential harms of their medications, but in many instances their physicians will be hard pressed to provide sound advice (see Wild and Biller-Andorno 2016). Such advice can only be available to pregnant women, if there are well-designed and executed clinical trials in pregnancy. A major barrier to such trials are current national and international research ethics laws, policies, guidelines, and practices that require the routine exclusion of pregnant women from clinical trials.

Without good clinical trial data on toxicity and dosing, there is the very real risk of under- or over-dosing. With under-dosing, the risk is that pregnant women and their developing foetuses will be exposed to the potential harms of the medication(s), without the corresponding potential benefits associated with addressing the underlying health problem. With over-dosing, the risk is that pregnant women and their developing foetuses will be exposed to greater potential harm than is required to manage the women's symptoms and achieve therapeutic benefit. Consider, for example, a medication that binds to receptors in a woman's body. Ideally, the appropriate dose of the medication would bind to these receptors without 'extra' drug free floating to potentially cause harm. With both under- and over-dosing, the harm-benefit ratio is skewed in a way that does not further the interests of pregnant women or their developing foetuses (see Little et al. 2016).

As it stands, each pregnant woman who takes (or stops taking) medications invariably finds herself in a clinical 'trial of one'. In this trial, as compared with a standard clinical trial, she is not being carefully monitored for adverse events, the medication she is taking is not being formally assessed for effectiveness, there may be no clear safety parameters, and no clear stopping rules. As well, there is no long-term follow-up of children exposed to medications during foetal development. This kind of 'trial and error' is not in the best interest of pregnant women, their developing foetuses or their future children (see Healy and Mangin 2016). For this reason, pregnant women should demand inclusion in relevant, well-designed clinical trials. This is the only way to ensure that fewer pregnant women, foetuses, and future children are not at risk of harm, resulting from the off-label use of medications – medications that come to market without reliable data for use in pregnancy (Macklin 2010).

2.3 Why Might Pregnant Women Not Want to Participate in Clinical Trials?

To this point, we have explained why physicians should be motivated to advocate for research in pregnancy. From this it follows that they should be motivated to contribute to the design of scientifically and ethically sound clinical research involving pregnant women and, as appropriate, to encourage their pregnant patients to enrol in these trials. We have also explained why pregnant women should want to participate in such clinical trials– so as to secure potential benefits for themselves, their developing foetuses, and their future children as well as secure benefits for pregnant women as a class. We nonetheless recognise that many pregnant women may not want to enrol in clinical trials, having been told for many years that research participation is a risky activity that can result in serious harm to the developing foetus. The thalidomide tragedy, where thousands of children were born worldwide with significant limb deformities, is often cited in this regard (see Langston 2016). As has been noted repeatedly, however, it is not research participation but rather the failure to test thalidomide in Phase I and Phase II clinical trials that explains the tragedy (Macklin 2010). Had there been such research, significantly fewer children would have suffered the harmful side-effects of the drug. The trial would have been stopped after one or a few adverse events, and the drug would never have been marketed to pregnant women.

Notwithstanding these facts, it is important to understand the legitimate concern of pregnant women for foetal well-being so that it can be addressed pro-actively, thereby contributing to the sea change required to make the inclusion of pregnant women in clinical trials the norm. Indeed, understanding why pregnant women may resist participating in clinical trials is critical to strategising about the best ways to explain the benefits of research participation.

2.4 How Might Pregnant Women Be Encouraged to Participate in Clinical Trials?

Imagine a world in which the routine exclusion of pregnant women from clinical trials is no longer the norm. That is, imagine a world in which manufacturers, regulators, and sponsors require fair, respectful, and responsible clinical research in pregnancy for medications that might reasonably be used by pregnant women; a world in which researchers are motivated to pursue appropriate research in pregnancy; and a world in which research ethics review committees are able and willing to approve scientifically and ethically responsible clinical trials in pregnancy. How might pregnant women invited to participate in such research respond? (see Wild and Biller-Andorno 2016; Ballantyne et al. 2016)

Recent data on the views of pregnant women about participating in H1N1 vaccine trials suggest that there are discrete circumstances in which at least some pregnant women perceive participation in a clinical trial as potentially safer than receiving so-called treatment in the standard clinical care setting (Lyerly et al. 2012). These are circumstances where the pregnant women and their developing foetuses are at significant risk of harm, knowledge about how to safely and effectively reduce the risk of harm is missing, the clinical trial promises careful monitoring, and is expected to produce generalizable knowledge. As these circumstances would apply to a number of pregnant populations, such as pregnant women with HIV (see Little et al. 2016) or pregnant women with diabetes, there is reason to think that in some circumstances some pregnant women might welcome the opportunity to participate in clinical trials.

Other pregnant women, however, might nonetheless be reluctant to participate in clinical trials because of perceived risk to their developing foetuses and future children. Consider, for example, a woman who suffers from significant inflammatory bowel disease. Her disease needs to be controlled in order for her to receive adequate nutrition, and subsequently nourish her foetus. Her treatment options include taking her usual medication(s) off-label and hopefully achieving an adequate dose in her body despite the fact that her body has significantly changed in pregnancy, *or* adjusting the dosage on her usual medication(s) in response to the physiological changes that accompany pregnancy, *or* not taking her usual medication(s). None of these options are ideal, and one or more of these options could potentially result in a growth-restricted foetus. Instead, the pregnant woman could be invited to participate in a Phase IV clinical trial where she would use her usual medication(s) in a managed way; the drug levels in her system would be in a therapeutic range; impact on herself and the developing foetus would be carefully monitored; and there would be long-term follow-up of the infant into childhood to look for potential long-term effects. Researchers following the child's progress would have access to information about other children who were exposed to the drug antenatally, and could look for patterns that persisted. Inclusion in such a trial would have clear benefits for the pregnant woman, her developing foetus, and other pregnant women who might benefit from the knowledge gained.

To be clear, off-label use of a medication is not safer (i.e., less risky) than use of the same medication within a clinical trial. In fact, the opposite is true insofar as a medication used within a clinical trial would be formally assessed for safety and effectiveness, the women and developing foetuses would be carefully monitored for adverse events, there would be clear safety parameters and stopping rules, and perhaps most importantly there should be long-term follow-up of infants exposed to medication during the trial. A similar point is made by Kristine Shields and Anne Drapkin Lyerly who note that: "Participation in an ethically designed Phase IV clinical study would be very much like treatment in clinical practice with the additional potential benefits of expanded informed consent, enhanced monitoring, and the patient's knowledge that she has contributed to the evidence base and has benefited other pregnant women" (Shields and Lyerly 2013).

Advocates of research in pregnancy (which should include physicians who treat pregnant women) need to help pregnant women (individually and as a class) to better understand the potential harms of using an over-the-counter or prescription medication in the absence of robust clinical trial evidence regarding toxicity and dosage. These potential harms exist whether the medication is used outside or within a clinical trial. To be very specific about this, if a medication is potentially harmful to the developing foetus (under any circumstances, or in specific dosages, or at particular developmental stages), this fact about the medication does not change because the medication is administered within a clinical trial. Moreover, by participating in a clinical trial it is possible to improve the harm-benefit ratio by securing some of the potential benefits of trial participation described above. These benefits are significant, and they are not available to pregnant women using an over-the-counter or prescription medication off-label.

It is our belief that pregnant women can be helped to overcome the therapeutic misperception that off-label use of a medication is a *bona fide* therapeutic intervention. Moreover, there is reason to believe that at least some pregnant women who come to understand and appreciate the point about medications and geography – namely, that the risk profile of a medication depends upon the medication, not whether it is provided outside or within a clinical trial – are going to want to secure the potential benefits of trial participation as a way to counterbalance the potential harms of a medication otherwise taken off-label. They will appreciate that their fears about the use of medications during pregnancy cannot be allayed by participating in a 'trial of one'.

Now clearly the issue will be different for generally healthy women using over-the-counter medications as compared with women who have an acute or chronic underlying health condition for which they are using prescription medications off-label. It will also be different for women who are pregnant for the first time and women who have experienced one or more pregnancies. As well, it will be different for women in their first trimester and women in their second or third trimester. The point is that pregnant women should be empowered to make reasonable choices for themselves as they weigh the potential harms and benefits of trial participation as compared with the potential harms and benefits of off-label use of a medication.

Having helped pregnant women to better understand the benefits of trial participation, it will be important to meaningfully engage them in identifying research priorities. This exercise can help researchers design and implement clinical trials that first and foremost will address the health priorities of women who are, or who anticipate becoming, pregnant. Establishing research priorities that are relevant and important to the target population will be a key factor motivating their participation. Physicians can assist in this task by clarifying where they most need robust clinical trial evidence in order for them to offer competent care to their pregnant patients.

For example, with clinical trials targeting pregnant women, there might be early attention given to the top ten over-the-counter medications used by pregnant women. Many of these medications, including pseudoephedrine and ondansetron, do not have clinical trial data confirming their safe and effective use in pregnancy. For many (if not most) of these medications, there is clinical trial data on their use in the general population and some retrospective safety data that could be used to design randomised controlled trials. The aims of such trials would be to learn how widely used medications are metabolised in pregnancy, and what their impact is on the health and well-being of pregnant women, their developing foetuses, and future children. Such information would enable physicians to provide pregnant patients with sound information in support of evidence-informed decision-making.

Another possible priority for routine research in pregnancy could be Phase IV clinical trials of prescription medications for chronic health conditions that are widely used by the general population but that could potentially threaten the health and well-being of women during pregnancy and, in turn, potentially threaten the health and well-being of their developing foetuses and future children. These would be clinical trials focused on the health needs of pregnant women who have experience with a medication prior to pregnancy, and who understand their bodies as an ecosystem where health challenges that affect them also affect their developing foetuses.

As with many aspects of health care, particularly in a western context, discrete medical symptoms are often addressed independently of the whole person within whom the symptoms manifest. This approach, while more simplistic (and thus more manageable), is deeply problematic and the problem is compounded when the person is pregnant. With a narrow focus on a pregnant woman's discrete medical symptoms one risks failing to properly attend to the physiological inter-connectedness of the pregnant woman and her foetus. It is important to take account of the fact that a threat to the pregnant woman's health is also a threat to the health of the foetus and future child(ren).

Consider, for example, a woman who suffers from moderate asthma. Her illness is controlled as long as she takes her asthma medications. Upon learning of her pregnancy, however, she adopts what she considers a cautious approach and immediately stops all of her medications, for fear of risk to her foetus. In so doing, she is not alone. Recent research suggests that about 30% of women with asthma will reduce or discontinue their asthma medications in the first trimester (Zetstra-van der Woude et al. 2013). This results in a significant worsening of her asthma, which is compounded by the changes in her respiratory physiology that naturally occur in pregnancy. These changes to her health ultimately result in the developing foetus being deprived of oxygen as it grows, resulting in a small for gestational age infant. In attempting to avoid the potential harms of the asthma medications, this woman would have inadvertently (and unintentionally) increased the harm to her developing foetus.

More generally, pregnant women with a chronic health condition, in consultation with their physicians, can guess at the best course of action with respect to drug use based on available evidence from case reports, adverse event registries, and so on. Alternatively, they can enrol in a clinical trial that aims to answer the research question: Should they continue their pre-pregnancy medication(s) regimen, continue their medication(s) but with a different dosage(s), or discontinue their medication(s)?

2.5 Conclusion

As noted at the outset, some pregnant women take over-the-counter and prescription medications during their pregnancy for which robust clinical trial data regarding use in pregnancy is lacking. These women may receive assurances from their physicians that the medications are 'likely safe even though they haven't been studied in pregnant women.' Other pregnant women do not take any over-the-counter and prescription medications fearing that the medications will harm their developing foetuses and future children. These women may receive assurances from their physicians that 'less is best' – a strategy used by some physicians to deal with their discomfort in prescribing under conditions of uncertainty.

This situation does not serve the best interests of pregnant women, their developing foetuses or future children, nor does it serve the interests of their treating physicians. Pregnant women are entitled to robust clinical trial evidence on the basis of which they can make evidence-informed decisions regarding their care. They can be helped to understand that current practice with respect to the use of off-label medications during pregnancy means that they are effectively participating in clinical 'trials of one', with very poor standards for recognising adverse events. If pregnant women were routinely included in clinical trials, the medications used would be formally assessed for safety and effectiveness, the women and developing foetuses would be carefully monitored for adverse events, there would be clear safety parameters and stopping rules, and perhaps most importantly there should be long-term follow-up of infants exposed to medications during pregnancy. While there are potential harms associated with participating in clinical trials, these can be counterbalanced by potential benefits.

In closing, we join others in insisting on the pressing need for fair, respectful, and responsible clinical trials in pregnancy to better understand and respond to the health needs of pregnant women, their developing foetuses, and their future children.

References

Anderka, M., A.A. Mitchell, C. Louik, M.M. Werler, S. Hernández-Diaz, S.A. Rasmussen, and National Birth Defects Prevention Study. 2012. Medications used to treat nausea and vomiting of pregnancy and the risk of selected birth defects. *Birth Defects Research, Part A: Clinical and Molecular Teratology* 94(1): 22–30.

Ballantyne, A., and W. Rogers. 2016. Pregnancy, vulnerability, and the risk of exploitation in clinical research. In *Clinical research involving pregnant women*, eds. F. Baylis and A. Ballantyne, 139–159. Cham: Springer.

Ballantyne, A., C. Barthow, and K. Wickens. 2016. Research into lifestyle changes in pregnancy. In *Clinical research involving pregnant women*, eds. F. Baylis and A. Ballantyne, 211–225. Cham: Springer.

Baylis, F. 2010. Pregnant women deserve better. *Nature* 465(7299): 689–690.

Baylis, F., and S. Halperin. 2012. Research involving pregnant women: Trials and tribulations. *Clinical Investigation* 2(2): 139–146.

Blehar, M.C., C. Spong, S.F. Grady, L. Sahin, and J.A. Clayton. 2013. Enrolling pregnant women: Issues in clinical research. *Women's Health Issues* 23(1): e39–e45.

Carlin, A., and Z. Alfirivic. 2008. Physiological changes of pregnancy and monitoring. *Best Practice & Research. Clinical Obstetrics & Gynaecology* 22(5): 801–823.

CIOMS (Council for International Organizations of Medical Sciences). 2002. *International ethical guidelines for biomedical research involving human subjects*. Geneva: World Health Organization. http://www.cioms.ch/publications/layout_guide2002.pdf. Accessed 3 Oct 2016.

FDA (US Food and Drug Administration). 2014. *Inside clinical research: Testing medical products in people*. http://www.fda.gov/Drugs/ResourcesForYou/Consumers/ucm143531.htm. Accessed 1 Apr 2016.

Healy, D., and D. Mangin. 2016. Does my bias look big in this? In *Clinical research involving pregnant women*, eds. F. Baylis and A. Ballantyne, 197–208. Cham: Springer.

Kaposy, C. 2016. Presumptive inclusion and legitimate exclusion criteria. In *Clinical research involving pregnant women*, eds. F. Baylis and A. Ballantyne, 51–62. Cham: Springer.

Kukla, R. 2016. Equipoise, uncertainty, and inductive risk in research involving pregnant women. In *Clinical research involving pregnant women*, eds. F. Baylis and A. Ballantyne, 179–196. Cham: Springer.

Langston, L. 2016. Better safe than sorry: Risk, stigma, and research during pregnancy. In *Clinical research involving pregnant women*, eds. F. Baylis and A. Ballantyne, 33–50. Cham: Springer.

Little, M.O., A.D. Lyerly, A.C. Mastroianni, and R.R. Faden. 2016. Ethics and research with pregnant women: Lessons from HIV/AIDS. In *Clinical research involving pregnant women*, eds. F. Baylis and A. Ballantyne, 227–246. Cham: Springer.

Lyerly, A.D., M.O. Little, and R. Faden. 2008. The second wave: Toward responsible inclusion of pregnant women in research. *International Journal of Feminist Approaches to Bioethics* 1(2): 5–22.

Lyerly, A.D., E.E. Namey, B. Gray, G. Swamy, and R. Faden. 2012. Women's views about participating in research while pregnant. *IRB: Ethics and Human Research* 34(4): 1–8.

Macklin, R. 2010. Enrolling pregnant women in biomedical research. *Lancet* 375(9715): 632–633.

Martin, J.A., B.E. Hamilton, S.J. Ventura, M.J.K. Osterman, E.C. Wilson, and T.J. Mathews. 2012. Births: Final data for 2010. *National Vital Statistics Reports* 61: 4–6.

Martin, J.A., B.E. Hamilton, M.J. Osterman, S.C. Curtin, and T.J. Matthews. 2015. Births: Final data for 2013. *National Vital Statistics Reports* 64(1): 1–65.

Miles, A., and M. Loughlin. 2011. Models in the balance: Evidence-based medicine versus evidence-informed individualized care. *Journal of Evaluation in Clinical Practice* 17(4): 531–536.

Mitchell, A.A., S.M. Gilboa, M.M. Werler, K.E. Kelley, C. Louik, S. Hernández Díaz, and National Birth Defects Prevention Study. 2011. Medication use during pregnancy, with particular focus on prescription drugs: 1976–2008. *American Journal of Obstetrics and Gynecology* 205(1): 51.e1–51.e8.

Pasternak, B., H. Svanström, and A. Hviid. 2013. Ondansetron in pregnancy and risk of adverse foetal outcomes. *The New England Journal of Medicine* 368(9): 814–823.

Sackett, D.L., W.M.C. Rosenberg, J.A. Muir Gray, R.B. Haynes, and W.S. Richardson. 1996. Evidence based medicine: What it is and what it isn't. *BMJ* 312(7023): 71–72.

Shields, K.E., and A.D. Lyerly. 2013. Exclusion of pregnant women from industry-sponsored clinical research. *Obstetrics and Gynecology* 122(5): 1077–1081.

Thall Bastow, B.D., and J.L. Holmes. 2016. Teratology and drug use during pregnancy. *Medscape*. http://emedicine.medscape.com/article/260725-overview#aw2aab6b3. Accessed 1 Apr 2016.

van der Zande, I.S.E., R. van der Graaf, J.L. Browne, and J.J.M. van Delden. 2016. Fair inclusion of pregnant women in clinical research: A systematic review of reported reasons for exclusion.

In *Clinical research involving pregnant women*, eds. F. Baylis and A. Ballantyne, 65–94. Cham: Springer.

Werler, M.M., A.A. Mitchell, S. Hernández-Diaz, and M.A. Honein. 2005. Use of over-the-counter medications during pregnancy. *American Journal of Obstetrics and Gynecology* 193(3 Pt 1): 771–777.

Wild, V., and N. Biller-Andorno. 2016. Pregnant women's views about participation in clinical research. In *Clinical research involving pregnant women*, eds. F. Baylis and A. Ballantyne, 119–136. Cham: Springer.

Zetstra-van der Woude, P.A., J.S. Vroegop, H.J. Bos, and L.T. de Jong-van den Berg. 2013. A population analysis of prescriptions for asthma medications during pregnancy. *The Journal of Allergy and Clinical Immunology* 131(3): 711–717.

Chapter 3
Better Safe Than Sorry: Risk, Stigma, and Research During Pregnancy

Lucy Langston

Abstract *Choosing to act within a precautionary paradigm is often the smart choice for pregnant women and for healthcare practitioners and researchers who interact with them. However, during pregnancy precaution is often conflated with inaction. This norm is identified in the literature as 'better safe than sorry/inaction is better than action.' I argue that the origin of this norm can be traced to the thalidomide and DES tragedies that mark the beginning of the stigmatisation of both prescription and over-the-counter pharmaceutical use during pregnancy. Conceptualising pharmaceutical use during pregnancy as 'stigmatised' is important because it helps explain the distorted perception of risk during pregnancy that arises from the norm of inaction. When reluctance to conduct pharmaceutical research during pregnancy is understood in terms of mistaken risk perception, then new tools of risk communication become available to support, critique and evaluate research during pregnancy.*

In the middle of the twentieth century, two major tragedies changed the way that medical science conceived of pharmaceutical use and risk, particularly in relation to pregnancy. It is estimated that between 8,000 and 12,000 children were prenatally affected by exposure to thalidomide during the 1950s (Knightley and Times of London 1979) and, similarly, a 40-fold increase in the risk of cancer has been recorded in women prenatally exposed to diethylstilbestrol (DES) in the 1940s, 1950s and 1960s (Swan 2000). These tragedies were not only a wakeup call regarding the potential dangers of pharmaceutical use during pregnancy, but were also central to the broader development of medical research regulation. The regulatory changes precipitated by the thalidomide and DES tragedies limited who could participate in research, expanded the mandate of regulators such as the US Food and Drug Administration (FDA), and, alongside other biomedical scandals, promoted the development of ever-more extensive and detailed regulations and guidelines for medical research (Dutton 1988).

L. Langston, MA (✉)
Department of Philosophy, McMaster University, Hamilton, Canada
e-mail: lucylangston@gmail.com

F. Baylis, A. Ballantyne (eds.), *Clinical Research Involving Pregnant Women*, Research Ethics Forum 3, DOI 10.1007/978-3-319-26512-4_3

In this chapter, I argue that the thalidomide and DES tragedies also created a stigma around pharmaceutical use during pregnancy. I believe that recognising this stigma and its origin is key to countering a problematic norm associated with pregnancy, whereby precaution has become conflated with inaction. This norm skews peoples' risk perception in favour of inaction and is key to understanding and improving research practices during pregnancy.

I first provide a historical overview of the tragedies. Second, I explore their regulatory legacy and how they contribute to ideas of risk during pregnancy via the notion of stigma. Third, I discuss the contemporary environment of pharmaceutical research during pregnancy and explore the roles of stigma, risk perception and the norm of inaction as precaution within this environment. Fourth and finally, I draw on the literature of risk communication developed for other stigmatised products and technologies to discuss how to improve pharmaceutical research during pregnancy, and combat the problematic norm of inaction as precaution.

3.1 Thalidomide

In 1956, thalidomide was introduced as a general sedative and promoted, in particular, for use as an antiemetic during periods of morning sickness. The pharmaceutical was advertised as extremely safe and impossible to overdose. Indeed, thalidomide was considered so safe that most countries approved it for over-the-counter sale.

Developed during World War II by the German pharmaceutical company Grunenthal, thalidomide (also known as Contergen or Distival) was marketed and distributed under licence worldwide. In countries where thalidomide was not approved by pharmaceutical regulators, such as New Zealand, the United States and Austria, pregnant women obtained limited amounts of thalidomide, within clinical trials, as free samples from physicians, and from overseas travel (van Boxtel et al. 2008). Of note, the information on the basis of which regulatory approval was given was drawn solely from Grunenthal's own claims about the safety and efficacy of thalidomide. Later scrutiny would not only show problems with the quality of their studies, but would also reveal a troubling pattern involving the suppression of reports of side-effects (Knightley and Times of London 1979).

The prescription of thalidomide to pregnant women came to an abrupt stop in 1962. Birth defects retrospectively linked to thalidomide had been reported as early as 1956; however, it took over 5 years for the causal link between thalidomide and the rise of phocomelia (severe limb malformation) to be identified (Knightley and Times of London 1979). Following thalidomide's identification as a teratogen that causes gross malformations and neuropathy, the pharmaceutical was withdrawn from sale in all markets within 12 months. Nonetheless, many sources estimate that thalidomide caused approximately 10,000 neonatal deaths and a significant number of miscarriages (Knightley and Times of London 1979; Silverman 2002). Beyond the significant neonatal mortality rate, worldwide estimates indicate that as a result of taking thalidomide, "~40,000 [pregnant women] developed peripheral neuropathy (numbing of the hands and/or feet); and, ~8,000–12,000 infants were born malformed, of these, ~5,000 survived beyond childhood" (Silverman 2002, 406). Today, the thalidomide tragedy and, to a lesser extent, the DES tragedy, remain the most widely noted examples of adverse reactions to pharmaceuticals, and both tragedies mark the beginning of an awareness of a new risk for pregnant women.

3.2 DES

While the use of DES preceded widespread use of thalidomide, the 20-year lag between ingestion and symptoms meant that its consequences would not be realised until after the thalidomide tragedy had captured worldwide attention.

> *Really? Yes…desPLEX to prevent abortion, miscarriage and premature labor. Recommended for routine prophylaxis in ALL pregnancies… bigger and stronger babies too.*

The above quotation first appeared in an advertisement by the Grant Chemical Company in the June 1957 issue of the *American Journal of Obstetrics and Gynecology* (Dutton 1988). First synthesised in 1938 as an oestrogen mimic, diethylstilbestrol or stilbestrol (DES) led to excitement in the medical community as the first cost-effective and potent synthetic oestrogen. A rapid research agenda was subsequently launched for a wide range of sex hormone disorders (Dutton 1988). By 1941, the benefits of DES were firmly established, with over 257 publications asserting its clinical effectiveness in conditions ranging from acne, gonorrhoea and cancer treatments, to lactation suppression and menopausal disorders (Davis 1940; MacBryde et al. 1940; Dutton 1988). At this time, the husband and wife team of Harvard researchers Olive and George Smith (along with colleague Priscilla White) developed a theory on the relationship between oestrogen and progesterone during pregnancy (Smith and Smith 1937). After confirming the theory in animal models, White successfully used DES to increase foetal survival rates from between 40 and 60 % to over 90 % in a small group of diabetic women (White et al. 1939; White and Hunt 1940, 1943; White 1945, 1949). Building on White's success, Smith alone and in partnership with her husband developed a large-scale trial in the general population, that appeared to find a range of foetal and maternal benefits from DES supplementation for a very broad range of risky pregnancies (Smith 1946; Watkins 1948[1]; Smith and Smith 1941, 1949). Prior to the Smiths' study, DES was often prescribed off-label to improve health during pregnancy. However, on the basis of their findings, in 1947 the FDA officially approved DES to prevent miscarriage (Dutton 1988).

In 1953, the Smiths' research was overturned when William Dieckmann failed to replicate their results in a much larger scale, blinded control study (Dieckmann et al. 1953). Dieckmann's study, however, did not find any harms associated with

[1] The author "Watkins" is Olive Waktins Smith and is identified as "Smith" in other publications cited in this chapter.

DES and it continued to be prescribed for general neonatal well-being (Dieckmann et al. 1953). This assessment of DES changed in 1970 when Herbst, a student of the Smiths, reported on a cluster of six cases of an extremely rare vaginal clear cell adenocarcinoma that occurred in adolescent girls around New England. Upon investigation, Herbst discovered that while none of the mothers of the 32 case control subjects had taken DES, all of the mothers of the young women in the subject group had taken DES during pregnancy. Subsequent work over the next few years – including follow-ups on all of the Smiths' patients and the establishment of a national registry in the United States for vaginal clear cell adenocarcinoma – confirmed the association between prenatal exposure to DES and cancer (Herbst et al. 1971; Herbst 1999). While there was no immediate widespread panic, knowledge and fear of this new form of pregnancy risk slowly spread through the wider population (Swan 2000).[2]

3.3 Regulation

Research regulations and guidelines changed worldwide in response to the thalidomide and DES tragedies. Although the United States was one of the few countries not to have approved thalidomide for sale, it nonetheless reacted quickly to the newly realised threat of pharmaceuticals to pregnancy.[3] As part of the review following the thalidomide tragedy, the US *Drug Efficacy Amendment* of 1962 shifted the burden of proof from regulators to manufacturers to prove both the safety and the efficacy of a pharmaceutical before approval for widespread use. This regulatory change greatly increased the power of the FDA. Similarly, in the United Kingdom, a Committee on the Safety of Drugs was formed in 1963, followed by a voluntary adverse pharmaceutical reaction reporting system in 1964. Similar legislation was also passed in Europe (van Boxtel et al. 2008).

In the 1970s, governance of pharmaceutical research would become even more sophisticated in response to other research ethics scandals such as the Tuskegee and Willowbrook experiments. Almost concurrently, the outcry around the legal struggles of thalidomide survivors for restitutions peaked, and the cancer risks associated with DES became public. This confluence of events created the perception of a

[2] Further adverse effects have subsequently been identified in both male and female, first and second generation, offspring of DES-treated women – including increased cancer risks, and a range of issues that make conceiving and carrying their own pregnancies more difficult (Swan 2000). Monitoring of, and research on, the health of children and grandchildren of DES patients continues today. As a result of this, new health issues continue to be identified and researched.

[3] The United States was one of the few countries to exclude thalidomide and insist on further testing. The FDA's concern was with the incidence of peripheral neuropathy in pregnant women prescribed thalidomide rather than with any potential foetal impacts. The FDA's progress in determining the rate of this side-effect was very slow and in the United States clinical research was still ongoing when the foetal malformations were first made public five years after the FDA was first approached to approve thalidomide (Archer 1979).

scientific process deeply in need of reform, particularly with regard to the rules for research involving vulnerable groups. Tightening of the rules governing clinical research are evident in the 1975 revision of the *Declaration of Helsinki*. This revision not only added a clause stressing that the interests of research participants should prevail over the interests of science and society, but also introduced an extra layer of oversight by an independent review committee to ensure the quality and ethics of all research involving humans (Shephard 1976). At the national level, in 1974, the United States formed the National Commission for the Protection of Human Subjects of Biomedical and Behavioral Research, which produced the *Belmont Report*, and also weighed in on the creation of the 1974 law for the Protection of Human Research Subjects (DHHS 2009). This law specifically included additional regulations for research involving pregnant women (subpart B added in 1975), and later prisoners (subpart C, in 1978) and still later children (subpart D, in 1983). Similarly, the 1977 FDA *Guidelines for Industry* required the exclusion of women of child bearing potential (i.e., all post-pubertal and premenopausal women) from participation in clinical research except at the latest stages of Phase III trials, and only once safety and efficacy were shown in humans and reproductive studies in animals were complete (DHHS 1977).

Taken together, these laws and guidelines increased the protections for human participants in clinical research, but they also left a negative legacy, particularly in US law, due to the grouping of pregnant women alongside prisoners and children under the label ‘vulnerable populations’ (DHHS 1977). The impact of this labelling was to associate pregnant women who, as a group, have a more complicated range of risks facing their participation in research, with prisoners and children – groups that have a reduced capacity to give informed consent for research participation (Levine et al. 2004). The issues that have arisen from labelling pregnant women as vulnerable and with reduced capacity to consent have only recently been examined (Macklin 2003; Coleman 2009; Wild 2012; Rogers and Lange 2013; Schonfeld 2013; see also Ballantyne and Rogers 2016; Johnson 2016). The categorisation of pregnant women as a vulnerable population (on the assumption that they cannot give informed consent to research participation) implies that we ought not to conduct research on them without greater safeguards. For this discussion, the most significant impact of this labelling of pregnant women is that it muddies the distinction between capacity for consent and potential exposure to greater risk.

Regulatory reaction to the thalidomide and DES tragedies was intended to protect vulnerable participants in medical research. However, it also resulted in the near total removal of fertile women from participation in clinical research (Levine et al. 2004). By the 1990s, the downsides of not testing pharmaceuticals and other types of medical interventions in women were becoming apparent (Mastroianni et al. 1994). Since then, practice has begun to shift and testing on women and other subpopulations has become more and more the standard (Levine et al. 2004; Foulkes et al. 2011). While most research guidelines still limit the participation of pregnant women in research, a few such guidelines have begun to presume the eligibility of pregnant women for participation in clinical trials, albeit as a specialised population. For instance, the 2002 Council for International Organizations of Medical

Sciences (CIOMS) guidelines for research involving humans specifically requires that pregnant women be presumed eligible for participation in research (CIOMS 2002, Guideline 17). This represents a complete reversal from previous iterations of CIOMS guidelines. For instance, the 1993 CIOMS guidelines state that "pregnant women should in no circumstances be the subjects of non-clinical research unless the research carries no more than minimal risk to the foetus or nursing infant and the object of the research is to obtain new knowledge about pregnancy or lactation" (1993, Guideline 11). A draft of the 2015 CIOMS revision clarifies and strengthens the 2002 position on research involving pregnant and lactating women.[4] Other recent research guidelines, such as the 2014 Canadian *Tri-Council Policy Statement, Ethical Conduct for Research Involving Humans* (*TCPS2*), go a step further and highlight the risk of exclusion due to "over[ly] protectionist attitudes or practices" and explicitly require a justification for any exclusion of pregnant women from research (Canada 2014). As well, a 2013 Canadian guidance document has been developed to assist clinical researchers and further facilitate the inclusion of pregnant women in clinical trials (Canada 2013). Similarly, as part of the 2020 National Institutes of Health (NIH) strategic plan, the US Office of Research on Women's Health recommends the inclusion of pregnant women in health research, and also includes guidance documents for clinical researchers to facilitate this end (Foulkes et al. 2011).

However, despite these regulatory shifts, in practice ethics review boards still regard pregnancy as "a near-automatic cause for exclusion" (Lyerly et al. 2008; see also Ells and Lyster 2016). Thus, the vast majority of research during pregnancy is limited to registries that track the use of specific pharmaceuticals during pregnancy, primarily to keep track of any potential long-term consequences for the foetus (Reiff-Eldridge et al. 2000; Meador et al. 2008; White et al. 2008). The early guidance documents and regulations (hereafter jointly referred to as guidelines) developed in response to the tragedies, indicate a strength of concern for the safety and well-being of pregnant women. However, the lack of uptake of the more recent revised guidelines – some of which have been in place for over a decade – point to a further issue that needs to be addressed in order for pregnant women to benefit from clinical research. The next section addresses the norm of precaution during pregnancy and examines how, in the wake of the thalidomide and DES tragedies, precaution became conflated with inaction and examines the role of this norm in improving research during pregnancy.

3.4 Risk

The role of industries in generating, shaping, and reinforcing norms, in addition to producing products, is often overlooked. The most obvious method of norm production within the pharmaceutical industry is via the advertising and marketing of

[4] Personal communication with Ruth Macklin April 2 2015.

pharmaceutical products – for instance, the type of advertising medium used, and who is or is not represented within the ad. But, as seen in the previous section on the regulation of research in pregnancy and the norm of inaction as precaution, norm production and shaping also occurs via guidelines, particularly those governing clinical trials. For example, it is normal in many countries for depression and anxiety to be treated primarily with pharmaceuticals instead of psychotherapy, or instead of both pharmaceuticals and psychotherapy (see Healy and Mangin 2016). While the marketing of pharmaceuticals to both clinicians and consumers has a role in producing a preference for pharmaceutical use as the norm, so do clinical trials, whose economic and regulatory systems are set up to make pharmaceutical research more attractive than research into the effectiveness of psychotherapies. Perhaps the tendency to avoid clinical research on pharmaceuticals during pregnancy makes pregnancy one of the few areas where the range of therapeutic options being tested in clinical research is not biased in favour of evaluating pharmaceuticals at the expense of other potential therapies (see Healy and Mangin 2016). Nevertheless, the shifting landscape of guidelines governing research during pregnancy over the last 50 years also points to the role of guidelines in producing research practices that ensure the best possible health outcomes during pregnancy.

Just as with any other aspect of life, evaluating risk and choosing to act with reasonable precaution is a smart choice for pregnant women and for healthcare practitioners and researchers who interact with them. However, because of the consequences of taking thalidomide and DES, during pregnancy, precaution and inaction have become conflated. Thalidomide and DES left a cultural legacy: the enduring belief among both lay people and health care professionals that taking pharmaceuticals during pregnancy is always a risky endeavour, and the prudent, lower risk option is to avoid taking pharmaceuticals wherever possible. The conflation of precaution and inaction has been identified in the literature as the norm, 'better safe than sorry/inaction is better than action'. This norm has a great influence on contemporary beliefs about the advisability of pharmaceutical use while pregnant (Lyerly et al. 2009). The continued lack of pharmaceutical research on pregnant women, in spite of guidelines emphasising the benefits of such research, is indicative of the strength of the cultural norm that inaction is better than action when it comes to using pharmaceuticals during pregnancy.

Inaction is not always a bad medical choice, but when the risks of inaction are constantly underestimated, inaction can work against best practice. A distorted perception of what is safe during pregnancy and, in particular, the tendency to regard inaction as safer than action, can lead researchers and health professionals, as well as pregnant women, to make inaccurate assumptions about the safest and best course of action during pregnancy. According to Lyerly and colleagues, this tendency towards distorted risk perception is widespread, and has "profoundly compromised the evidence base for medical decision making in pregnancy" (Lyerly et al. 2007, 983). For instance, the consequences of uncontrolled asthma, depression, or anxiety during pregnancy can be far riskier for both maternal and foetal health than continuing many medications. Similarly, in situations of acute trauma where maternal and foetal death is a strong possibility, overemphasising the risks of radiological imaging despite studies showing minimal foetal risk could increase harm to pregnant women and their foetuses (Lyerly et al. 2007). It follows that the tendency to view inaction as safer than action (through a lens of distorted risk perception) is something that needs to be countered.

The thalidomide and DES tragedies mark the beginning of the modern iteration of risk perception during pregnancy. This newfound awareness of the potentially vulnerable nature of the pregnant body collided with a long-developing belief in maternal responsibility for good foetal outcomes. In this view, women were expected to manage a range of external and internal risk factors, including appetites and emotions, while also ensuring their experiences while pregnant were positive (Kukla 2005). However, prior to these tragedies, there was no expectation that pregnant women should restrict or change their intake of food, drink, and other consumables such as pharmaceuticals from the regular patterns of consumption that applied to non-pregnant women. For instance, when German clinicians investigating the causes of the phocomelia outbreak asked women what they had ingested during their pregnancies, many affected women failed to mention having taken Contergan (as thalidomide was branded in Germany). When these women were later questioned about why they had not mentioned taking Contergan, many felt the pharmaceutical was "too innocent to mention on the questionnaire" (Taussig 1962, 842). At the time of the thalidomide tragedy, it was also assumed that the placenta, and thus the foetus, was impervious to any pharmaceuticals ingested by the pregnant woman – unless the pharmaceutical actually resulted in death, or was a known abortifacient. Despite findings to the contrary from a few animal studies as early as the 1940s, knowledge that pharmaceuticals could cross the placental membrane was not widely known by researchers who worked with pregnant women (Greek et al. 2011).

Beyond creating an instant heightening of caution with regard to pharmaceutical use during pregnancy, the thalidomide tragedy identified a new risk to be managed during pregnancy: consumables. Since the 1960s, the perception of risk associated with consumption during pregnancy has strengthened, and the list of risky products has increased (Armstrong 2003). In an environment where a new product – food, drink, pharmaceutical – seems to be added to the list of risky products every other week, avoidance is perceived as the safe and sensible option. Today, pregnancy is a time of hyper-vigilance for women who are faced with many prohibitions: *don't do this, don't eat or drink that.* While many recommendations are beneficial to foetal well-being – particularly those that encourage moderate intake or avoidance of foodstuffs prone to foodborne pathogens – when hyper-vigilance extends to a blanket avoidance of pharmaceuticals, the well-being of both the pregnant woman and the foetus may be jeopardised. Of particular concern in North America is the fact that one in three women, either of their own accord or on the advice of a healthcare provider, stop taking anti-depressives or anti-anxiety pharmaceuticals upon becoming pregnant – despite best practice recommendations to the contrary (Lyerly et al. 2009; see also Healy and Mangin 2016).

While the mid-twentieth century pharmaceutical tragedies associated with thalidomide and DES were not solely responsible for producing the narrative of risky consumption during pregnancy, they strongly contributed to the rise of the norm of inaction as precaution (Beck 1992). This norm, however, is at odds with the continued need to treat pregnant women – not just for conditions associated with of pregnancy, but for everyday illnesses of general life. Thus, while the world of clinical research and the guidelines governing such research began to restrict research on pregnant women in an effort to protect them from the newly apparent risks of using pharmaceuticals, the rates of prescription to pregnant women remained high throughout the 1970s and 1980s (Doering and Stewart 1978; Bonati et al. 1990; Donati et al. 2000; Egen-Lappe and Hasford 2004). By the early 1990s, this contradiction between research and treatment practices became apparent. As well, the unique pharmacokinetic features of pregnancy, which complicate any extrapolation of therapeutic dosage for pregnant women from studies on non-pregnant populations, came to be recognised. Nevertheless, the narrative of risk avoidance during pregnancy – inaction as precaution – continued to contribute to the low rate of research during pregnancy.

3.5 Stigma

Stigma associated with the use of pharmaceuticals during pregnancy is part of the legacy of thalidomide and DES. Indeed, viewing the thalidomide and DES tragedies in terms of stigma around pharmaceutical use can help us understand how precaution during pregnancy became conflated with inaction as the safest option (see Little et al. 2016). Understanding how this harmful norm of inaction arose, and comparing the stigma of pharmaceutical use during pregnancy with other stigmatised technologies during pregnancy, may help in developing strategies to overcome some of the reluctance to conduct and participate in pharmaceutical research during pregnancy.

The predominant theoretical model of stigma arises out of the work of Gregory et al. (2001). They argue that perceptions of risk and stigma are closely linked, and they propose a model of stigma whereby a stigma associated with a person, product, technology, or place can distort risk perception in the wake of a catastrophe. This model of stigma emphasises social as well as psychological aspects of stigma that are often overlooked by those focusing on risk communication.

According to Gregory and colleagues there are nine criteria common to phenomena that develop a stigma. All nine criteria are applicable to the thalidomide and DES tragedies. The first criterion is that a stigma is "something that is to be shunned or avoided because it overturns or destroys a positive condition; what was or should be something good is now marked as blemished or tainted" (Gregory et al. 2001, 3). Historically, for an object or person to be stigmatised, it must first be considered a hazard. However, contemporary stigmatisation is more than simple hazard identification, it also involves the benign or good suddenly turning menacing or harmful. Pharmaceuticals are given to improve a person's well-being, but – as in the cases of thalidomide and DES – they sometimes cause harm, the opposite of the intended result. Thus, the thalidomide and DES tragedies meet the first criterion required for a technology or product to develop a stigma: something intended for benefit instead causes damage.

Closely related to the first criterion for stigma is the second, where the abnormal nature of a precipitating event violates or overturns the standard of what is right or natural. The injury or death of innocent people, as in both the thalidomide and DES tragedies, is identified by Gregory and colleagues as an example of the violation of what is right and natural (Gregory et al. 2001).

The production and creation of a visual mark, in particular negative imagery, is a third common factor in the production of stigma. The birth defects associated with thalidomide are a particularly apt illustration of the role of negative imagery in stigma production. The malformations associated with thalidomide are especially vivid, and include the rare and very memorable stunted limb malformation known as phocomelia, which literally translates as 'seal flippers' (Silverman 2002). While DES causes no obvious visual marker and thus does not fulfil this criterion it does not matter for the overall argument: according to Gregory and colleagues, not all criteria must be met in order for there to be stigma (Gregory et al. 2001).

Gregory and colleagues' fourth criterion is that there must be "some critical event, accident or report of a hazardous condition" that "sends a strong signal of abnormal risk" (Gregory et al. 2001, 4). The thalidomide and DES tragedies are critical events in the histories of both pregnancy and pharmaceuticals that still resonate strongly in the cultural memory. For example, phocomelia remains strongly associated with thalidomide in contemporary culture (Fraser 2005; von Glasow 2008). Another reason for the continued cultural significance of both thalidomide and DES is the presence of highly organised national and international survivor networks whose representatives regularly interact with the press weighing-in on social and ethical aspects of bio-scientific debates (Irish Thalidomide Survivors Society 2010; Thalidomide Society 2013; DES Action Groups 2014; The Thalidomide Trust 2015). These networks also remain directly in the biomedical consciousness because of the ongoing health needs of both thalidomide and DES survivors, and because of the DES cohort studies to which almost all offspring of identified DES lineages belong.

The fifth criterion proposed by Gregory and colleagues is that the perception of how a hazard is being managed can contribute to whether a stigma is created. One factor that can contribute to the perception of hazard management is the existence of pre-existing distrust. In a society already primed towards distrust, people judge more harshly and give less leeway when "concerns about competence, conflicts of interest or a failure to apply proper values and precautions" arise (Gregory et al. 2001, 5). The thalidomide and DES tragedies arose in a historical moment that was already primed towards distrust. There were the atrocities conducted by Nazi and Japanese scientists during World War II. As well, in North America, World War II was used to justify greater risks with research subjects with little consideration for the vulnerable status of targeted research populations – institutionalised children and adults, conscientious objectors and soldiers.

With specific reference to thalidomide, it became increasingly apparent during court cases that the manufacturer of thalidomide, Grunenthal, had suppressed knowledge of harmful side-effects. As early as 1959, Grunenthal had received internal warnings from staff about the safety of thalidomide, but did not act on them. When the side-effects of thalidomide were first reported publicly, Grunenthal not only consistently denied these findings, but also tried to discredit the physicians and prevent their articles from being published in the medical literature (Knightley and Times of London 1979). These actions by Grunenthal, the decades of dispute over compensation and apologies involving both Grunenthal and other pharmaceutical companies that were licensed to manufacture thalidomide, and the secrecy around the legal proceedings heightened public concerns about government and healthcare competence and increased distrust in the pharmaceutical companies involved. Today, as new evidence about the efforts of pharmaceutical companies to avoid responsibility and deny the victims justice comes to light, the production of stigma remains strong (Evans 2014).

Gregory and colleagues' sixth criterion is that products which become stigmatised often have had an unequal distribution of harm across populations and/or geographic areas. Thalidomide only affected women who were pregnant, and the children they carried while ingesting this pharmaceutical. Among those who consumed thalidomide there was further uneven distribution of harm insofar as only those who took thalidomide during a specific 10 day window during their pregnancy suffered harmful consequences. In addition to the unequal distribution of harms across populations, there was unequal geographic impact. The consequences of thalidomide occurred primarily in just a few jurisdictions – Germany, the United Kingdom, Canada, and Australia. Similarly, DES had an uneven impact across populations and geographic areas. Only some of the children of women who consumed DES have developed cancers and/or had reproductive difficulties – perpetuating the perception of uneven distribution of harm across populations. Also, DES was more widely prescribed in the United States – where the initial research for its use during pregnancy was conducted – than elsewhere including Australia, New Zealand, Canada, and Europe.

The seventh factor common among those products that develop a stigma is that the initial precipitating event have an impact that is "unbounded in the sense that its magnitude and persistence over time is not well known" (Gregory et al. 2001, 5). Not only is there significant variation in the estimates of the number of people affected by thalidomide, but a wide range of symptoms are possible. Within survivor networks and in popular discussion, there were also ongoing fears of thalidomide having harmful effects into a second generation. It is only in the last decade that the teratogenic mechanism of thalidomide has been identified and these fears have begun to dissipate (Vargesson 2009). Unlike thalidomide, where concerns about second generation impact have been laid to rest, higher rates of cancers and reproductive issues caused by DES have been identified in the second generation (Blatt et al. 2003; Brouwers et al. 2006). Furthermore, the major impact of DES is an increased likelihood of cancer across people's entire lives. Thus, even more so than thalidomide, the consequences of DES can be perceived as unbounded in magnitude and scope. With thalidomide, the harmful effects were visited on one generation and these were evident at the time of birth, or shortly after. In comparison, DES daughters are in their 60s, and still getting cancer at increased rates in comparison to the general population (Swan 2000).

An eighth common feature of stigma identified by Gregory and colleagues is involuntary exposure. While both pharmaceuticals were intentionally given to pregnant women, this was done under the mistaken belief that pharmaceuticals could not cross the placenta and affect the foetus. Thus, foetal exposure was involuntary. Dreadful consequences is the ninth and final feature common to stigmatised technologies or products. The consequences of both thalidomide and DES – death, disfigurement, and cancers – are dreadful outcomes.

In summary, the incidents involving thalidomide and DES fit well within the model of stigma proposed by Gregory and colleagues. They involved products that were intended to benefit, but instead caused harm; violated what was right and natural; included critical hazardous events; included strikingly memorable negative imagery; created the perception of a failure of hazard management; were unequally distributed across populations and geography; had unbounded magnitude and persistence; produced involuntary foetal exposure; and, had consequences that were dreadful.

3.6 Risk Communication

Recognising the stigma around pharmaceutical consumption produced by the thalidomide and DES tragedies is important for current efforts to increase the quality and quantity of research during pregnancy. This is because, when a product or technology is stigmatised, it always produces a distorted perception of the risks around that product or technology thus explaining the rise of the problematic norm of inaction as precaution with regards to pregnancy (Gregory et al. 2001). In applying the model of stigma proposed by Gregory and colleagues to the use of pharmaceuticals during pregnancy, I have situated pregnancy and pharmaceuticals within a wider literature of risk communication that focuses on the effects of stigma on risk perception.

Given the applicability of the stigma model to pharmaceutical use during pregnancy, understanding risk perception and developing effective communication tools is central to improving pharmaceutical research during pregnancy. Integrating the findings from the wider literature on risk perception could save us reinventing the wheel (Pidgeon et al. 2003; Aakko 2004; Breakwell 2000, 2007; Fischhoff 2009). Efforts to educate a range of stakeholders about the balance of risks and potential benefits of various technologies and products have been investigated and evaluated; those hoping to increase research during pregnancy could draw on these findings (Bak 2001; Saba and Messina 2003; Gaskell et al. 2004; Lee et al. 2005; Whitfield et al. 2009). Further, a careful review of the communication tools developed to mitigate risk misperception with pharmaceutical consumption (e.g., Tylenol) and chemical scares (e.g., Alar) may be particularly useful (Rosen 1990; Mitchell 2001).

According to Slovic there are four areas of risk communication where the adverse effects of stigma can be minimised: preventing stigmatising events; reducing perceived risk; reducing the social amplification of stigmatising messages; and reducing the impact of stigma (Slovic 2000). In seeking to address the stigma of pharmaceutical use during pregnancy, there are two reasons to focus on the second and third areas of risk communication. First, preventing stigmatising events (the first area of risk communication) is overly challenging. Second, decreasing the impacts of stigma (the fourth area of risk communication) is the likely result of reducing perceived risk and reducing the social amplification of stigmatising messages (the second and third areas of risk communication) (see Baylis and MacQuarrie 2016; Kukla 2016).

As regards the second area of risk communication, with pharmaceutical use during pregnancy, the issue is not so much reducing perceived risk as improving the accuracy of risk perception so as to correctly represent the risks of both action and inaction. In a more detailed discussion, Kasperson identifies two stages at which the amplification or attenuation of perceived risk can occur: first, via the transfer of information about particular risks; second, in society's response mechanisms, such as how we deal with the social and economic consequences of a hazardous product, technology or event (Kasperson et al. 1988).

Not only do the particular risks of pharmaceutical research during pregnancy need to be clearly communicated but, in planning for future hazards arising from pharmaceutical use and research during pregnancy, clear economic and social response policies need to be developed. This may require broader considerations than the adverse event planning currently required in clinical research. For instance, we could evaluate both historical and ongoing responses to relevant adverse events such as the thalidomide and DES tragedies to examine whether further mitigation work can be done to reduce the stigma around pharmaceutical use during pregnancy, overcome the norm of inaction as a safer option, and thus improve the accuracy of risk perception. Accordingly, education of three key stakeholder groups should be the next step. Drawing on the broader stigma and risk literature, ethics review boards, pregnant women and researchers should all be educated about how risk is often misperceived during pregnancy, and how this misperception often results in the (false) belief that inaction is safer than action. The specific consequences of the tendency towards inaction with regards to clinical research during pregnancy should be clearly articulated.

Educating these stakeholder groups about risk perception is a necessary step in improving the quality and quantity of research during pregnancy. Simply communicating accurate risk information to the parties involved in research during pregnancy will not by itself overcome the tendency to regard inaction as the safer option during pregnancy. This is true because researchers and health professionals themselves are affected by the norm of inaction as precaution (see Baylis and MacQuarrie 2016). Moreover, it is important to be mindful of the fact that, as Flynn and colleagues advise: "Scientific rationality by itself will not substitute [in risk communication with the public]; it is neither the final answer to addressing public concerns nor the only consideration for guiding public policy and the management of hazardous technologies" (Flynn et al. 2001, 308).

3.7 Conclusion

Low rates of pharmaceutical use by women during pregnancy, in part a consequence of limited clinical research in pregnancy, can be detrimental to the health of pregnant women. Identifying and separating out the origin and force of the problematic norm of inaction helps explain why the challenges of involving pregnant women in clinical research have only recently become a serious concern. The effects of disordered risk perception also help to explain the continued lack of progress, despite broad attempts within healthcare to address the issue. The thalidomide and DES tragedies were an important precipitating factor in the formation of a stigma around pharmaceutical use during pregnancy, and these tragedies continue to distort risk perception about what can safely be consumed during pregnancy. In healthcare, the perception of safety is almost as important as actual safety. If people think a technology or product is unsafe, then they simply will not risk engagement, interaction, or ingestion of the technology or product. This was aptly demonstrated with the plunge in apple sales that occurred in the United States in 1989 in response to the scare about the colour-enhancing chemical Alar as a potential cause cancer. Despite scientific evidence to the contrary, the perceived risk of cancer affected the behaviour of consumers (Rosen 1990).

Understanding pharmaceutical use during pregnancy as being stigmatised, helps us to better understand and target, in a more nuanced manner, the continued resistance to having pregnant women participate in clinical research. Second, considering the thalidomide and DES tragedies in terms of stigma, also aids in understanding how the norm of inaction as precaution has become entrenched over time. Understanding that our perception of risk during pregnancy is distorted to favour inaction, underlines the need for effective risk communication strategies to counteract this tendency. Finally, situating pharmaceutical use during pregnancy within a model of stigma creates links to other technologies and products, about which people also have significantly skewed perceptions of risk, and helps us to understand the formation of the contemporary social and psychological dynamics at play with respect to risk perception and pregnancy.

References

Aakko, E. 2004. Risk communication, risk perception, and public health. *Wisconsin Medical Journal* 103(1): 25–27.

Armstrong, E.M. 2003. *Conceiving risk, bearing responsibility: Fetal alcohol syndrome and the diagnosis of moral disorder*. Baltimore: Johns Hopkins University Press.

Bak, H.J. 2001. Education and public attitudes toward science: Implications for the "deficit model" of education and support for science and technology. *Social Science Quarterly* 82(4): 779–795.

Ballantyne, A., and W. Rogers. 2016. Pregnancy, vulnerability, and the risk of exploitation in clinical research. In *Clinical research involving pregnant women*, eds. F. Baylis and A. Ballantyne, 139–159. Cham: Springer.

Baylis, F., and R. MacQuarrie. 2016. Why physicians and women should want pregnant women included in clinical trials. In *Clinical research involving pregnant women*, eds. F. Baylis and A. Ballantyne, 17–31. Cham: Springer.

Beck, U. 1992. *Risk society: Towards a new modernity*, 1st ed. London/Newbury Park: SAGE.

Blatt, J., L. Van Le, T. Weiner, and S. Sailer. 2003. Ovarian carcinoma in an adolescent with transgenerational exposure to Diethylstilbestrol. *Journal of Pediatric Hematology* 25(8): 635–636.

Bonati, M., R. Bortolus, F. Marchetti, M. Romero, and G. Tognoni. 1990. Drug use in pregnancy: An overview of epidemiological (drug utilization) studies. *European Journal of Clinical Pharmacology* 38(4): 325–328.

Breakwell, G.M. 2000. Risk communication: Factors affecting impact. *British Medical Bulletin* 56(1): 110–120.

Breakwell, G.M. 2007. *The psychology of risk*. Cambridge: Cambridge University Press.

Brouwers, M.M., W.F.J. Feitz, L.A.J. Roelofs, L.A.L.M. Kiemeney, R.P.E. de Gier, and N. Roeleveld. 2006. Hypospadias: A transgenerational effect of diethylstilbestrol? *Human Reproduction* 21(3): 666–669.

Canada (Canadian Institutes of Health Research, Natural Sciences and Engineering Council of Canada, and Social Sciences and Humanities Research Council of Canada). 2014. *Tri-Council Policy Statement: Ethical conduct for research involving humans (TCPS2)*. http://www.pre.ethics.gc.ca/pdf/eng/tcps2-2014/TCPS_2_FINAL_Web.pdf. Accessed 3 Oct 2016.

CIOMS (Council for International Organizations of Medical Sciences). 1993. *International ethical guidelines for biomedical research involving human subjects*. Geneva: World Health Organization. http://www.codex.uu.se/texts/international.html. Accessed 23 Apr 2015.

CIOMS (Council for International Organizations of Medical Sciences). 2002. *International ethical guidelines for biomedical research involving human subjects*. Geneva: World Health Organization. http://www.cioms.ch/publications/layout_guide2002.pdf. Accessed 3 Oct 2016.

Coleman, C.H. 2009. Vulnerability as a regulatory category in human subject research. *The Journal of Law, Medicine & Ethics* 37(1): 12–18.

Davis, M.E. 1940. A clinical study of stilbestrol. *American Journal of Obstetrics & Gynecology* 39(6): 938–953.

DES Action Groups. 2014. DES action groups. *Diethylstilbestrol DES: Journal of a DES daughter*. http://diethylstilbestrol.co.uk/des-action-groups/. Accessed 12 Apr 2015.

DHEW (US Department of Health, Education and Welfare, Public Health Service, Food and Drug Administration). 1977. *General considerations for the clinical evaluation of drugs* (HEW/FDA-77-3040).

DHHS (US Department of Health and Human Services). 2009. *Code of Federal Regulations: Title 45, Part 46, Protection of Human Subjects*. http://www.hhs.gov/ohrp/regulations-and-policy/regulations/45-cfr-46/index.html. Accessed 3 Oct 2016.

Dieckmann, W.J., M.E. Davis, L.M. Rynkiewicz, and R.E. Pottinger. 1953. Does the administration of diethylstilbestrol during pregnancy have therapeutic value? *American Journal of Obstetrics and Gynecology* 66(5): 1062–1081.

Doering, P., and D. Stewart. 1978. The extent and character of drug consumption during pregnancy. *JAMA* 239(9): 843–846.

Donati, S., G. Baglio, A. Spinelli, and M.E. Grandolfo. 2000. Drug use in pregnancy among Italian women. *European Journal of Clinical Pharmacology* 56(4): 323–328.

Dutton, D.B. 1988. *Worse than the disease: Pitfalls of medical progress*. Cambridge: Cambridge University Press.

Egen-Lappe, V., and J. Hasford. 2004. Drug prescription in pregnancy: Analysis of a large statutory sickness fund population. *European Journal of Clinical Pharmacology* 60(9): 659–666.

Ells, C., and C. Lyster. 2016. Research ethics review of drug trials targeting medical conditions of pregnant women. In *Clinical research involving pregnant women*, eds. F. Baylis and A. Ballantyne, 95–118. Cham: Springer.

Evans, H. 2014. Thalidomide: How men who blighted lives of thousands evaded justice. *The Guardian*, November 14.

Fischhoff, B. 2009. Risk perception and communication. In *Oxford textbook of public health, Volume 2: The methods of public health*, eds. R. Detels, R. Beaglehole, M.A. Lansing, and M. Gulliford, 940–953. Oxford: Oxford University Press.

Flynn, J., P. Slovic, and H. Kunreuther (eds.). 2001. *Risk, media, and stigma: Understanding public challenges to modern science and technology*. New York: Earthscan.

Foulkes, M.A., C. Grady, C.Y. Spong, A. Bates, and J.A. Clayton. 2011. Clinical research enrolling pregnant women: A workshop summary. *Journal of Women's Health* 20(10): 1429–1432.

Fraser, M. 2005. *Thalidomide!! A Musical.*

Gaskell, G., N. Allum, W. Wagner, N. Kronberger, H. Torgersen, J. Hampel, and J. Bardes. 2004. GM foods and the misperception of risk perception. *Risk Analysis* 24(1): 185–194.

Greek, R., N. Shanks, and M.J. Rice. 2011. The history and implications of testing thalidomide on animals. *The Journal of Philosophy, Science & Law* 11. http://jpsl.org/archives/history-and-implications-testing-thalidomide-animals/. Accessed 20 Apr 2015.

Gregory, R., J. Flynn, and P. Slovic. 2001. Technological stigma. In *Risk, media, and stigma: Understanding public challenges to modern science and technology*, eds. J. Flynn, H. Kunreuther, and P. Slovic, 3–8. London: Earthscan.

Health Canada. 2013. *Considerations for inclusion of women in clinical trials and analysis of sex differences*. http://www.hc-sc.gc.ca/dhp-mps/prodpharma/applic-demande/guide-ld/clini/womct_femec-eng.php. Accessed 12 May 2015.

Healy, D., and D. Mangin. 2016. Does my bias look big in this? In *Clinical research involving pregnant women*, eds. F. Baylis and A. Ballantyne, 197–208. Cham: Springer.

Herbst, A.L. 1999. Diethylstilbestrol and adenocarcinoma of the vagina. *American Journal of Obstetrics and Gynecology* 181(6): 1576–1578.

Herbst, A.L., H. Ulfelder, and D.C. Poskanzer. 1971. Adenocarcinoma of the vagina. *New England Journal of Medicine* 284(15): 878–881.

Irish Thalidomide Survivors Society. 2010. Irish Thalidomide Survivors Society. http://irishthalidomidesurvivorssociety.webs.com/. Accessed 12 Apr 2015.

Johnson, L.S.M. 2016. When hypothetical vulnerability becomes actual: Research participation and the autonomy of pregnant women. In *Clinical research involving pregnant women*, eds. F. Baylis and A. Ballantyne, 161–178. Cham: Springer.

Kasperson, R.E., O. Renn, P. Slovic, H.S. Brown, J. Emel, R. Goble, J.X. Kasperson, and S. Ratick. 1988. The social amplification of risk: A conceptual framework. *Risk Analysis* 8(2): 177–187.

Knightley, P., and The Sunday Times of London. 1979. *Suffer the children: The story of thalidomide*. London: Viking.

Kukla, R. 2005. *Mass hysteria: Medicine, culture, and mothers' bodies*. Oxford: Rowman & Littlefield Publishers.

Kukla, R. 2016. Equipoise, uncertainty, and inductive risk in research involving pregnant women. In *Clinical research involving pregnant women*, eds. F. Baylis and A. Ballantyne, 179–196. Cham: Springer.

Lee, C.-J., D.A. Scheufele, and B.V. Lewenstein. 2005. Public attitudes toward emerging technologies examining the interactive effects of cognitions and affect on public attitudes toward nanotechnology. *Science Communication* 27(2): 240–267.

Levine, C., R. Faden, C. Grady, D. Hammerschmidt, L. Eckenwiler, and J. Sugarman. 2004. The limitations of "vulnerability" as a protection for human research participants. *The American Journal of Bioethics* 4(3): 44–49.

Little, M.O., A.D. Lyerly, A.C. Mastroianni, and R.R. Faden. 2016. Ethics and research with pregnant women: Lessons from HIV/AIDS. In *Clinical research involving pregnant women*, eds. F. Baylis and A. Ballantyne, 227–246. Cham: Springer.

Lyerly, A.D., L.M. Mitchell, E.M. Armstrong, L.H. Harris, R. Kukla, M. Kuppermann, and M.O. Little. 2007. Risks, values, and decision making surrounding pregnancy. *Obstetrics & Gynecology* 109(4): 979–984.

Lyerly, A.D., M.O. Little, and R. Faden. 2008. The second wave: Toward responsible inclusion of pregnant women in research. *International Journal of Feminist Approaches to Bioethics* 1(2): 5–22.

Lyerly, A.D., L.M. Mitchell, E.M. Armstrong, L.H. Harris, R. Kukla, M. Kuppermann, and M.O. Little. 2009. Risk and the pregnant body. *Hastings Center Report* 39(6): 34–42.

MacBryde, C.M., H. Freedman, E. Loeffel, and D. Castrodale. 1940. The synthetic estrogen stilbestrol: Clinical and experimental studies. *Journal of the American Medical Association* 115(6): 440–443.

Macklin, R. 2003. Bioethics, vulnerability, and protection. *Bioethics* 17(5–6): 472–486.

Mastroianni, A., R. Faden, and D. Federman (eds.). 1994. *Women and health research: Ethical and legal issues of including women in clinical studies, Volume 1*. Washington, DC: National Academy Press.

Meador, K.J., P.B. Pennell, C.L. Harden, J.C. Gordon, T. Tomson, P.W. Kaplan, G.L. Holmes, J.A. French, W.A. Hauser, P.G. Wells, J.A. Cramer, and The HOPE Work Group. 2008. Pregnancy registries in epilepsy: A consensus statement on health outcomes. *Neurology* 71(14): 1109–1117.

Mitchell, M. 2001. Impact of external parties on brand name capital: The 1982 Tylenol poisonings and subsequent cases. In *Risk, media, and stigma: Understanding public challenges to modern science and technology*, eds. J. Flynn, P. Slovic, and H. Kunreuther, 203–217. New York: Earthscan.

Pidgeon, N., R.E. Kasperson, and P. Slovic (eds.). 2003. *The social amplification of risk*. Cambridge: Cambridge University Press.

Reiff-Eldridge, R., C.R. Heffner, S.A. Ephross, P.S. Tennis, A.D. White, and E.B. Andrews. 2000. Monitoring pregnancy outcomes after prenatal drug exposure through prospective pregnancy registries: A pharmaceutical company commitment. *American Journal of Obstetrics and Gynecology* 182(1 Pt 1): 159–163.

Rogers, W., and M.M. Lange. 2013. Rethinking the vulnerability of minority populations in research. *American Journal of Public Health* 103(12): 2141–2146.

Rosen, J.D. 1990. Much ado about Alar. *Issues in Science and Technology* 7(1): 85–90.

Saba, A., and F. Messina. 2003. Attitudes towards organic foods and risk/benefit perception associated with pesticides. *Food Quality and Preference* 14(8): 637–645.

Schonfeld, T. 2013. The perils of protection: Vulnerability and women in clinical research. *Theoretical Medicine and Bioethics* 34(3): 189–206.

Shephard, D.A. 1976. The 1975 Declaration of Helsinki and consent. *Canadian Medical Association Journal* 115(12): 1191–1192.

Silverman, W.A. 2002. The schizophrenic career of a "monster drug.". *Pediatrics* 110(2 Pt 1): 404–406.

Slovic, P. 2000. Informing and educating the public about risk. In *The perception of risk*, ed. P. Slovic, 182–198. Abingdon/New York: Earthscan/Routledge.

Smith, O.W. 1946. Increased excretion of pregnanediol in pregnancy from diethylstilbesterol with special reference to the prevention of late pregnancy accidents. *Obstetrical & Gynecological Survey* 1(4): 478–479.

Smith, O.W., and G.V.S. Smith. 1937. Prolan and estrin in the serum and urine of diabetic and nondiabetic women during pregnancy, with especial reference to late pregnancy toxemia. *American Journal of Obstetrics & Gynecology* 33(3): 365–379.

Smith, G.V.S., and O.W. Smith. 1941. Estrogen and progestin metabolism in pregnancy. III. The effect of hormone administration in pre-eclampsia. *The Journal of Clinical Endocrinology & Metabolism* 1(6): 477–484.

Smith, O.W., and G.V.S. Smith. 1949. The influence of diethylstilbestrol on the progress and outcome of pregnancy as based on a comparison of treated with untreated primigravidas. *American Journal of Obstetrics & Gynecology* 58(5): 994–1009.

Swan, S.H. 2000. Intrauterine exposure to diethylstilbestrol: Long-term effects in humans. *APMIS* 108(12): 793–804.

Taussig, H.B. 1962. A study of the German outbreak of phocomelia. *Obstetrical and Gynecological Survey* 17(6): 840–844.

Thalidomide Society. 2013. Thalidomide Society. http://www.thalidomidesociety.org/. Accessed 12 Apr 2015.

The Thalidomide Trust. 2015. The Thalidomide Trust. https://www.thalidomidetrust.org/. Accessed 13 Apr 2015.

Van Boxtel, C.J., B. Santoso, and I.R. Edwards (eds.). 2008. *Drug benefits and risks: International textbook of clinical pharmacology*, 2nd ed. Amsterdam: Ios Press.

Vargesson, N. 2009. Thalidomide-induced limb defects: Resolving a 50-year-old puzzle. *BioEssays* 31(12): 1327–1336.

Von Glasow, N. 2008. *Nobody's perfect*. Palladio Films.

Watkins, O. 1948. Diethylstilbestrol in the prevention and treatment of complications of pregnancy. *American Journal of Obstetrics & Gynecology* 56(5): 821–834.

White, P. 1945. Pregnancy complicating diabetes. *Journal of the American Medical Association* 128(3): 181–182.

White, P. 1949. Pregnancy complicating diabetes. *The American Journal of Medicine, Symposium on Diabetes Mellitus* 7(5): 609–616.

White, P., and H. Hunt. 1940. Prediction and prevention of pregnancy accidents in diabetes. *Journal of the American Medical Association* 115(24): 2039–2040.

White, P., and H. Hunt. 1943. Pregnancy complicating diabetes: A report of clinical results. *The Journal of Clinical Endocrinology & Metabolism* 3(9): 500–511.

White, P., R. Titus, E. Joslin, and H. Hunt. 1939. Prediction and prevention of late pregnancy accidents in diabetes. *American Journal of Medical Sciences* 198(4): 482–491.

White, N.J., R.M. McGready, and F.H. Nosten. 2008. New medicines for tropical diseases in pregnancy: Catch-22. *PLoS Medicine* 5(6): e133.

Whitfield, S.C., E.A. Rosa, A. Dan, and T. Dietz. 2009. The future of nuclear power: Value orientations and risk perception. *Risk Analysis* 29(3): 425–437.

Wild, V. 2012. How are pregnant women vulnerable research participants? *International Journal of Feminist Approaches to Bioethics* 5(2): 82–104.

Chapter 4
Presumptive Inclusion and Legitimate Exclusion Criteria

Chris Kaposy

Abstract *This chapter presents an ethics framework for decisions about whether to exclude pregnant women from a clinical research trial. It begins by articulating several background assumptions about the care of pregnant women in the clinical setting and the involvement of pregnant women in clinical research. The uncontroversial truth of these background assumptions supports the idea that pregnant women should be presumed to be included in clinical research, and that their exclusion requires justification. After making the case for the presumptive inclusion of pregnant women, I outline the ethics framework for the legitimate exclusion of pregnant women from clinical research. This framework consists of nine factors that researchers and research ethics committees should consider when deciding whether to exclude pregnant women. Details about research ethics committee review, the nature of risks in pregnancy, the balance between risk and potential benefit, and the context of clinical care are addressed by the framework.*

This chapter outlines an ethics framework for decision-making about the exclusion of pregnant women from clinical research. I provide a brief argument for the inclusion of pregnant women in clinical research as a default position and then articulate criteria that should be considered when departing from this starting presumption. The framework is informed by a series of background assumptions about health care decisions faced by pregnant women, and about their involvement in clinical research. For the most part, I take these assumptions to be uncontroversial, though two of the assumptions require some clarification. This brief review of my starting assumptions is meant to support the claim that there should be a default position in favour of the inclusion of pregnant women in clinical research. The final section of the chapter presents a complex set of criteria that can assist in decisions about when it is justifiable to exclude pregnant women from clinical research.

C. Kaposy, PhD (✉)
Faculty of Medicine, Memorial University, St. John's, Canada
e-mail: christopher.kaposy@med.mun.ca

F. Baylis, A. Ballantyne (eds.), *Clinical Research Involving Pregnant Women*, Research Ethics Forum 3, DOI 10.1007/978-3-319-26512-4_4

4.1 Background Assumptions

Seven background assumptions inform the proposed ethics framework for decision-making about research involving pregnant women. Many of these assumptions are shared by other authors in this book (see Baylis and Ballantyne 2016), and my articulation of these assumptions is inspired by previous work on this issue by these authors. If one holds these assumptions to be true, it is clear that the exclusion of pregnant women from clinical research requires justification. Some of these background assumptions concern pregnant women in the clinical setting; others concern the involvement of pregnant women in research (Baylis 2013).

Assumptions About the Clinical Setting

1. Clinicians should practice evidence-informed decision-making.
2. Pregnant women are capable of making decisions about their health and well-being.
3. Pregnant women are as entitled as any other patient populations to information and professional advice on the basis of which to make decisions about their health.
4. Pregnant women care about their foetuses and future children (Baylis 2012).

Assumptions About Clinical Research

5. By definition clinical research is a potentially risky activity because it involves the unknown.
6. Risks of harm can often be better managed (and potentially diminished) within rather than outside a clinical trial.
7. Some clinical research is too risky to involve pregnant women, or to involve certain classes of pregnant women (i.e., pregnant women at certain gestational stages).

In my view, these background assumptions are non-controversial, though admittedly two claims warrant further explanation. One such claim is that clinicians should practice evidence-informed decision making. Evidence-informed decisions require evidence provided by research. Within the evidence-informed paradigm, the highest quality information is provided by clinical trials (see Healy and Mangin 2016). Because of the fears associated with including pregnant women in clinical trials, clinicians are often forced to rely on lower-quality information when treating pregnant women, such as information from pre-clinical data, case reports, and the retrospective analysis of data. But these sources of information are not the evidence-informed standard for other patient populations and should not be the standard for pregnant patients (see Healy and Mangin 2016).

The second claim requiring further explanation is that risk of harm can often be better managed within a clinical trial than outside a clinical trial. Exclusion from clinical research does not always achieve the goal of protecting the foetus from harm. Exclusion may modify the risk posed to the foetus, but does not eliminate the risk and in some instances may even increase the risk. For instance, exclusion may expose the foetus to risks associated with non-treatment, or risks associated with treatment in a less-controlled clinical context (see Baylis and MacQuarrie 2016).

4.2 The Default Position

The background assumptions that I have sketched above suggest that, on occasion, there can be scientifically and ethically valid reasons to exclude pregnant women from some clinical research, but that these exclusions should occur *only* when there are good reasons. Insofar as the background assumptions are not idiosyncratic, this conclusion should not be controversial. Nonetheless, pregnant women are under-represented as participants in clinical research. Lyerly, Little, and Faden make the striking claim that "only a dozen medications are approved by the US Food and Drug Administration (FDA) for use during pregnancy" (Lyerly et al. 2008, 7). This under-representation is dangerous for the health of pregnant women, their foetuses, and their future children (Lyerly et al. 2008).

One reason for the problem of under-representation is an overly protectionist mind-set that assumes the involvement of pregnant women in clinical research requires justification. This way of thinking is reflected, for instance, in the US *Common Rule* which places limits on the inclusion of pregnant women in human subjects research (DHHS 2009, 45 CFR 46 Subpart B). Inclusion requires that the research meets ten conditions. This regulatory hurdle can deter researchers from including pregnant women in their studies (Lyerly et al. 2008). A protectionist mind-set is also reflected in work by other authors who specifically address the issue of inclusion and exclusion criteria for pregnant women in research (for example: Chervenak and McCullough 2011; Strong 2011). These authors clearly prioritise the need to protect foetuses and pregnant women from potential research-related harms without recognising the harm done by neglecting research involving this population.

As a remedy to the protectionist mind-set, many authors have recommended that the justificatory burden should be shifted from inclusion to exclusion (Lyerly et al. 2008; Kaposy and Baylis 2011). That is, the default position should be that pregnant women will be included in research unless there is justification for excluding them. The justificatory burden is placed on the exclusion of pregnant women in the *International Ethical Guidelines for Biomedical Research Involving Human Subjects* published by the Council for International Organizations of Medical Sciences (CIOMS). According to this Guideline, "Pregnant women should be presumed to be eligible for participation in biomedical research" (CIOMS 2002, 74).[1]

While Canada's research ethics guidelines have not fully embraced the presumed eligibility of pregnant women, they are nonetheless alert to the dangers of protectionism. Canada's *Tri-Council Policy Statement*, second edition (*TCPS2*) states that

[1] See Macklin (2010) for a discussion of the CIOMS guideline on pregnancy and ambiguities found within this guideline.

pregnant women "shall not be inappropriately excluded from research" (2014, Article 4.3). Exclusion requires "a valid reason" (2014, Article 4.3). *TCPS2* does not articulate any specific exclusion criteria aside from this, but states that ethics review committees should consider foreseeable risks and potential benefits of inclusion and exclusion from research for the pregnant woman, her foetus, and the infant who may result from the pregnancy.

Presented below is a list of criteria that should inform exclusion decisions. This list can be used as a guideline for deciding whether a clinical research study meets the justificatory burden for the exclusion of pregnant women. As Lyerly and her colleagues state, "There are many trials in which that burden may be met" (2008, 18). Some trials are indeed too risky to involve pregnant women. But exclusions should be based on evidence and considered decisions, rather than convenient avoidance of a difficult standard of inclusion. These criteria can guide decisions that result in research that is safe and valuable for pregnant women, which will enable evidence-based care for pregnant women.

Aside from well-founded exclusion criteria, another way to promote safety in clinical research is through trial design. For instance, clinical research trials could build in increased periodic data analysis to detect any early signs of safety failures or lack of efficacy (Kaposy and Lafferty 2012). Another proposal is that Phase I trials involving pregnant women could begin concurrently with Phase III trials of the same intervention that involve the general population (Baylis 2010; Baylis and Halperin 2012). Alternatively, Phase I trials involving pregnant women could be embedded in standard Phase II or Phase III trials with additional safety monitoring for the pregnant research participants (Baylis 2010; Baylis and Halperin 2012). Since this chapter deals only with exclusion criteria, I do not investigate these other elements of trial design. I note, however, that more work needs to be done on research design that promotes the safety of research involving pregnant women.

4.3 Criteria Relevant for Exclusion Decisions

In 2013, Health Canada published a guidance document *Considerations for Inclusion of Women in Clinical Trials and Analysis of Sex Differences*, which supports the *TCPS2* guideline discussed in the previous section. The Health Canada guidance document addresses the inclusion of pregnant women in clinical trials. It states that,

> A decision to enrol pregnant women in a specific trial should be individualized and based on a careful risk/benefit assessment taking into consideration: the nature and severity of the disease; the availability and results of previous nonclinical data on pregnant and non-pregnant animals, and results from clinical data; the availability of alternative therapy/therapies and knowledge about their associated risks; the stage of pregnancy in relation to overall development of the foetus, especially regarding foetal brain development; and the potential for harm to the woman, the foetus or child. (Health Canada 2013)

This guidance document places the burden of justification on the inclusion of pregnant women in clinical trials. It does not adopt the opposite default position of presuming pregnant women eligible for participation in clinical research and requiring justification for their exclusion. Aside from this problem, the document provides a helpful list of factors that should be taken into consideration in decision-making about the exclusion of pregnant women from clinical research. The list is helpful for the formulation of legitimate exclusion criteria since it captures a number of factors that help identify risk, and that account for the nature of risk. Health Canada's list also takes into consideration contextual details about treatment alternatives for pregnant women. Other relevant factors that should be added to this list include: the pregnant woman's choice regarding continuing or terminating the pregnancy (Strong 2012); the risk posed to the individual pregnant woman of *not* participating in the proposed clinical trial; and the likelihood that the drug or intervention under study will be used off-label by pregnant women.

I will elucidate and justify each factor identified on the Health Canada list as well as these additional three factors. These criteria can be used as a frame of reference for identifying instances of legitimate exclusion of pregnant women from clinical research.

Individualised Review

The decision to exclude pregnant women from clinical research should be trial specific. Pregnant women should not be excluded from specific clinical research simply because they are pregnant. Since the uniform exclusion of pregnant women is unjustified, research ethics committees must make decisions about the acceptability of the exclusion of pregnant women based on the harm-benefit profile specific to the proposed clinical trial. Though this sort of review is already common practice within research ethics committees, individualised review entails recognisable challenges. Research ethics committees are faced with multi-dimensional uncertainty about the risks associated with many interventions in pregnancy (see Ells and Lyster 2016).

First of all there is no consensus on the threshold against which different risks in pregnancy can be measured. To give an example of such a threshold in another area of research, in nontherapeutic studies involving children there is the 'minimal risk' standard articulated in a number of research ethics guidelines. For example in Canada's *TCPS2*, children are included among those who may lack capacity to consent for themselves. The participation of this group is generally limited to research entailing only minimal risk (2014, Article 4.6), which is defined as research whose probability and magnitude of possible harms are "no greater than those encountered by participants in those aspects of their everyday life that relate to the research" (Canada 2014, 22). In the United States, the US *Common Rule* (DHHS 2009, 45 CFR 46.102(i)) provides a similar definition of 'minimal risk'. Nontherapeutic paediatric research runs afoul of the *Common Rule* when it exceeds minimal risk, or in some cases a minor increase over minimal risk.

Research ethics committees do not have a similar authoritative threshold to which they can refer when reviewing exclusion decisions in clinical research involving pregnant women. For example, the CIOMS (2002) guideline offers no standard for acceptable risk, and leaves the determination of acceptable risk largely in the hands of pregnant research participants through the informed consent process. In contrast, commentators such as Strong (2011) and Chervenak and McCullough (2011) articulate much stricter standards of acceptable risk.

Secondly, there is a lack of good data about the possible harms of various interventions in pregnancy and the probability of these harms. For example, consider the possible harms associated with an allergy skin test for the purposes of research. Rid et al. (2010) find six potential harms identified in the literature that are associated with allergy skin testing. These potential harms range from skin prick pain to various degrees of allergic reaction and death (Rid et al. 2010). The more catastrophic harms have lower probability. The lack of similar harm data about interventions during pregnancy is caused by the historical avoidance of research in pregnancy (Lyerly et al. 2008). In a study of all drugs approved by the US FDA between 2000 and 2010, researchers found that for 168 of the 172 drugs (97.7 %) teratogenic risk in human pregnancy was 'undetermined' (Adam et al. 2011).

Because of these multiple uncertainties, it would be impossible to develop a formal algorithm for determining whether a research study should exclude pregnant women. Instead, we rely heavily on the judgement of research ethics committees. Because of the uncertainties about the risk threshold and the lack of data about harms in pregnancy, there is legitimate reason to be concerned about the validity of research ethics committee review (Rid et al. 2010). However, the general guidelines provided in this chapter can help diminish the wholesale reliance on the idiosyncratic judgements of research ethics committees.

Nature and Severity of the Disease

The assessment of whether to exclude pregnant women should take the nature and severity of the disease into consideration. If a specific research study concerns an intervention for a disease or condition that commonly affects women, then there is good reason not to exclude pregnant women from participation in the study. Any disease or condition that affects women of reproductive age – such as hypertension, diabetes, and depression – could affect pregnant women. This reason for not excluding pregnant women is weightier if the disease in question has severe effects on pregnant women. Clinicians require evidence on the basis of which pregnant women can be treated. Pregnant women are as entitled to such evidence-based care as any other patients. The exclusion of pregnant women from relevant research studies would deny them and their clinicians the evidence needed for safe and effective treatment. By the same token, if the research concerns an intervention for a disease or condition that typically does not affect pregnant women (like prostate cancer, or

Alzheimer's disease), this may be grounds for excluding this population from participation.

Previous Nonclinical Data on Animals and Results from Clinical Data

The availability and results of previous nonclinical data on pregnant and nonpregnant animals and results from clinical data should factor into the assessment of risk versus potential benefit. Though research is, by definition, a risky activity, there may be ways to diminish risk by attending to the effects of a particular intervention on animals or from other data. National and international research ethics guidelines usually recommend a reliance on previous animal and nonpregnant human studies in order to define the risks associated with research that will likely involve pregnant human participants (CIOMS 2002; DHHS 2009). If these previous studies indicate that the research would be potentially harmful to the pregnant woman, foetus, or future infant, the guidelines typically recommend that pregnant women should be excluded. In particular, if prior studies with pregnant animals or clinical studies with humans indicate a risk of teratogenicity, mutagenicity, or miscarriage of the foetus, or serious health problems for the pregnant woman herself, then this population can typically be excluded from the research.[2]

One possible exception is when the potential research participants who are pregnant are suffering from a very serious or potentially terminal condition. Illness of this nature would typically also imperil the viability of the foetus. In such a scenario, pregnant women need not be excluded from research because of risk to the foetus. It would be unjustified to exclude pregnant women from research because of a pregnancy they would be likely to lose anyway, or if their own life is in danger.

Availability of Alternative Therapy

The availability of an intervention that is the standard of care, and the knowledge about the associated risks of such a standard of care, are relevant for determining whether pregnant women should be excluded. In general, clinical research involving humans is warranted when there is (1) no standard of care for treating the condition in question other than the intervention being studied, or (2) the standard of care is unsatisfactory because of side-effects, access issues, cost, or other reasons,

[2] Note that this dependence on pre-clinical animal studies requires that female animals be used in pre-clinical research. The inclusion of female animals has actually lagged behind the inclusion of female humans in clinical studies (Clayton and Collins 2014). In the US, the NIH is attempting to rectify this harmful deficiency of pre-clinical animal studies through policy changes (Clayton and Collins 2014).

or (3) there is clinical equipoise (defined as honest clinical and stakeholder disagreement about the effectiveness and value of two or more available treatment options) (Freedman 1987).[3] When the condition affects pregnant women, this rationale applies to research that would involve them.

If there is already an acceptable standard of care, and there is good evidence of safety and effectiveness from previous research involving pregnant women to support the standard of care, then this may be justification for excluding pregnant women from a study of an intervention that is not the standard of care.

Stage of Pregnancy in Relation to Overall Development of the Foetus

The risk versus potential benefit assessment should take into account the stage of pregnancy in relation to overall development of the foetus, especially regarding foetal brain development. Some drugs or interventions might pose risks during the development of particular organs or systems but no other risk at later developmental stages. It might be possible to involve pregnant women in such research at later developmental stages in order to minimise risk. The relevant question to pose in the context of some research is not whether pregnant women in general should be excluded, but whether certain sub-groups of pregnant women should be excluded based on factors such as gestational age, while others are included.

Risks to the Pregnant Woman, the Foetus, or Child

Exclusion decisions should consider the risks to the pregnant woman, the foetus, or child. Risks can be short-term or long-term. There may be research-related risks for the pregnant woman herself or the foetus such as miscarriage or premature labour, or longer-term health risks for the child once born. Each of these categories should be taken into account when assessing prior animal and human data for judging whether exclusion is justified (see Kukla 2016).

As I have argued above, there should be an initial presumption in favour of including pregnant women in clinical research absent good reason for excluding them. Data from nonclinical research with pregnant animals and clinical data with pregnant women could provide risk information about health effects on the pregnant woman, foetus, or the future child. Without this kind of data justifying exclusion, pregnant women normally should be included in clinical research subject to their own harm/benefit calculation during the informed consent process.

[3] On the subject of clinical equipoise, Rebecca Kukla (2016) defends a nuanced understanding of equipoise that takes factors other than safety and effectiveness as relevant.

Choice to Continue or Terminate the Pregnancy

The pregnant woman's choice regarding continuing or terminating the pregnancy may be a relevant factor in determining whether they should be excluded. In some cases, pregnant women need not be excluded from research that is risky to the foetus, such as when the pregnant woman has made a firm decision to terminate her pregnancy. Such research is most justified when the foetus has been diagnosed with a condition that is invariably fatal, or in situations in which the research process itself involves the termination of the pregnancy (Strong 2012). Such research is more controversial when there is a possibility that the pregnant woman could change her mind and decide to keep the pregnancy after being involved in clinical research that poses health risks to the foetus or the child if born (Strong 2012; see Harris 2016).

Risks to the Individual Pregnant Woman of Not Participating in Clinical Research

It is important to consider the risks to individual pregnant women of not participating in proposed clinical research. Excluding pregnant women from particular clinical research might deny them the possible benefits of an experimental intervention (Shivakumar et al. 2011). The risks of untreated illness could be great enough to outweigh the risk posed to the foetus by research participation, especially since untreated illness itself poses health risks to the foetus (see Baylis and MacQuarrie 2016).

If we assume that pregnant women are capable of making choices about their own health care, and that pregnant women care about their foetuses, then there should be a strong presumption in favour of allowing pregnant women to give informed consent to research interventions that might benefit them. The routine exclusion of pregnant women from clinical research denies them this opportunity. Furthermore, risks do not disappear when pregnant women are excluded from research. As discussed earlier, such exclusions might drive pregnant women into seeking care that they need in a less-controlled clinical context, or into avoiding care altogether.

It might be difficult to operationalise this consideration in the context of research ethics committee oversight. Such committees might not have information about treatment options outside of the research context. But the presumption of inclusion means that investigators do not have to justify the inclusion of pregnant women – instead, they have to justify exclusion. Investigators should be required to provide written justification for the exclusion of pregnant women on standard research ethics review forms. These justifications should delineate how the potential research harms outweigh the potential research benefits to the pregnant woman, which would force investigators to consider how exclusion would affect these women.

Likelihood That the Intervention Will Be Used by Pregnant Women Even Without Research Evidence

The likelihood that the drug or intervention in the study will be used or needed by the population of pregnant women even without research evidence should affect deliberations about whether to exclude pregnant women. Many classes of drugs are used by pregnant women without research evidence demonstrating safety, and effectiveness (Lyerly et al. 2008; Baylis and Kaposy 2010). If the drug or intervention under study is likely to be used (or is being widely used already) in this population, then there is little justification to exclude pregnant women from studies of the drug or intervention, unless there are reliable prior indications of foetal or maternal risk incommensurate with the likely benefits to the pregnant woman. A similar argument applies if the intervention (such as a vaccine) is already being commonly used among pregnant women to promote the health of the foetus.

An exclusion of pregnant women from studies of interventions needed by pregnant women and likely to be used by them denies clinicians the ability to make evidence-informed decisions and denies pregnant women access to evidence-based care. It is better to expose a small number of pregnant women to the risks of research in a controlled research environment, when more women would otherwise be at risk in a clinical context in which safety and effectiveness are not known.

Some may argue that those who conduct research are not responsible for clinical care standards outside of the research context – that is, that ethical responsibilities in research extend only to research participants. This argument is difficult to accept, however, because the research enterprise is not a closed system. The goal of clinical research is to find cures or improve care in the clinical context. Therefore researchers and their funders are responsible for unjustified exclusions from research that affect the quality of care available to patients, including pregnant women.

4.4 Conclusion

The illegitimate presumptive exclusion of pregnant women from clinical research means that such women are treated off-label and subject to risk in a context where there is limited knowledge directing decision-making in the clinical setting. The inclusion of pregnant women in research is an ethical imperative. As I have argued above, the evidence-informed decision-making standard for pregnant women should be the same as for other patient populations. If the general patient population has access to better evidence for their care than pre-clinical data, case reports, and the retrospective analysis of data, then pregnant women should have access to better evidence as well.

Clinical research takes place in a cultural context in which pregnant women are expected to refrain from all sorts of activities that are perceived as risky. For example, pregnant women are regularly advised to avoid eating sushi and cookie dough, to refrain from scooping the cat's litter box, sitting in the bathtub too long, sleeping in the wrong position, and so forth (Lyerly et al. 2009). Pregnant women who smoke, use street drugs, or drink alcohol are treated as social pariahs. The foetal protectionist impulse behind excluding pregnant women from clinical research is a symptom of this larger cultural context. At its root, the protective impulse is an expression of the fact that people value the health of pregnant women, their foetuses, and their future infants. The downstream effects of foetal protectionism, however, show this impulse to be a perverse and counter-productive expression of value. If one values the health of pregnant women, foetal health, and child health, then research participation is necessary. The treatment of illnesses in these groups requires medical knowledge generated by clinical research.

In many cases, pregnant women can choose for themselves to participate (or not) in a clinical research study when they weigh the harms and benefits for themselves, their foetuses and their future children, and when they look at their available options. If these options include off-label treatment with less-supervised risks versus participation in a trial with data collection for knowledge production and the oversight of risks, the exercise of autonomous choice may lead pregnant women to consent to trial participation. When there is no scientifically and ethically sound reason to exclude pregnant women from such trials, they should be allowed this choice.

Acknowledgments I would like to thank Françoise Baylis for her substantial contributions to this chapter and Angela Ballantyne for her help with revisions.

References

Adam, M.P., J.E. Polifka, and J.M. Friedman. 2011. Evolving knowledge of the teratogenicity of medications in human pregnancy. *American Journal of Medical Genetics Part C: Seminars in Medical Genetics* 157C(3): 175–182.

Baylis, F. 2010. Pregnant women deserve better. *Nature* 465(7299): 689–690.

Baylis, F. 2012. *Research ethics: The obligation to include pregnant women in research*. Science in the city, Bourns Lecture in Bioethics, McMaster University, Hamilton, Ontario, 14 February 2012. https://www.youtube.com/watch?v=XTSRzf60Hlc

Baylis, F. 2013. *Missed trials: Research involving pregnant women*. People's Health Matters, Memorial University, St. John's, Newfoundland, 8 March 2013.

Baylis, F., and A. Ballantyne. 2016. Missed trials, future opportunities. In *Clinical research involving pregnant women*, eds. F. Baylis and A. Ballantyne, 1–13. Cham: Springer.

Baylis, F., and S. Halperin. 2012. Research involving pregnant women: Trials and tribulations. *Clinical Investigation* 2(2): 139–146.

Baylis, F., and C. Kaposy. 2010. Wanted: Inclusive guidelines for research involving pregnant women. *Journal of Obstetrics and Gynaecology Canada* 32(5): 473–476.

Baylis, F., and R. MacQuarrie. 2016. Why physicians and women should want pregnant women included in clinical trials. In *Clinical research involving pregnant women*, eds. F. Baylis and A. Ballantyne, 17–31. Cham: Springer.

Canada (Canadian Institutes of Health Research, Natural Sciences and Engineering Council of Canada, and Social Sciences and Humanities Research Council of Canada). 2014. *Tri-Council Policy Statement: Ethical conduct for research involving humans (TCPS2)*. http://www.pre.ethics.gc.ca/pdf/eng/tcps2-2014/TCPS_2_FINAL_Web.pdf. Accessed 3 Oct 2016.

Chervenak, F.A., and F.B. McCullough. 2011. An ethically justified framework for clinical investigation to benefit pregnant and foetal patients. *American Journal of Bioethics* 11(5): 39–49.

CIOMS (Council for International Organizations of Medical Sciences). 2002. *International ethical guidelines for biomedical research involving human subjects*. Geneva: World Health Organization. http://www.cioms.ch/publications/layout_guide2002.pdf. Accessed 3 Oct 2016.

Clayton, J.A., and F.S. Collins. 2014. Policy: NIH to balance sex in cell and animal studies. *Nature* 509(7500): 282–283.

DHHS (US Department of Health and Human Services). 2009. *Code of Federal Regulations: Title 45, Part 46, Protection of Human Subjects*. http://www.hhs.gov/ohrp/regulations-and-policy/regulations/45-cfr-46/index.html. Accessed 3 Oct 2016.

Ells, C., and C. Lyster. 2016. Research ethics review of drug trials targeting medical conditions of pregnant women. In *Clinical research involving pregnant women*, eds. F. Baylis and A. Ballantyne, 95–118. Cham: Springer.

Freedman, B. 1987. Equipoise and the ethics of clinical research. *New England Journal of Medicine* 317(3): 141–145.

Harris, L.H. 2016. Clinical research involving pregnant women seeking abortion services: United States perspectives. In *Clinical research involving pregnant women*, eds. F. Baylis and A. Ballantyne, 265–284. Cham: Springer.

Health Canada. 2013. *Considerations for inclusion of women in clinical trials and analysis of sex differences*. http://www.hc-sc.gc.ca/dhp-mps/prodpharma/applic-demande/guide-ld/clini/womct_femec-eng.php. Accessed 15 Jan 2015.

Healy, D., and D. Mangin. 2016. Does my bias look big in this? In *Clinical research involving pregnant women*, eds. F. Baylis and A. Ballantyne, 197–208. Cham: Springer.

Kaposy, C., and F. Baylis. 2011. The Common Rule, pregnant women, and research: No need to "rescue" that which should be revised. *American Journal of Bioethics* 11(5): 60–62.

Kaposy, C., and L. Lafferty. 2012. Overcoming liability concerns in vaccine trials involving pregnant women. *Accountability in Research* 19(3): 156–174.

Kukla, R. 2016. Equipoise, uncertainty, and inductive risk in research involving pregnant women. In *Clinical research involving pregnant women*, eds. F. Baylis and A. Ballantyne, 179–196. Cham: Springer.

Lyerly, A.D., M.O. Little, and R. Faden. 2008. The second wave: Toward responsible inclusion of pregnant women in research. *The International Journal of Feminist Approaches to Bioethics* 1(2): 5–22.

Lyerly, A.D., L.M. Mitchell, E.M. Armstrong, L.H. Harris, R. Kukla, M. Kuppermann, and M.O. Little. 2009. Risk and the pregnant body. *The Hastings Center Report* 39(6): 34–42.

Macklin, R. 2010. Enrolling pregnant women in biomedical research. *The Lancet* 375: 632–633.

Rid, A., E.J. Emanuel, and D. Wendler. 2010. Evaluating the risks of clinical research. *Journal of the American Medical Association* 304(13): 1472–1479.

Shivakumar, G., S. Inrig, and J.Z. Sadler. 2011. Community, constituency, and morbidity: Applying Chervenak and McCullough's criteria. *American Journal of Bioethics* 11(5): 57–59.

Strong, C. 2011. How should risks and benefits be balanced in research involving pregnant women and foetuses? *IRB: Ethics and Human Research* 33(6): 1–5.

Strong, C. 2012. Abortion decisions as inclusion and exclusion criteria in research involving pregnant women and foetuses. *Journal of Medical Ethics* 38(1): 43–47.

Part II

Chapter 5
Fair Inclusion of Pregnant Women in Clinical Research: A Systematic Review of Reported Reasons for Exclusion

Indira S.E. van der Zande, Rieke van der Graaf, Joyce L. Browne, and Johannes J.M. van Delden

Abstract *This empirical chapter provides a systematic review of literature relevant to the inclusion of pregnant women in clinical trials. In particular, it addresses barriers to fair inclusion identified within the literature. The 31 articles reviewed discuss the exclusion of pregnant women from clinical trials. Reasons given for such exclusion were grouped under several themes, including: foetal safety, collective memory or social controversies, liability, regulations, research ethics committee interpretations, research design, willingness to participate and consent. The discussion reviews arguments in the literature for how many of these barriers to fair inclusion can be surmounted. The authors find that barriers to fair inclusion of pregnant women in clinical research interact. While there are practical solutions for surmounting some barriers, others require further discussion.*

In the last decade, fair inclusion of pregnant women in clinical research has been widely promoted (Lyerly et al. 2007; Little et al. 2009; Baylis 2010). This is motivated by the need to produce evidence-based knowledge concerning medications that are prescribed to women during pregnancy for both obstetric and non-obstetric illnesses (Shields and Lyerly 2013). Currently, the percentage of pregnant women who take medications – for which there is not substantial data on safety, efficacy, and foetal risk evaluation – may be as high as 84–99 % (EMA 2005; Haas et al. 2011; Lupattelli et al. 2014; Noah 2014). While protection of the foetus is commonly cited as a reason for the exclusion of pregnant women from research, maternal as well as foetal well-being can be promoted by more frequent inclusion of pregnant women in

I.S.E. van der Zande, MA, MSc (✉) • R. van der Graaf, PhD
J.J.M. van Delden, MD, PhD
Julius Center for Health Sciences and Primary Care, Department of Medical Humanities, University Medical Center Utrecht, Utrecht, The Netherlands
e-mail: i.s.e.vanderzande@umcutrecht.nl

J.L. Browne, MD, MSc
Julius Center for Health Sciences and Primary Care, Department of Epidemiology, University Medical Center Utrecht, Utrecht, The Netherlands

F. Baylis, A. Ballantyne (eds.), *Clinical Research Involving Pregnant Women*, Research Ethics Forum 3, DOI 10.1007/978-3-319-26512-4_5

clinical research as this may provide more information on prevention and treatment options (Zajicek and Giacoia 2007; Little et al. 2009; Frew et al. 2014). Lack of a sound evidence base leads to suboptimal care or even under-treatment of pregnant women (see Baylis and MacQuarrie 2016). Poorly treated asthma, for example, places pregnant women at higher risk of hypertension, preeclampsia, and uterine haemorrhage. As well, asthma is associated with foetal growth restriction, premature birth, and low birth weight (Little et al. 2009). In contrast, when asthma is well-controlled by medication, maternal and perinatal outcomes are as good as comparable groups without asthma (Tan and Thomson 2000). Creating a more solid evidence base can lead to consensus in treatment guidelines, and ultimately result in better health outcomes for pregnant women and their foetuses.

The research community has not ignored the call for fair inclusion of pregnant women in clinical research, and there have been various efforts to take on the challenge. In the United States, the Office of Research on Women's Health (ORWH) of the Department of Health and Human Services (DHHS) has endorsed the view that pregnant women are to be presumed eligible for participation in clinical research (1994). This view was later adopted by the Council for International Organizations of Medical Sciences (CIOMS) in its *Ethical Guidelines for Biomedical Research Involving Human Subjects* (2002). And in 2009, the Second Wave Initiative was launched – a collaborative academic initiative to find ethically and scientifically responsible means to increase the knowledge base for the treatment of pregnant women with medical illness (Lyerly et al. 2007; Second Wave Initiative 2016).

Despite multiple attempts to challenge the status quo, the under-representation and exclusion of pregnant women in clinical research remains common practice (Cain et al. 2000; Shields and Lyerly 2013). Many people have hypothesised about the reasons for the current situation, often with a focus on the diethylstilboestrol (DES) and thalidomide tragedies. From 1938 to 1971, DES was prescribed to an estimated 1.5–3 million women during pregnancy to prevent miscarriage. Only in 1971 was it realised that the drug did not prevent miscarriage and was linked to several adverse complications for the offspring, including vaginal and cervical carcinomas in young women, and malformation of reproductive organs in both male and female children (Swan 2000; Allesee and Gallagher 2011). In the late 1950s, thalidomide was prescribed for nausea during pregnancy without prior testing in pregnant women, and resulted in unforeseen teratogenic effects with severe birth defects in over 10,000 children (Macklin 2010). These tragic events had a great impact on the research community, even though neither tragedy involved clinical research (see Langston 2016). Although the memory of the events that took place over 40 years ago likely contributes to the exclusion of pregnant women from clinical research today, additional barriers to fair inclusion may be at play.

Understanding the barriers to fair inclusion of pregnant women in clinical research (i.e., understanding the putative reasons for the exclusion of pregnant women from clinical research), and the way in which these barriers intersect is important relative to the goal of promoting fair inclusion. With this systematic review, we first identify the barriers to fair inclusion. We then briefly discuss those barriers that, in our estimation, can easily be addressed. Other barriers to fair inclusion, such as those that relate to the level of acceptable research risk for pregnant women, and the protection owed to alleged vulnerable populations are not easily addressed, and for this reason are not discussed in this chapter (see Kukla 2016).

5.1 Design

We conducted a systematic review of reasons for the exclusion of pregnant women loosely based on the review of reasons as developed by Strech and Sofaer (2012) and the thematic synthesis methods for the categorisation of reasons (Barnett and Thomas 2009). Sofaer and Strech incorporate the Preferred Reporting Items for Systematic Reviews and Meta-Analyses (PRISMA) statement and allow for analytical strategies that are typically used in qualitative research (Moher et al. 2009). Instead of the comprehensive approach for categorisation of reasons we relied on the thematic synthesis of Barnett et al., because it is helpful to identify key themes among different article types.

Search Strategy

A search of PubMed, EMBASE, and Philosopher's Index was conducted to identify relevant articles in May 2015. These databases were selected for their comprehensive coverage of biomedical and ethics research journals and articles. Additional articles were retrieved through cross-referencing by way of manually searching the reference lists. A broad search strategy that included the following keywords was applied: ((pregnan* OR expecting wom*) AND (research)) AND ((challeng* OR reason* OR motivation* OR view* OR decision*OR attitude* OR willing* OR consideration* OR concern* OR barrier* OR issue*) AND (exclu*)). Table 5.1 contains the detailed search strategy.

Study Selection and Inclusion Criteria

One researcher (IvdZ) independently reviewed all titles and/or abstracts to select articles eligible for review, while a second reviewer (JB) subsequently checked a sample from the PubMed results for consistency (n=55; 8%). Articles were included in which the exclusion of pregnant women from clinical research was a specific topic or aspect discussed, determined on the basis of references to the topic in either the title or the abstract. Articles from which it was apparent from the title that the content was out of the research question's scope, were excluded. When this could not be determined based on the title, the abstract was consulted. We excluded articles that were not in English, only reported on primary research reports of trials, or did not include pregnant research participants.

Data Extraction and Analysis

Our first strategy was to collect the contextual data of the included articles, such as the aim and scope, the country of origin, and the article type. We categorised each article as a: (i) systematic review, (ii) qualitative analysis, (iii) case study/ies, (iv) reasoned opinion, or (v) consensus document, where a reasoned opinion is an article written in an argumentative style and a consensus document is an article of the same type issued by an organisation or institution. We then collated all of the reasons for the exclusion of pregnant women from clinical research mentioned in the articles, and categorised them into themes determined by consensus within our study team.

5.2 Results

Search and Selection

After removing duplicate references, we screened 669 unique references on the basis of the title and the abstract. Subsequently, 63 articles were assessed in full text, of which 38 met the inclusion criteria. After further assessment for eligibility, seven articles were excluded because they did not provide specific reasons for the exclusion of pregnant women from clinical research. Consequently, 31 articles were included in the final review (see Table 5.2, the PRISMA flow diagram).

Study Characteristics

Table 5.3 summarises the characteristics of included articles and the reasons given in these articles for the exclusion of pregnant women. The majority of the articles originate from North America, especially from the United States (n = 22). Most of the articles in the review are reasoned opinions (n = 22). Others are systematic reviews (n = 4), consensus documents (n = 3), a case study (n = 1) and a qualitative analysis (n = 1).

Synthesis of the Reasons for Exclusion

Table 5.4 provides an overview of reported reasons for exclusion of pregnant women in clinical research as identified in the articles. This includes articles that explored a single reason for exclusion (for example, foetal protection is the only reason for exclusion in: Schonfeld and Gordon 2005; Beran 2006 and Goldkind et al. 2010), as well as articles that mentioned multiple reasons. There was considerable consistency among the reported reasons and we were able to identify nine discrete themes: foetal safety (n=22), liability (n=15), regulations/wording (n=15), research design (n=13), institutional review board (IRB) interpretation (n=9), collective memory/social controversies (n=7), willingness to participate (n=7), vulnerability (n=4), and consent (n=4). We then clustered closely related themes into four groups (see Table 5.5). These groups are briefly described below.

Foetal Safety, Collective Memory/Social Controversies, and Liability

Protecting the foetus from harm was most frequently cited as a reason for exclusion (n=22). Since medications can cross the placenta, this can affect the foetus with possibly profound implications, leading to reluctance on the part of many to expose the foetus to clinical research (Lyerly et al. 2007; Frew et al. 2014). One article mentioned that concern about research involving pregnant women was complicated by social controversies in the United States that influenced the research community, such as the abortion debate in the 1970s which highlighted maternal-foetal conflicts (Alexander in ORWH 2011). More frequently, however, there was mention of the collective memory of several historical tragedies, primarily DES and thalidomide (n=6). In addition to the direct catastrophic health outcomes with these drugs, there were a large number of liability claims resulting in huge financial losses for manufacturers. Many stakeholders, including manufacturers, research ethics review committees, sponsors and researchers, were mentioned as among those worried about legal liability claims (n=15), possibly explaining the number of times liability was identified as a reason for exclusion.

Regulations/Wording and IRB Interpretation

Research regulations and guidelines (hereafter collectively referred to as regulations)[1] governing clinical research with pregnant women are among the reasons given for the exclusion of pregnant women (n=15). A difference can be found between reasons relating to the actual content and meaning of the regulations (n=9) and reasons relating to the wording, i.e. the comprehensiveness and phrasing of the regulations (n=6). International and national guidelines provided a mixed picture. For example, the *Declaration of Helsinki* (applicable in most European countries), does not make any reference to research involving pregnant women (World Medical Association 2013). Meanwhile, according to the CIOMS guidelines, pregnant

[1] Because most of the articles originate from the United States where there are regulations, the term 'regulations' is used to refer to both regulations and guidelines except when there is specific reference to an identifiable guideline.

women are presumed eligible for research participation (CIOMS 2002). According to the *Common Rule* in the United States, pregnant women are a vulnerable population. In Canada, the *Tri-Council Policy Statement* (*TCPS2)* says "researchers and REBs [Research Ethics Boards] shall take into account foreseeable risks and potential benefits for the woman and her embryo, fetus or infant" (Canada 2014). Furthermore, according to some authors, if clinical research in pregnant women is mentioned in regulations, the wording is sometimes restrictive and sometimes vague (Levine 2011; Macklin 2010; Foulkes et al. 2011). Some authors also noted that since most research regulations do not require researchers to justify the exclusion pregnant women, it is simpler not to include pregnant women in clinical research (Lyerly et al. 2007; Baylis and Kaposy 2010).

While research regulations establish relevant norms, research ethics review committees are entrusted to apply them. Since the articles included in this review are mostly about the United States, the comments below focus on Institutional Review Boards (IRBs). Several authors pointed out that IRBs and their members vary in their interpretation of the federal regulations and relevant international guidelines, or interpret these in a conservative or overcautious manner (n=4) (Levine 2011: i.e. the Secretary Advisory Committee on Human Subject Research; IOM). In addition, some authors noted a tendency for IRBs to almost automatically exclude pregnant women or impose criteria limiting participation in clinical research (n=4). The practice of exclusion occurred even in studies where there were no additional risks for the foetus and there were costs associated with the exclusion for both the women and their foetuses. One article identified IRBs as the gatekeepers for access to research (Lyerly et al. 2008). Lastly, one article mentioned documentation required by IRBs as a bureaucratic barrier to research involving pregnant women without providing further explanation.

Research Design

There are unique design challenges with research involving pregnant women that are not experienced with the design of research involving other populations. Among these challenges are research set-up, the recruitment of research participants, and the use of placebo-controlled designs (n=4). Another reason cited for the routine exclusion of pregnant women from research was their alleged physiological complexity. Physiological changes that occur during pregnancy can potentially alter a drug's pharmacokinetics and pharmacodynamics, and make clinical research within this population more difficult (n=4). As an example, pregnancy research may require greater numbers of research participants across gestational ages to clearly identify and define optimal treatment regimens (Blehar et al. 2013). A third reason found to contribute to the difficulty of designing clinical research with pregnant women related to the costs of research (n=3). As with all population-specific research, data can only be gathered through additional research, which makes conducting research in pregnancy more expensive. Since pregnant women only make up a small percentage of the population likely to use a certain medication, and possibly only take the medication during the nine months of pregnancy, pharmaceutical companies lack financial incentives to investigate the safety of drugs in pregnant women or develop post-marketing studies (Theiler 2009). Lastly, one article mentioned that pregnant women were excluded from research either due to the low prevalence of the condition under study in pregnant women (n = 1) or because the researchers wanted to control for risk (n = 1). Regarding the latter, upon asking the researchers about their motivations, the authors found that researchers excluded pregnant women because the drug included in the research had not been approved for use in pregnant women (because of a lack of safety data).

Willingness to Participate, Vulnerability, and Consent

The last three themes that emerged were willingness to participate (n = 7), vulnerability (n = 4), and consent (n = 4). Concerns relating to the recruitment of participants for clinical research had to do with both the presumed unwillingness of pregnant women to participate in research (n = 5), as well as the unwillingness of clinicians to enrol them in research (n = 2). It appeared that clinicians' willingness to promote research to their pregnant patients was hampered by a lack of resources and time constraints, among other things (Brandon et al. 2014; Madan et al. 2014).

As concerns the themes of vulnerability and consent, the exclusion of pregnant women from clinical research resulted from their classification as a vulnerable population in regulations (n = 2). Nevertheless, some articles mentioned that the concept of vulnerability had shifted over time, and noted that vulnerability was primarily a historical reason for exclusion (n = 2). Finally, the fact that a foetus is legally unable to consent was cited as another potential reason for exclusion (n = 3).

5.3 Discussion

Foetal Safety, Collective Memory/Social Controversies, and Liability

Unsurprisingly, the most frequently mentioned reason for the exclusion of pregnant women from research related to the potential harm to foetuses. Although the 1970s social controversy surrounding the abortion debate has lessened in most jurisdictions (except perhaps the United States), our collective memory of the DES and thalidomide tragedies remains, and this has had an impact on clinical research (see Langston 2016). Changing the current perception of research as an unacceptably risky activity may be particularly difficult. Nevertheless, highlighting dangers to the foetus from routine interventions for which safety evidence is lacking could be an effective way to address this barrier. In addition, advancements in research technologies may contribute to decreasing certain risks for foetuses which in turn might shift the assumption that the best way to protect the foetus from harm is to exclude pregnant women from clinical research. To illustrate this point, placental perfusion experiments can be used to predict placental drug transfer and could facilitate the assessment of the risks and potential benefits of drug therapy in pregnancy in the pre-clinical phases of research (Myren et al. 2007; Hutson et al. 2011). An underlying unresolved ethical issue is what counts as an acceptable level of risk for the foetus in clinical research with pregnant women. Certainly, foetal safety should always be considered when conducting research with pregnant women; however, with realistic assessment of the risks, this barrier to research participation need not be as solid as it is often portrayed.

With regards to clinical research in the United States there is fear regarding potential liability claims since all who are involved in the design of clinical research in pregnant women could potentially be sued under tort law if foetuses are injured as a result of research participation (Merkatz 1998; Clayton in Mastroianni et al. 1999; Levine 2011). The risk of legal liability notwithstanding, the likelihood that anyone would be held liable is actually fairly low (which explains the limited existing litigation) (Merton 1993; Clayton 1999; Frew et al. 2014). Demonstrating the predicted low occurrence of tort liability claims could be a first step towards overcoming this obstacle to research participation. There could be other solutions, however. For example, Chris Kaposy and Lorraine Lafferty note that in both the United States with the 2011 H1N1 influenza vaccine trial, and in Canada with the Pertussis Maternal Immunization Study, the manufacturers were protected from tort liability. In the United States this was through the US Federal Government's *Public Readiness and Vaccine Trials Involving Pregnant Women Emergency Preparedness Act*. In Canada, liability was shifted to the research institute and its insurance providers (Kaposy and Lafferty 2012). The authors propose that such strategies could be extended to enhance further clinical research with pregnant women. Additionally, Wendy Mariner proposes that in some instances tort law could be avoided by introducing a compensation system where responsibility for research injuries is shifted from the manufacturer to society as a whole. She argues that, since research participants take on risks for society's sake, in return society has a moral obligation to compensate those who are harmed (Mariner 1999). In short, even though legally liability is a very real concern, there might be ways to work around this barrier.

Regulations/Wording and IRB Interpretation

While liability concerns may influence IRBs (and research ethics review committees more generally) in their overcautious interpretation of regulations (Levine 2011; Sahin 2011; Blehar et al. 2013; Shields and Lyerly 2013), providing training and guidance for IRB members on the harms of exclusion and possible liability risks because of *exclusion* might help to increase fair inclusion (Merton 1993; Flannery and Greenberg in Mastroianni et al. 1999; Lyerly et al. 2007; Blehar et al.

2013). Further clarification of the regulations and explications of certain wording is needed in order to facilitate practical implementation. For example, despite the fact that several ethicists have tried to define minimal risk (Binik et al. 2011; Strong 2011), the wording remains ambiguous and there is no consensus on how to weigh risks and potential benefits. A workable notion of what constitutes an acceptable risk for the foetus or the pregnant woman is needed. Currently, as a result of ambiguity concerning acceptable research risks, pregnant women have been excluded from clinical research that did not pose any risks or where risks were negligible. Consider, for example, observational research or research on physiologic processes involving FDA approved drugs already used by pregnant women (i.e., Cain et al. 2000; Little et al. 2009; Westreich et al. 2013; see also Ells and Lyster 2016).

Another proposal concerns a change in the language of regulations: not leaving inclusion of pregnant women in clinical research optional, but instead requiring a justification for the exclusion of pregnant women from clinical research (see Kaposy 2016). Such a formulation would not only take away the perception that pregnancy is always a reason for exclusion, or that pregnant women should simply be ignored in clinical research, but would also ensure that potential benefits are distributed more equally (Lyerly et al. 2007; Little et al. 2009; Baylis 2010). The idea of requiring justification for the exclusion of pregnant women from research is grounded in a notion of justice as equity or as corrective justice (van der Graaf et al. 2013).

Justice as equity calls for equal treatment and precludes exclusion for arbitrary reasons. On this view, pregnant women should be included in clinical research in the same way as other populations. According to corrective justice, justifying exclusion is essential to restore differences between trial populations. To illustrate, the lack of research on the pharmacokinetics and pharmacodynamics of medications in pregnancy has a negative impact on the health of pregnant women and their foetuses which results in class injustice for this particular group. Considered from a corrective justice point of view, one could, for example, require the prioritisation of pregnant women in clinical research until they are more equally represented. However, there are methodological limitations to the routine inclusion of pregnant women in clinical research. For instance, when it is unknown whether an intervention's effect differs between pregnant women and non-pregnant women, the inclusion of pregnant women in a clinical trial in which this intervention is tested should have a favourable harm-benefit ratio that is either proportional or substantial in order to be methodologically meaningful. Thus, justice as equity and corrective justice will not necessarily make inclusion of pregnant women as a study population more fair (van der Graaf et al. 2013).

Research Design

There is an urgent need for clinical research on safety, efficacy, and dosing of medications that pregnant women take either due to chronic medical conditions, or because of acute pregnancy problems. As long as the risks for the pregnant woman and her foetus are acceptable, which is implied by the use of specific medications in clinical treatment, it is imperative that there be research in the population that is actually taking the medications, i.e., in pregnant women (Zajicek and Giacoia 2007). Besides, numerous studies with pregnant women in randomised clinical trials and observational studies demonstrate that the perceived barriers of costs and physiological complexity in clinical research with pregnant women can be overcome (Foulkes et al. 2011). In addition, innovative research designs, such as specialised cohort registries, may be able to strike a favourable balance between minimising the risks and burdens of research procedures and interventions, while maintaining scientific validity (Baylis 2010). Moreover, the inclusion of pregnant women in Phase IV clinical trials could increase the knowledge base on the risks and potential benefits of certain medications (Briggs et al. 2015). Systematically collecting data from post-marketing studies is the least that can be done to enhance evidence-based medicine for pregnant women (Little et al. 2009; Baylis 2010). Next, in order to determine whether the recruitment of pregnant women and the motivation of clinicians to enrol them in clinical research constitute an actual barrier for which possible solutions might be found, more information is needed. This can only be established by adding to the existing empirical data on the views of pregnant women and clinicians (see Wild and Biller-Adorno 2016).

Willingness to Participate, Vulnerability, and Consent

Vulnerability is generally on the agenda in relation to the exclusion of pregnant women from clinical research (DHHS 2009; Foulkes et al. 2011; Blehar et al. 2013). For a long time, the concept has been connected to the capacity to give informed consent and to the anticipated exposure to potential harm. Obviously, pregnant women are capable of decision-making and not automatically vulnerable in this aspect (Levine et al. 2004; ACOG 2007; Hurst 2008; Luna and Van der Poel 2013). However, potential exposure to the harms of research cannot be negated and risks must be taken into account when talking about vulnerability. As such, vulnerability may play a more implicit role, primarily conceived of as risks for the foetus which we found to be the most frequently cited reason for exclusion. Since we can express risks through risk-benefit assessments, the classification of pregnant women as a vulnerable group might no longer be needed. Indeed, some authors have challenged the utility of traditional uses of the concept of vulnerability and argued that it needs to be reconceptualised in order to regain its usefulness as a concept in relation to pregnant women in clinical research (Macklin 2003; Levine et al. 2004; Hurst 2008; Luna 2009; Schroeder and Gefenas 2009; see also Ballantyne and Rogers 2016). This may explain why vulnerability was only mentioned four times in our review of the literature.

Although pregnant women are capable of giving informed consent, foetuses are not and this inability to give consent was another area where barriers to fair inclusion were mentioned (Godlovitch 2003; Lyerly et al. 2008). This relates in particular to the moral status of the foetus. With research involving pregnant women, the interests of the pregnant woman might be in conflict with the interests of the foetus, and whose interests should prevail depends on whether the foetus has independent moral status. There is scarce literature on conflicting maternal-foetal interests in clinical research. Some regard the foetus as a patient (McCullough and Chervenak 2011), while others consider the foetus a research participant (Lyerly et al. 2011). In addition, some regard the interests of the pregnant woman and the foetus as distinct (Macklin 1990; Chervenak and McCullough 2011; Lyerly et al. 2011), whereas others label the conflict between the pregnant woman and the foetus as a false dichotomy since research participation can benefit both the pregnant woman and the foetus (Blehar et al. 2013). To better understand and evaluate the potential conflict between pregnant women and their foetuses, the moral status of the foetus needs to be clarified, indicating the duties that various stakeholders have.

5.4 Limitations

This systematic review has some limitations. First, there is no tool available to perform an adequate quality assessment of the different reasons for exclusion. Therefore, we were unable to determine whether the most frequently mentioned reasons for the exclusion of pregnant women from clinical research correspond to the strongest arguments in support of this view. In addition, the systematic review primarily included articles from North America, probably depicting a narrow scope of the issues. We tried to increase the number of articles by authors outside of North America by conducting a small search on English-language articles authored by Europeans, however, we were not successful in identifying any additional sources. Finally, since we were specifically looking for articles in which the reasons for exclusion of pregnant women from clinical research was a major subject (and not a mere mention as part of exclusion criteria in a trial), we chose to exclude a large number of articles based on title and abstract. In part, our ability to do so is a reflection of a broad search strategy in which the term 'research' as a keyword was included instead of more narrow synonyms like 'study' or 'trial' or 'method*' in the two biomedical databases (PubMed and EMBASE) that we searched. As such, it is possible that we might have excluded relevant articles.

5.5 Conclusions

The systematic review of reasons for the exclusion of pregnant women from clinical research indicates that there are a number of interacting barriers hindering the fair inclusion of pregnant women. These include issues surrounding foetal safety, collective memory/social controversies and liability; ambiguity regarding regulations/wording and IRB interpretation; the unique challenges of research design; and questions concerning the willingness to participate, vulnerability and consent. While there are practical solutions to some of these barriers, there are also a number of barriers that need further discussion. In particular, barriers associated with claims/concerns about acceptable levels of risks, and claims about vulnerability of pregnant women remain important ethical challenges (see Ballantyne and Rogers 2016).

Appendix

Table 5.1 Searches

PUBMED		
Date of search: May 10th, 2015		
Search	Terms	Hits
1	((((((((((challeng*[Title/Abstract]) OR reason*[Title/Abstract]) OR motivation*[Title/Abstract]) OR view*[Title/Abstract]) OR decision*[Title/Abstract]) OR attitude*[Title/Abstract]) OR willing*[Title/Abstract]) OR consideration*[Title/Abstract]) OR concern*[Title/Abstract]) OR barrier*[Title/Abstract]) OR issue*[Title/Abstract]	2,312,595
2	exclu*[Title/Abstract]	380,158
3	1 AND 2	60,360
4	((pregnan*[Title/Abstract]) OR expecting wom*[Title/Abstract] AND research* [Title/Abstract]))	19,792
5	3 AND 4	387
EMBASE		
Date of search: May 18th, 2015		
Search	Terms	Hits
1	(challenge*:ab,ti OR reason*:ab,ti OR motivation*:ab,ti OR view*:ab,ti OR decision*:ab,ti OR willing*:ab,ti OR attitude*:ab,ti OR consideration*:ab,ti OR concern*:ab,ti OR barrier*:ab,ti OR issue*:ti,ab)	2,792,242
2	Exclu*:ti,ab	530,749
3	1 AND 2	90,193
4	((pregnan*:ti,ab OR expecting wom*:ti,ab) AND research*:ti,ab)	14,891
5	3 AND 4	365
Philosophers index		
Date of search: May 20th, 2015		
Search	Terms	Hits
1	(research* or trial* or stud*).mp. [mp=abstract, title, heading word]	57,827
2	pregnan*.mp. [mp=abstract, title, heading word]	695
3	1 and 2	138
4	limit 3 to (english)	117

Table 5.2 PRISMA 2009 flow diagram

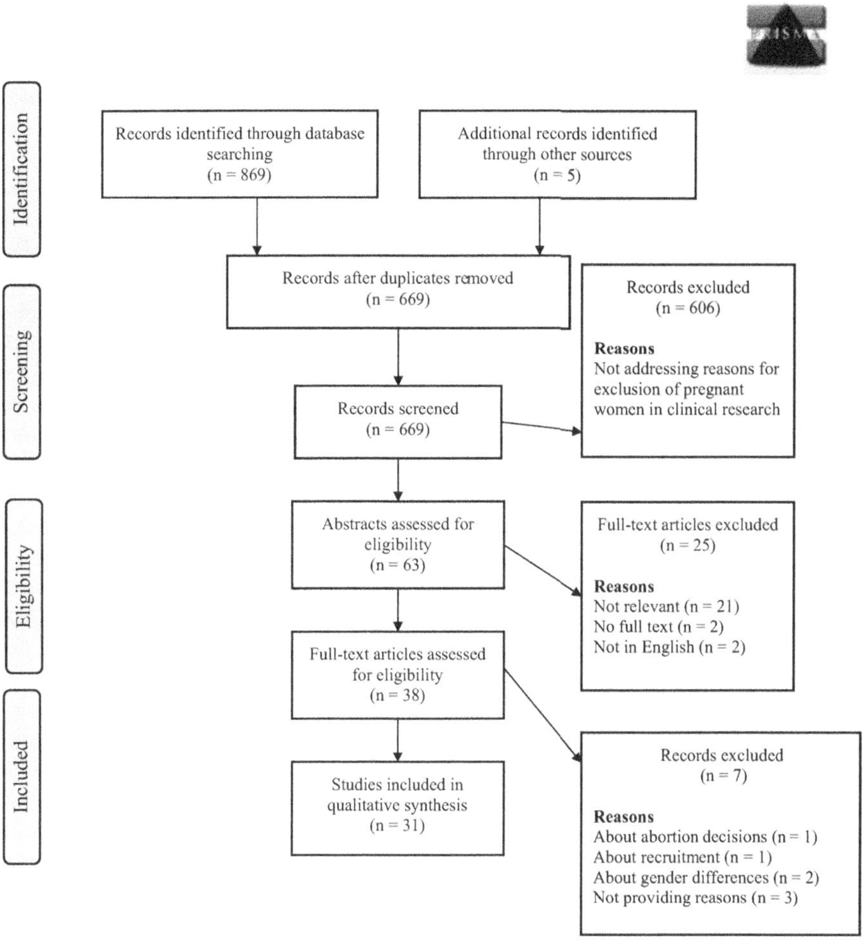

From: Moher D., A. Liberati, J. Tetzlaff, D.G. Altman, and The PRISMA Group. 2009. Preferred Reporting Items for Systematic Reviews and Meta-Analyses: The PRISMA Statement. PLoS Med. 6(6): e1000097. doi:10.1371/journal.pmed1000097. For more information, visit www.prisma-statement.org

Table 5.3 Summary of papers selected for the review

	References	Country	Paper type	Scope of paper	Aim of paper	Reported reason(s) for exclusion[a]
1	ACOG (2007)	USA	Consensus document	Clinical research	To update previous committee opinion on earlier publication on pregnant women in research	Pregnant women were viewed as a vulnerable population and there was concern that trial participation would result in harm to the foetus and increased liability risk for researchers
2	Alexander (2011)	USA	Reasoned opinion	Clinical research	To provide a historical overview of the Federal regulations (45 CFR 46 Subpart B) and discuss the revision	Fears of legal liability and social controversies (abortion debate in the USA) have led to exclusion
3	Allesee and Gallagher (2011)	USA	Reasoned opinion	Clinical research	To examine the ethics of the exclusion or inclusion of pregnant women in clinical trials	It is too dangerous for the baby if a pregnant woman participates; liability considerations of manufacturers and IRBs; IRB concerns with inherent and unknown danger to the foetus; reticence manufacturers due to historical tragedies
4	Battino (2001)	Italy	Reasoned opinion	Antiepileptic drug trials	To promote the inclusion of inclusion women in AED trials	Historically pregnant women were considered as a vulnerable population in relation to foetus
5	Baylis (2010)	CAN	Reasoned opinion	Clinical research	To argue that routine exclusion is unethical and unscientific and regulators must mandate change	Pregnant women are routinely excluded because of the harm the intervention might do to the developing foetus

6	Baylis and Kaposy (2010)	CAN	Reasoned opinion	Guidelines	To illustrate the negative ethical implications of excluding pregnant women in guidelines	The Tri-Council Policy Statement takes into account potential harms instead of having to give reasons for exclusion; existing fear of exposing foetuses to substances of unknown teratogenicity
7	Beran (2006)	AUS	Reasoned opinion	Anti-epileptic medications	To provide arguments for inclusion of AEM trials based on a review of existing information	Non-maleficence supports exclusion due to potential teratogenicity and protection of the potential offspring remains mandatory from the perspective of the non-malevolence principle
8	Blehar et al. (2013)	USA	Reasoned opinion	Clinical research	To further analyse three major interrelated scientific and science regulatory issues that emerged as important concerns at the 2010 ORWH meeting	Fear of harm to the foetus, threat of legal liability, concern about the complicated physiology of pregnant women, uncertainty whether pregnant women would be willing to participate, regulations which classify pregnant women as a vulnerable population and vague and restrictive wording of regulations which IRBs in turn interpret conservatively for pregnant subjects
9	Brandon et al. (2014)	USA	Qualitative analysis	Perinatal mental health research	To describe existing perspectives held by investigators (n = 15) and IRB members (n = 6) on acceptable perinatal mental health research	Semi-structured interviews indicate study design/methodology; safety concerns (risk mother, foetus, minimal risk interpretation); participant selection and recruitment and autonomy (competency informed consent, need for parental consent)

(continued)

Table 5.3 (continued)

	References	Country	Paper type	Scope of paper	Aim of paper	Reported reason(s) for exclusion[a]
10	Cain et al. (2000)	USA	Systematic literature review	Clinical protocols	To survey the contraceptive requirements in clinical trial protocols which were presented to an IRB (n = 410)	"[T]he near complete exclusion of pregnant women found in this review illustrates the impact that legal and economic concerns of sponsors about fetal outcome have (beyond issues of teratogenicity)"
11	Foulkes et al. (2011)	USA	Consensus document	Clinical research	To summarise a NIH workshop on barriers and opportunities enrolling pregnant women	Pregnant women are physiologically complicated, they will complicate our research, and they would be difficult to recruit. There are also challenges concerning legal liability and regulatory obscurity
12	Frew et al. (2014)	USA	Systematic literature review	Clinical research	To synthesise the body of literature that details challenges, facilitators, and best practices toward recruitment and retention through the socioecological model of health promotion	Public scandals, federal guidelines, IRB provisions and liability issues. In addition there are community/social level factors (i.e. willingness clinicians) and individual level factors (i.e. willingness participants)

13	Godlovitch (2003)	CAN	Reasoned opinion	Clinical research	To explore some of the ethical issues, the implicit ageism and sexism in the exclusion of seniors.	Babies are legally incapable of consenting, and pharmaceutical companies might not want potential exposure to later legal claims
14	Goldkind et al. (2010)	USA	Reasoned opinion	H1N1 influenza pandemic	To argue that it is not only permissible but also imperative that pregnant women be judiciously included in research	The effort to protect the foetus from research related risks is generally a reason for exclusion
15	Hurlimann et al. (2011)	CAN	Systematic literature review	Nutrigenetics clinical research	To examine inclusion in nutrigenetics clinical research and its scientific and ethical challenges	Challenges are the obtainment and quality of informed consent of vulnerable populations (i.e. ethical questions in terms of benefit and protection of future autonomy children), the notion of minimal risk and acceptable burden in research with vulnerable populations is controversial
16	Kaposy and Lafferty (2012)	CAN	Reasoned opinion	Vaccine trials	To argue that vaccine research in pregnancy can be ethical and to explore methods for overcoming vaccine manufacturers' fear of liability	Vaccine manufacturers' fear of liability is a primary force behind the exclusion

(continued)

Table 5.3 (continued)

	References	Country	Paper type	Scope of paper	Aim of paper	Reported reason(s) for exclusion[a]
17	Kornblum (1994)	USA	Reasoned opinion	Clinical research	To discuss whether pregnant women should be research subjects	'During a recent exchange, some NIEHS [National Institute of Environmental Health Sciences] scientist argued that existing federal rules against experimenting on pregnant women are beneficial to protect a fetus and prevent exploitation."
18	Levine (2011)	USA	Reasoned opinion	IRBs and policies	To review the development of policies and discuss current guidelines	Apparent IRB reasons: Subpart B of the Federal Code presents challenges, guidelines are restrictive, conservatism, exposure to legal liability, influenced by frightening history, and a documentation focus
19	Little et al. (2009)	USA	Reasoned opinion	Clinical research	To outline reasons for inclusion and outline possible steps to advance responsible inclusion	Researchers/IRBs view pregnancy as near-automatic cause of exclusion; no (legal) incentives to design studies; studies are more costly; there is a concern for the fetus, resistance due to thalidomide tragedy; vagueness leads to excluding interpretations

20	Lyerly (2011)	USA	Reasoned opinion	Clinical research	To provide a brief background of the need for research with pregnant women	Historical reasons for exclusion: complicated physiologies of pregnant women; the need to protect women and foetuses from potential risks of the drug; thalidomide which led to an almost universal exclusion of pregnant women from research.
21	Lyerly et al. (2008)	USA	Reasoned opinion	Clinical research	To outline resulting knowledge gaps and their costs and then highlight the ethical obligation to confront the challenges of including pregnant women in clinical research	Challenges concerning exclusion: safety of medication for the foetus; capacity to consent; cultural anxiety to place any risk on the foetus for sake of woman; liability concerns; IRBs are the gatekeepers of access to research; no legislation concerning justification of exclusion
22	Lyerly et al. (2007)	USA	Reasoned opinion	Medical decision-making around pregnancy	To outline three patterns in risk perception and reasoning that affect medical decision-making around pregnancy	Many researchers and IRBs continue to regard pregnancy as a virtually automatic cause for exclusion due to the tendency to notice risks of intervening to the exclusion of risks of not intervening

(continued)

Table 5.3 (continued)

	References	Country	Paper type	Scope of paper	Aim of paper	Reported reason(s) for exclusion[a]
23	Macklin (2010)	USA	Reasoned opinion	Biomedical research	To provide a perspective on the enrolment of pregnant women in biomedical research	Exclusion exist because researchers took the wrong message from thalidomide, ethical guidelines provide a mixed picture, and foetal safety concerns
24	Madan et al. (2014)	AUS	Case study/ies	Randomised trial methods	To explore difficulties in recruitment to improve future trial participation using a case-study of an RCT on cervical ripening	Clinicians are gatekeeping "(patients are not to be approached for the trial)" because of time constraints, lack of resources and a lack of equipoise
25	Merkatz (1998)	USA	Reasoned opinion	Clinical research	To outline a historical overview of exclusion and discuss reasons for current policy changes	Research practice shows that protection of the foetus remains an essential priority in research and that liability fears are still germane in present day research
26	Merton (1993)	USA	Reasoned opinion	Biomedical research	To catalogue ways in which women have been disadvantaged by exclusion; recent developments to redress them; dissect underlying rationales	Researcher's commitment to quality science (physiology and expenses); moral duty to avoid infliction of foetal harm; dictates of the law (government regulations) and tort phobia: the risk of liability to the offspring

27	Noah (2014)	USA	Reasoned opinion	Clinical research	To describe the current status of inclusion and discuss the FDA-related regulatory barriers to collecting safety and efficacy information	Current regulations discourage inclusion in a "misguided attempt to avoid challenging ethical issues, possible injuries to research participants and fetuses, and potential liability."
28	Sahin (2011)	USA	Consensus document	Clinical research	To present the FDA Perspective on the ethical barriers of conducting research in pregnant women in relation to recent publications	Pregnant women are an understudied population due to the ethical and medical-legal considerations of harming the foetus and in the effort to protect the foetus, institutional review boards' interpretation of the regulation may be overly cautious.
29	Schonfeld and Gordon (2005)	USA	Reasoned opinion	Contraception in research	To give a policy suggestion on contraception requirements in research	"[m]any investigators and sponsors are reluctant to include women of childbearing potential in research because of the risk of fetal harm should subject become pregnant during the study."

(continued)

Table 5.3 (continued)

	References	Country	Paper type	Scope of paper	Aim of paper	Reported reason(s) for exclusion[a]
30	Shields and Lyerly (2013)	USA	Systematic literature review	Phase VI trials	To measure the current exclusion of pregnant women from industry sponsored clinical trials as a baseline for future comparison	Condition has low prevalence in pregnant women; not enough safety data/not approved in pregnant women; considered vulnerable by FDA; pregnancy is always an exclusion criteria. In addition they hypothesise that the desire not to do harm, over interpretation of federal guidelines and the presumption that women might not be willing to participate are reasons as well
31	Theiler (2009)	USA	Reasoned opinion	Antimicrobial therapy	To provide a perspective on evidence-based antimicrobial therapy in pregnancy	Concerns for foetal well-being; current drug approval mechanisms (X labelling without tests in humans); legal environment; profit driven drug pipeline: post marketing studies are expensive and there are liability issues

[a]N.B.: authors often report on reasons for exclusion but do not actually promote or endorse these reasons

Table 5.4 Overview of reported reasons for exclusion

General theme	Article[a]
Foetal safety (n = 22)	
Concern that trial participation would result in harm to the foetus[a]	1
Too dangerous for the baby if a pregnant woman participates	3
IRB concerns with inherent and unknown danger to the foetus	3
The harm the intervention might do to the developing foetus	5
Existing fear of exposing foetuses to substances of unknown teratogenicity	6
Non-maleficence supports exclusion due to potential teratogenicity	7
Protection of the potential offspring remains mandatory	7
Fear of harm to the foetus	8
Protect the foetus from research related risks	14
Existing federal rules against experimenting on pregnant women are beneficial to protect a foetus and prevent exploitation	17
Concern for the foetus	19
The need to protect women and foetuses from potential risks of the drug[a]	20
Worries about the safety of medication for the foetus	21
Cultural anxiety to place any risk on the foetus for sake of woman	21
Foetal safety concerns	23
Protection of the foetus remains an essential priority in research	25
Moral duty to avoid infliction of foetal harm	26
Ethical and medical-legal considerations of harming the foetus	28
Effort to protect the foetus	28
Risk of foetal harm	29
Desire not to do harm	30
Concerns for foetal well-being	31
Liability (n = 15)	
Increased liability risk for researchers[a]	1
Fears of legal liability	2
Liability considerations of manufacturers and IRBs	3
Threat of legal liability	8
Legal concerns of sponsors about foetal outcome	10
Economic concerns of sponsors about foetal outcome	10
Challenges concerning legal liability	11
Liability issues	12
Pharmaceutical companies might not want potential exposure to later legal claims	13
Vaccine manufacturers' fear of liability	16
IRB exposure to legal liability	18
Liability concerns	21
Liability fears are still germane	25
Tort phobia: the risk of liability to the offspring	26
Liability issues	31

(continued)

Table 5.4 (continued)

General theme	Article[a]
Regulations/Wording (n = 15)	
Tri-Council Policy Statement takes into account potential harms instead of having to give reasons for exclusion	6
Existing regulations are somewhat ambiguous	8
Safety concerns (risk mother, foetus, minimal risk interpretation)	9
Language of minimal risk relative to the foetus is unclear (Subart B Federal Code)	11
Federal guidelines	12
Notion of minimal risk/acceptable burden for vulnerable populations is controversial	15
Subpart B of the Federal Code presents challenges	18
Guidelines are restrictive	18
Vagueness leads to excluding interpretations	19
There is no legislation concerning justification of exclusion	21
Ethical guidelines provide a mixed picture	23
Dictates of the law (government regulations)	26
The current regulations discourage inclusion in a "misguided attempt to avoid challenging ethical issues, possible injuries to research participants and foetuses, and potential liability."	27
Current drug approval mechanisms (X labelling without tests in humans)	31
Legal environment	31
Research design (n = 13)	
Concerns about the complicated physiology of pregnant women	8
Study design/methodology	9
Pregnant women are physiologically complicated	11
Pregnant women complicate our research	11
There are no (legal) incentives to design studies	19
Studies are more costly	19
Complicated physiologies of pregnant women[a]	20
Researcher's commitment to quality science (physiology)	26
Condition has low prevalence in pregnant women	30
There is not enough safety data/not approved in pregnant women	30
Pregnancy is always an exclusion criteria	30
Pharmaceutical companies have little incentive to investigate the safety of drugs	31
Profit driven drug pipeline: postmarketing studies are expensive	31
IRB interpretation (n = 9)	
Vague and restrictive wording of regulations which IRBs in turn interpret conservatively for pregnant subjects	8
IRB provisions (stringent regulation of protocols)	12
IRB conservatism	18
IRB documentation focus	18
Researchers/IRBs view pregnancy as near-automatic cause of exclusion	19
IRBs are the gatekeepers of access to research	21

(continued)

Table 5.4 (continued)

General theme	Article[a]
Researchers and IRBs continue to regard pregnancy as a virtually automatic cause for exclusion due to the tendency to notice risks of intervening to the exclusion of risks of not intervening	22
Institutional review boards' interpretation of the regulation may be overly cautious	28
Overinterpretation of federal guidelines	30
Willingness to participate (n = 7)	
Uncertainty whether pregnant women would be willing to participate	8
Participant selection and recruitment	9
Pregnant women would be difficult to recruit	11
Individual level factors (i.e. willingness participants)	12
Community/social level factors (i.e. willingness clinicians)	12
Clinicians are gatekeeping ("patients are not to be approached for the trial") "because of time constraints, lack of resources and a lack of equipoise	24
There is a presumption that women might not be willing to participate	30
Collective memory/Social controversies (n = 7)	
Social controversies have led to exclusion[a]	2
Reticence of manufacturers due to historical tragedies	3
Public scandals	12
IRBs are influenced by frightening history in the field	18
Resistance due to thalidomide tragedy	19
Thalidomide led to an almost universal exclusion of pregnant women from research[a]	20
Researchers took the wrong message from thalidomide	23
Consent (n = 4)	
Autonomy (competency informed consent, possible need for parental consent)	9
Babies are legally incapable of consenting	13
Challenges are the obtainment and quality of informed consent of vulnerable populations (i.e. ethical questions in terms of benefit and protection of future autonomy children)	15
Capacity to consent	21
Vulnerability (n = 4)	
Pregnant women were viewed as a vulnerable population[a]	1
Pregnant women were considered as a vulnerable population in relation to foetus[a]	4
Regulations which classify pregnant women as a vulnerable population	8
Pregnant women are considered vulnerable by FDA	30

[a]Mentioned as an earlier existing reason which might have lost its relevance

Table 5.5 Grouping of reported reasons for exclusion

General theme	Article[a]
Group 1	
Foetal safety (n = 22)	1, 3, 3, 5, 6, 7, 7, 8, 14, 17, 19, 20, 21, 21, 23, 25, 26, 28, 28, 29, 30, 31
Collective memory/Social controversies (n = 7)	
Social controversies	2
Reticence of manufacturers due to historical tragedies	3, 12, 18, 19, 20, 23
Liability (n = 15)	
Concerns liability (general)	2, 8, 11, 12, 21, 25, 26, 31
Liability considerations of researchers, manufacturers, sponsors and IRBs	1, 3, 10, 10, 13, 16, 18
Group 2	
Regulations/Wording (n = 15)	
Regulations	6, 8, 12, 18, 21, 26, 27, 31, 31
Formulation ambiguity	9, 11, 15, 18, 19, 23
IRB interpretation (n = 9)	
Conservatism/caution in interpretation	8, 18, 28, 30
Pregnancy as near-automatic cause of exclusion	12, 19, 21, 22
IRB documentation focus	18
Group 3	
Research design (n = 13)	
Study design/methodology	9, 11, 19, 30
Complicated physiology	8, 11, 20, 26
Studies are more costly	19, 31, 31
Low prevalence of a condition	30
Control for risk	30
Group 4	
Willingness to participate (n = 7)	
(Presumed) unwillingness pregnant women	8, 12, 30
Recruitment difficulties	9, 11
Unwillingness clinicians	12, 24
Vulnerability (n = 4)	
Pregnant women were viewed as a vulnerable population	1, 4, 8, 30
Consent (n = 4)	
Capacity to consent foetus	9, 13, 21
Obtainment/quality consent vulnerable populations	15

[a]Several articles mentioned multiple reasons, in that case the number of the article is repeated

References

ACOG (American College of Obstetricians and Gynecologists). 2007. ACOG committee opinion no. 377: Research involving women. *Obstetrics and Gynecology* 110(3): 731–736 [At the time of writing, ACOG committee opinion no. 377 was in effect. It has since been withdrawn and replaced with American College of Obstetricians and Gynecologists. 2015. ACOG committee opinion no. 646: Ethical considerations for including women as research participants. *Obstetrics and Gynecology* 126: e100–e107].

Alexander, D. 2011. Global health issues and US regulation 45 CFR 46 Subpart B. In *ORWH workshop: Enrolling pregnant women: Issues in clinical research. An ORWH 20th Anniversary event*, ed. ORWH, 57–61. Bethesda: National Institutes of Health.

Allesee, L., and M. Gallagher. 2011. Pregnancy and protection: The ethics of limiting a pregnant woman's participation in clinical trials. *Journal of Clinical Research & Bioethics* 2(108).

Ballantyne, A., and W. Rogers. 2016. Pregnancy, vulnerability, and the risk of exploitation in clinical research. In *Clinical research involving pregnant women*, eds. F. Baylis and A. Ballantyne, 139–159. Cham: Springer.

Barnett-Page, E., and J. Thomas. 2009. Methods for the synthesis of qualitative research: A critical review. *BMC Medical Research Methodology* 9: 59.

Battino, D. 2001. Trials in women. *Epilepsy Research* 45(1–3): 141–145.

Baylis, F. 2010. Pregnant women deserve better. *Nature* 465: 689–690.

Baylis, F., and C. Kaposy. 2010. Wanted: Inclusive guidelines for research involving pregnant women. *Journal of Obstetrics and Gynaecology Canada* 32(5): 473–476.

Baylis, F., and R. MacQuarrie. 2016. Why physicians and women should want pregnant women included in clinical trials. In *Clinical research involving pregnant women*, eds. F. Baylis and A. Ballantyne, 17–31. Cham: Springer.

Beran, R.G. 2006. The ethics of excluding women who become pregnant while participating in clinical trials of anti-epileptic medications. *Seizure* 15(8): 563–570.

Binik, A., C. Weijer, and M. Sheehan. 2011. Minimal risk remains an open question. *American Journal of Bioethics* 11(6): 25–27.

Blehar, M.C., C. Spong, C. Grady, S.F. Goldkind, L. Sahin, and J.A. Clayton. 2013. Enrolling pregnant women: Issues in clinical research. *Women's Health Issues: Official Publication of the Jacobs Institute of Women's Health* 23(1): e39–e45.

Brandon, A.R., G. Shivakumar, S.J. Inrig, J.Z. Sadler, and S.J. Craddock Lee. 2014. Ethical challenges in designing, conducting, and reporting research to improve the mental health of pregnant women: The voices of investigators and IRB members. *AJOB Empirical Bioethics* 5(2): 25–43.

Briggs, G.G., and J.E. Polifka, K.L. Wisner, E. Gervais, R.K. Miller, A. Berard, et al. 2015. Should pregnant women be included in Phase IV clinical drug trials? *American Journal of Obstetrics and Gynecology* 231(6): 810–815.

Cain, J., J. Lowell, L. Thorndyke, and R. Localio. 2000. Contraceptive requirements for clinical research. *Obstetrics and Gynecology* 95(6 Pt 1): 861–866.

Canada (Canadian Institutes of Health Research, Natural Sciences and Engineering Council of Canada, and Social Sciences and Humanities Research Council of Canada). 2014. *Tri-Council Policy Statement: Ethical conduct for research involving humans (TCPS2)*. http://www.pre.ethics.gc.ca/pdf/eng/tcps2-2014/TCPS_2_FINAL_Web.pdf. Accessed 3 Oct 2016.

Chervenak, F.A., and L.B. McCullough. 2011. An ethically justified framework for clinical investigation to benefit pregnant and fetal patients. *American Journal of Bioethics* 11: 39–49.

CIOMS (Council for International Organizations of Medical Sciences). 2002. *International ethical guidelines for biomedical research involving human subjects*. Geneva: World Health Organization. http://www.cioms.ch/publications/layout_guide2002.pdf. Accessed 3 Oct 2016.

Clayton, E.W. 1999. Liability exposure when offspring are injured because of their parents' participation in clinical trials. In *Women and health research: Ethical and legal issues of including women in clinical studies, volume 2, workshop and commissioned papers*, eds. A.C. Mastroianni, R. Faden, and D. Federman, 103–112. Washington, DC: National Academies Press.

DHHS (US Department of Health and Human Services). 2009. *Code of Federal Regulations: Title 45, Part 46, Protection of Human Subjects*. http://www.hhs.gov/ohrp/regulations-and-policy/regulations/45-cfr-46/index.html. Accessed 3 Oct 2016.

Ells, C., and C. Lyster. 2016. Research ethics review of drug trials targeting medical conditions of pregnant women. In *Clinical research involving pregnant women*, eds. F. Baylis and A. Ballantyne, 95–118. Cham: Springer.

EMA (European Medicines Agency). 2005. *Guideline of the exposure to medicinal products during pregnancy: Need for post-authorisation data*. London. http://www.ema.europa.eu/docs/en_GB/document_library/Regulatory_and_procedural_guideline/2009/11/WC500011303.pdf. Accessed 17 June 2015.

Flannery, E., and S. Greenberg. 1999. Liability exposure for exclusion and inclusion of women as subjects in clinical studies. In *Women and health research: Ethical and legal issues of including women in clinical studies, volume 2, workshop and commissioned papers*, eds. A.C. Mastroianni, R. Faden, and D. Federman, 91–102. Washington, DC: National Academies Press.

Foulkes, M.A., C. Grady, C.Y. Spong, A. Bates, and J.A. Clayton. 2011. Clinical research enrolling pregnant women: A workshop summary. *Journal of Women's Health* 20(10): 1429–1432.

Frew, P.M., D.S. Saint-Victor, M.B. Isaacs, S. Kim, G.K. Swamy, J.S. Sheffield, et al. 2014. Recruitment and retention of pregnant women into clinical research trials: An overview of challenges, facilitators, and best practices. *Clinical Infectious Diseases: An Official Publication of the Infectious Diseases Society of America* 59(Suppl 7): S400–S407.

Godlovitch, G. 2003. Age discrimination in trials and treatment: Old dogs and new tricks. *Monash Bioethics Review* 22(3): 66–77.

Goldkind, S.F., L. Sahin, and B. Gallauresi. 2010. Enrolling pregnant women in research – lessons from the H1N1 influenza pandemic. *New England Journal of Medicine* 362(24): 2241–2243.

Haas, D.M., B. Gallauresi, K. Shields, D. Zeitlin, S.M. Clark, M.F. Hebert, et al. 2011. Pharmacotherapy and pregnancy: Highlights from the third international conference for individualized pharmacotherapy in pregnancy. *Clinical and Translational Science* 4(3): 204–209.

Hurlimann, T., R. Stenne, V. Menuz, and B. Godard. 2011. Inclusion and exclusion in nutrigenetics clinical research: Ethical and scientific challenges. *Journal of Nutrigenetics and Nutrigenomics* 4(6): 322–343.

Hurst, S.A. 2008. Vulnerability in research and health care: Describing the elephant in the room? *Bioethics* 22(4): 191–202.

Hutson, J.R., F. Garcia-Bournissen, A. Davis, and G. Koren. 2011. The human placental perfusion model: A systematic review and development of a model to predict in vivo transfer of therapeutic drugs. *Clinical Pharmacology and Therapeutics* 90(1): 67–76.

Kaposy, C. 2016. Presumptive inclusion and legitimate exclusion criteria. In *Clinical research involving pregnant women*, eds. F. Baylis and A. Ballantyne, 51–62. Cham: Springer.

Kaposy, C., and L. Lafferty. 2012. Overcoming liability concerns in vaccine trials involving pregnant women. *Accountability in Research* 19(3): 156–174.

Kornblum, A. 1994. Trial and error: should pregnant women be research subjects? *Environmental Health Perspective* 102(9): 752–753.

Kukla, R. 2016. Equipoise, uncertainty, and inductive risk in research involving pregnant women. In *Clinical research involving pregnant women*, eds. F. Baylis and A. Ballantyne, 179–196. Cham: Springer.

Langston, L. 2016. Better safe than sorry: Risk, stigma, and research during pregnancy. In *Clinical research involving pregnant women*, eds. F. Baylis and A. Ballantyne, 33–50. Cham: Springer.

Levine, R.J. 2011. IRB perspective on inclusion of pregnant women in clinical research. In *ORWH workshop: Enrolling pregnant women: Issues in clinical research. An ORWH 20th Anniversary Event*, ed. ORWH, 37–43. Bethesda: National Institutes of Health.

Levine, C., R. Faden, C. Grady, D. Hammerschmidt, L. Eckenwiler, and J. Sugarman. 2004. The limitations of "vulnerability" as a protection for human research participants. *The American Journal of Bioethics* 4(3): 44–49.

Little, M.O., A.D. Lyerly, and R.R. Faden. 2009. Pregnant women & medical research: A moral imperative. *Bioethics Forum* 2(2): 60–65.

Luna, F. 2009. Elucidating the concept of vulnerability: Layers not labels. *International Journal of Feminist Approaches to Bioethics* 2(1): 121–139.

Luna, F., and S. van der Poel. 2013. Not the usual suspects: Addressing layers of vulnerability. *Bioethics* 27(6): 325–332.

Lupattelli, A., O. Spigset, M.J. Twigg, K. Zagorodnikova, A.C. Mårdby, and M.E. Moretti. 2014. Medication use in pregnancy: A cross-sectional, multinational web-based study. *BMJ Open* 4(2): e004365.

Lyerly, A.D. 2011. Direct benefit to pregnant women. In *ORWH workshop: enrolling pregnant women: issues in clinical research. An ORWH 20th anniversary event*, ed. ORWH, 57–61. Bethesda: National Institutes of Health.

Lyerly, A.D., L.M. Mitchell, E.M. Armstrong, L.H. Harris, R. Kukla, M. Kuppermann, and M.O. Little. 2007. Risks, values, and decision making surrounding pregnancy. *Obstetrics and Gynecology* 109(4): 979–984.

Lyerly, A.D., M.O. Little, and R.R. Faden. 2008. The second wave: Toward responsible inclusion of pregnant women in research. *International Journal of Feminist Approaches to Bioethics* 1(2): 5–22.

Lyerly, A.D., M.O. Little, and R.R. Faden. 2011. Reframing the framework: Toward fair inclusion of pregnant women as participants in research. *American Journal of Bioethics* 11(5): 50–52.

Macklin, R. 1990. Maternal-fetal conflict: An ethical analysis. *Women's Health Issues* 1(1): 28–30.

Macklin, R. 2003. Bioethics, vulnerability and protection. *Bioethics* 17(5–6): 472–486.

Macklin, R. 2010. The art of medicine: Enrolling pregnant women in biomedical research. *Lancet* 375: 632–633.

Madan, A., S. Tracy, R. Reid, and A. Henry. 2014. Recruitment difficulties in obstetric trials: A case study and review. *The Australian & New Zealand Journal of Obstetrics & Gynaecology* 54(6): 546–552.

Mariner, W.K. 1999. Compensation for Research Injuries. In *Women and health research: Ethical and legal issues of including women in clinical studies: Volume 2: Workshop and commissioned papers*, eds. A. Matroianni, R. Faden, and D. Federman. Washington, DC: National Academies Press.

Mastroianni, A.C., R. Faden, and D. Federman (eds.). 1999. *Women and health research: Ethical and legal issues of including women in clinical studies, volume 2, workshop and commissioned papers*. Washington, DC: National Academies Press.

McCullough, L.B., and F.A. Chervenak. 2011. The fetus as a patient and the ethics of human subjects research: Response to commentaries on "An ethically justified framework for clinical investigation to benefit pregnant and fetal patients.". *American Journal of Bioethics* 11(5): W3–W7.

Merkatz, R.B. 1998. Inclusion of women in clinical trials: A historical overview of scientific, ethical, and legal issues. *Journal of Obstetric, Gynecologic, and Neonatal Nursing* 27(1): 78–84.

Merton, V. 1993. The exclusion of pregnant, pregnable, and once-pregnable people (a.k.a. women) from biomedical research. *American Journal of Law and Medicine* 19(4): 369–451.

Moher, D., A. Liberati, J. Tetzlaff, D.G. Altman, and The PRISMA Group. 2009. Preferred reporting items for systematic reviews and meta-analyses: The PRISMA statement. *PLoS Medicine* 6(7): e1000097.

Myren, M., T. Mose, L. Mathiesen, and L.E. Knudsen. 2007. The human placenta – An alternative for studying foetal exposure. *Toxicology in Vitro* 21(7): 1332–1340.

Noah, B.A. 2014. The inclusion of pregnant women in clinical research. *St. Louis University Journal of Health Law and Policy* 7: 353–388.

ORWH (United States. Department of Health and Human Services, Public Health Service, National Institutes of Health, Office of Research on Women's Health). 2011. *Enrolling pregnant women: Issues in clinical research*. Bethesda: National Institutes of Health.
Sahin, L. 2011. The FDA perspective on the ethical barriers of conducting research in pregnant women. *AJOB Primary Research* 2(1): 13–14.
Schonfeld, T., and B. Gordon. 2005. Contraception in research: A policy suggestion. *IRB* 27(2): 15–20.
Schroeder, D., and E. Gefenas. 2009. Vulnerability: Too vague and too broad? *Cambridge Quarterly of Healthcare Ethics* 18(2): 113–121.
Second Wave Initiative. 2009. *The Second Wave Initiative: Toward the responsible inclusion of pregnant women in medical research*. http://secondwaveinitiative.org/. Accessed 17 June 2015.
Second Wave Initiative. 2016. http://secondwaveinitiative.org/. Accessed 7 Apr 2016.
Shields, K.E., and A.D. Lyerly. 2013. Exclusion of pregnant women from industry-sponsored clinical trials. *Obstetrics and Gynecology* 122(5): 1077–1081.
Strech, D., and N. Sofaer. 2012. How to write a systematic review of reasons. *Journal of Medical Ethics* 38(2): 121–126.
Strong, C. 2011. Minimal risk in research involving pregnant women and fetuses. *The Journal of Law, Medicine & Ethics* 39(3): 529–538.
Swan, S.H. 2000. Intrauterine exposure to diethylstilbestrol: Long-term effects in humans. *APMIS* 108(12): 793–804.
Tan, K.S., and N.C. Thomson. 2000. Asthma in pregnancy. *American Journal of Medicine* 109(9): 727–733.
Theiler, R.N. 2009. Evidence-based antimicrobial therapy in pregnancy: Long overdue. *Clinical Pharmacology and Therapeutics* 86(3): 237–238.
Van der Graaf, R., R.H.H. Groenwold, S. Kalkman, D.E. Grobbee, and J.J.M. van Delden. 2013. From justifying inclusion to justifying exclusion of study populations: Strengths and limitations. *World Medical Journal* 59(5): 192–197.
Westreich, D., M. Rosenberg, S. Schwartz, and G. Swamy. 2013. Representation of women and pregnant women in HIV research: A limited systematic review. *PloS One* 8(8): e73398.
Wild, V., and N. Biller-Andorno. 2016. Pregnant women's views about participation in clinical research. In *Clinical research involving pregnant women*, eds. F. Baylis and A. Ballantyne, 119–136. Cham: Springer.
World Medical Association. 2013. *WMA Declaration of Helsinki: Ethical Principles for Medical Research Involving Human Subjects*. http://www.wma.net/en/30publications/10policies/b3/. Accessed 3 Oct 2016.
Zajicek, A., and G.P. Giacoia. 2007. Obstetric clinical pharmacology: Coming of age. *Clinical Pharmacology and Therapeutics* 81(4): 481–482.

Chapter 6
Research Ethics Review of Drug Trials Targeting Medical Conditions of Pregnant Women

Carolyn Ells and Caroline Lyster

Abstract *In this chapter we examine ways in which research ethics committees can appropriately conduct ethics review of clinically important trials pertaining to the management of medical conditions of pregnant women. Given the well-documented variability of research ethics committees' decision-making, it is reasonable to predict variability among committees regarding their reviews and decisions of research involving pregnant women. At least some of this variability is due to a lack of sufficient guidance on the part of national and international research guidelines, which results in a reluctance to approve clinical research involving pregnant women. After summarising the problems inherent in the relevant guidelines, we propose additional considerations and recommendations to guide research ethics committees, researchers, trial sponsors, and funding agencies in the review and oversight of such research.*

The bulk of the literature on research involving pregnant women tends to be narrowly focused on why that research ought to be done (Lyerly et al. 2008), and the frameworks that can be used to justify such research (Chervenak and McCullough 2011). Less has been written on the practical steps that might be implemented to facilitate the design and ethics review of clinical trials that enrol, and pertain to the medical conditions of, pregnant women. While part of the issue is undoubtedly that

C. Ells, RRT, PhD (✉)
Biomedical Ethics Unit, Department of Medicine, McGill University, Montreal, Canada

Lady Davis Institute for Medical Research, Montreal, Canada
e-mail: carolyn.ells@mcgill.ca

C. Lyster, MA
Clinical Research Unit, Alberta Diabetes Institute, University of Alberta, Edmonton, Canada

F. Baylis, A. Ballantyne (eds.), *Clinical Research Involving Pregnant Women*, Research Ethics Forum 3, DOI 10.1007/978-3-319-26512-4_6

research ethics committees[1] see few clinical trials that propose to include pregnant participants, for a variety of reasons, research ethics committees may discourage or decline proposals of this nature. Yet, given their essential role in the process of research leading to the availability of regulated medications that can be safely prescribed to those who could benefit from them, we believe that research ethics committees have an important role in fostering the conduct of high quality clinical trials focused on treating the chronic conditions of women during pregnancy.

As we discovered through searching the literature and in our own communications with research ethics committees, it appears that most committees lack specific policies or other guidance to facilitate the review of research involving pregnant women. The ethics review of such research is guided mainly by the major national and international guidelines, which offer insufficient guidance with respect to many aspects of this kind of research. In the wake of the thalidomide disaster, fear of doing harm has also led to a protectionist stance concerning foetal involvement (Levine 2010, 37; see also Langston 2016), and research ethics committees also appear particularly concerned with legal liability, even though litigation against individual Research Ethics Boards (REBs) is rare in practice (Levine 2010, 42). This combination of factors seems to have resulted in reluctance on the part of research ethics committees to approve clinical trials that would involve pregnant participants.

In this chapter, we offer analysis and recommendations for researchers and research ethics committees to consider. Our purpose is twofold. First, we aim to facilitate research ethics review of high quality, safe, and ethical research involving pregnant women by further specifying what ethics review of such research should entail. This work is specifically directed at research ethics committees and bodies involved with developing guidelines for research ethics review, though researchers, trial sponsors and funding agencies may gain from what we have to say. Second, we aim to encourage research ethics committees to ask why pregnant women have not been included in the research they review, and to question whether this exclusion is justified.

We limit our focus to clinical trials of investigational drugs (including investigation of drugs used off-label during pregnancy) that address clinically important questions pertaining to the management of medical conditions of pregnant women, such as lupus, high blood pressure, and mental illness. In some cases these conditions will be pre-existing and possibly exacerbated by pregnancy. In other cases they will arise during pregnancy. The management of these conditions is complicated by the medical complexities of pregnancy and insufficient evidence-based management strategies. Although we focus on this subset of clinical trials, some of our considerations and recommendations may be relevant to a broader set of research questions involving pregnant women and a broader set of methods used to answer them.

---

[1] This is the term commonly used in Europe and elsewhere. It is equivalent to Institutional Review Board (IRB) and Research Ethics Board (REB) used in the United States and Canada, respectively.

The chapter proceeds as follows: First, we provide a short summary of our cursory search for policies and procedures on ethics review of clinical research involving pregnant participants created by research ethics committees. We then move on to a discussion of the major national and international guidelines, and discuss how their guidance is, for the most part, insufficient. Finally, we outline how research ethics committees can combat these insufficiencies through policy changes.

6.1 Local Research Ethics Committee Policies

To augment our literature search and inform our analysis, we sought to review policies and procedures that research ethics committees may have created locally to supplement existing guidelines for research involving pregnant women. To this end, in the summer of 2014 we sent a query to research ethics committees at major Canadian research institutions that might reasonably be expected to have experience with research involving pregnant women. We also posted our query to the listservs of the Canadian Association of Research Ethics Boards (CAREB) and the International Network on Feminist Approaches to Bioethics (FAB). In all these queries we asked research ethics committees whether they had specific policies or procedures in place for the review of clinical research involving pregnant women, and to share any such documents with us.

We received only 11 replies, mainly indicating that the respondent's research ethics committee did not have specific policies or procedures to govern the review of research involving pregnant women. Responders from Canada indicated that their review process followed the *Tri-Council Policy Statement: Ethical Conduct for Research Involving Humans* (TCPS2 2014) and International Council for Harmonisation *(ICH) Harmonised Tripartite Guideline: Guideline for Good Clinical Practice E6* (ICH E6 1996), while those responding from the United States followed the *Code of Federal Regulations* (DHHS 2009, 45 CFR 46 Subpart B). Some respondents indicated that research proposals involving pregnant women might face more 'intense scrutiny' due to the pregnancy issue, though they were not specific as to what that additional scrutiny might entail.[2] Given the emphasis that various guidelines place on reducing potential harm to the foetus, it would seem that the intense scrutiny referred to by respondents would be in regards to risks, and possibly the need to involve a greater number of, or more expert, reviewers than would otherwise be required. Other responses made direct mention of ensuring that harm to the foetus was minimised, with some aiming to ensure that no harm occurred to either the pregnant woman or the foetus.

---

[2] In Canada, the *TCPS2* is the most widely applicable policy guidance for research ethics review. It allows for, and uses the language of, greater or lesser 'scrutiny' (in accordance with the degree of foreseeable risks) in its proportionate approach to research ethics review (see, for example, Chapter 1C).

While most respondents indicated an absence of additional policies or procedures pertaining to research involving pregnant women, we did find that some institutions had internal guidelines. The *IWK Health Centre – Research Ethics Board Review Criteria* (*IWK Review*), for instance, includes two points of particular interest. The first point, which is not specific to research in pregnant participants, is that risks regarding reproduction, lactation and foetal development must be explained in relevant protocols; these protocols must contain a plan for follow-up with any unintended pregnancies; and the consent forms must clearly explain to female participants why it is necessary to follow those pregnancies (IWK 2013, 2). While the review criteria require protocols to state clearly acceptable methods of contraception that participants of childbearing potential must use, and the length of time they need to use this, research ethics committee members are not prompted to question whether the exclusion of pregnant women from research is ever justified. The second point of interest in the *IWK Review* has to do with data storage: in the specific case of studies involving pregnant women, records must be stored for a minimum of 10 years past the age of majority (which is 18), i.e., for at least 28 years. This long-term storage of records enables a potential return to the data for further examination or analysis if new relevant information becomes available. It also facilitates follow-up with participants and their children for safety or other reasons, provided the code key is kept that links their identity to the data and participants have consented to potential follow-up.

The Southern Illinois University research ethics guidelines contain one mention of pregnancy requiring the consent document to include a statement that the treatment or procedure being investigated may involve currently unforeseeable risks to the subject, or to the embryo or foetus if the subject is or may become pregnant (Office of Sponsored Projects Administration 2010, 11). Given the general lack of knowledge about the teratogenicity and mutagenicity of Food and Drug Administration (FDA) approved medications (Lo and Friedman 2002, 468) the inclusion of a statement to this effect is important, and ensures that the consent of participants who are or may become pregnant during the study is informed in this regard. It also may offer protection from legal action for the researchers, and the research institution, if participants suffer harm that was not explicitly mentioned in the consent document.

Our query to research ethics committees found no evidence of sophisticated in-house guidelines specifically regulating the review of clinical trials including pregnant women.

6.2 National and International Guidelines

Ethics review of clinical research involving pregnant women, first and foremost, is governed by national and international guidelines. Typically, these are based on widely accepted research ethics principles including respect for persons, concern for welfare, and justice. We chose a total of nine Canadian, American, European, and international guidelines,[3] and examined what each said about research involving pregnant women. Our analysis revealed areas where guidance may not be sufficient, and where individual research ethics committees and institutions may want to augment that guidance with policies of their own to better ensure a high quality research ethics review. Pregnancy also complicates other areas of the guidelines not specific to research in this group, especially those involving study design and informed consent.

Below, we summarise the major issues that became apparent through our review and analysis. Following that, we discuss these issues further and propose recommendations for how individual institutions and research ethics committees might mitigate these issues through policy development.

Pregnancy and Vulnerability

Whether pregnant women are considered vulnerable (or in a vulnerable situation) for the purposes of research ethics review has important implications. Additional protections and restrictions apply for research participants who are deemed vulnerable (or in a vulnerable situation) due to their (presumed) increased risk of exploitation or other harm by participating in research, sometimes with no corresponding potential for benefit. Notably, of the nine major national and international research ethics guidelines we reviewed *only one*, the American *Code of Federal Regulations*, explicitly identifies pregnant women as vulnerable with respect to potential research participation (DHHS 2009, 21 CFR 56 Subpart C and 45 CFR 46 Subpart A). While this position is an outlier among the guidelines we reviewed, it is consistent with a common assumption that pregnant women are vulnerable and in need of special protection in research (Blehar et al. 2013, e40; Wild 2012, 83). We find such attribution problematic and support the position that pregnant women should *not* be considered a vulnerable class of potential research participants for the purposes of research ethics. Due to limitations of space, we refer readers to other chapters in this volume (see Ballantyne and Rogers 2016; Johnson 2016) for analyses of pregnancy and vulnerability in and for research guidelines, and will turn next to the complicated issue of assessing risks and potential benefits of participation in clinical trials during pregnancy.

---

[3] *TCPS2*; *Code of Federal Regulations*; *Directive 2001/20/EC of the European Parliament and of the Council*; *Additional Protocol to the Convention on Human Rights and Medicine, concerning Biomedical Research* (Council of Europe); *Declaration of Helsinki* (WMA); *Standards and Operational Guidance for Ethics Review of Health-Related Research with Human Participants* (WHO); *General Considerations for Clinical Trials* (ICH); *Guideline for Good Clinical Practice* (ICH); and *International Ethical Guidelines for Biomedical Research Involving Human Subjects* (CIOMS).

Assessing Risks and Potential Benefits

The interests of pregnant women are not always opposed to those of the foetuses they carry. Yet some may be tempted to make an overly simplistic calculation whereby the foetus, on the one hand, risks being exposed to a teratogen, having its development impacted in some other way, or not being carried to term if the drug under investigation were to cause a miscarriage. The woman, on the other hand, stands to reap the health benefits associated with improvements to her medical care.

This simple kind of analysis appears to be what the *Code of Federal Regulations* has in mind, as criterion (b) states explicitly that any risks to the foetus must be balanced by direct benefits to the pregnant woman (DHHS 2009, 45 CFR 46.204; see Table 6.1); as Schonfeld (2013) points out, no mention is made of risks to the pregnant woman (190). Likewise, as Margaret Little (2010) has noted, the focus of many guidelines tends to be on the risks associated with participation in research, which are primarily risks to the foetus, while no mention is made of the risks to the pregnant woman when the research is not performed (24).

Pregnant women face serious medical conditions, such as high blood pressure, heart disease, diabetes, lupus, and cancer, and all the medicines used to treat these conditions in pregnant women are being used off-label. Indeed, in the United States there are only 12 medications approved by the FDA for use in pregnancy, and those are used to prevent premature labour or treat labour pains (Little 2010, 23). The Canadian Pharmacists Association has provided its members with a reference table for the treatment of certain conditions during pregnancy, but this table does not address dosage changes that might be required given the physiological changes associated with pregnancy (Diav-Citrin and Koren 2011, 1725–1728). This lack of guidance means that the standard of care for pregnant women with medical conditions superimposed upon pregnancy is woefully inadequate. Medical conditions are either not treated, managed without medications, or managed with off-label use of medications (i.e., medications used for indications that have not been subjected to regulatory evaluation and approval). This far from ideal standard of care makes the risk/benefit assessment more complex, as the pregnant woman is exposed to the risk of poor quality medical management when research is not done. Unlike the *Code of Federal Regulations*, the Canadian *TCPS2* (Article 4.3; see Table 6.2) takes this additional risk into account, and directs research ethics committees to not only consider the risks and potential benefits associated with participation in research, but also to take into consideration the harm that could befall pregnant women if the research is not done.

Another type of oversimplification can be seen in the way that guidelines seem to treat the pregnant woman and the foetus as two separate research participants who have conflicting interests. As Verina Wild (2012; see also Wild and Biller-Andorno 2016) has argued convincingly, however, this two-person model of the foetus as an individual, separate from the pregnant woman, is an artificial construct and a consequence of technological innovation, namely visualisation, in medicine (89). The pregnant woman and foetus are intimately connected, and just as the pregnant woman stands to benefit from the research being done, so too does the foetus. Benefits associated with the woman's core functioning – improvements in her blood pressure, her blood glucose levels, or her mental health – will, to some degree, benefit the foetus, even though the foetus may potentially be exposed to risks, teratogenic or otherwise.

However, admitting that there is an intimate relationship between the pregnant woman and the foetus complicates the assessment of risks and potential benefits, an assessment made even more complex because the exact risks to the foetus are, in many cases, unknown. Lo and Friedman (2002) determined that the teratogenic risk in pregnancy was still undetermined in 91 % of drugs approved by the FDA between 1980 and 2000 (468). What's more, within the other 9 % of drugs only a portion – 16 medications or classes of medications – have been proven as teratogenic: the others are suspected teratogens, but drug manufacturers may have legal rather than scientific reasons for assigning this designation to a product (Diav-Citrin and Koren 2011, 1721–24).

Teratogenic or Mutagenic Effects

When considering the risks that the foetus might be exposed to during a clinical trial, research ethics committees are likely to place a great deal of weight on the potential for severe birth defects that could have a devastating impact on the lives of children. Many will recall the early use of thalidomide. Children of women who took this drug during pregnancy were born with significant impairments, including stunted limbs that affected their functional capacities. The fear – if not dread – no doubt lingers, among research ethics committees, researchers, and the pharmaceutical industry that a research ethics committee may approve research on a medication that will unknowingly contribute to the disabling of children who were exposed to the product in utero for research purposes (see Langston 2016). Strictly speaking, however, this line of thinking is not correct, as the scope of the thalidomide disaster could have been prevented by taking a more cautious approach to drug development research, which would have included pre-clinical research in pregnant animals, and small-scale clinical research involving pregnant participants.

It is now well-known that medications can cross the placental barrier, and so the *Code of Federal Regulations* requires that research proposals involving pregnant women include data from animal studies in order to determine the teratogenicity/mutagenicity of the drug being studied (DHHS 2009, 45 CFR 46.204; see Table 6.1), and the Council for the International Organizations of Medical Sciences (CIOMS) *International Ethical Guidelines for Biomedical Research Involving Human Subjects* (CIOMS 2002; see Table 6.3) outlines a similar requirement. Such requirements are prudent as animal studies of this sort provide crucial evidence to inform the risk assessment of offering research participation to pregnant women. Regulatory agencies could go further however to require animal studies to determine the teratogenicity/mutagenicity of drugs on a routine basis (at least for drugs that address medical conditions that pregnant women commonly experience). Public reporting of such research would open up the potential for more (and more informed) clinical trials addressing medical conditions that pregnant women experience.

While some teratogenic or mutagenic effects might be immediately apparent upon the birth of a child, others may only become apparent as the child ages. To account for this, other guidelines, including the CIOMS guidelines (see Table 6.3) and *ICH E8* (see Table 6.4) recommend that research proposals involving pregnant women include a plan for follow-up, so that the health of both the woman and her child may be monitored – a requirement that could have inspired the IWK guideline mentioned earlier.

The problem with these recommendations, however, is that the guidelines do not elaborate on the form that this follow-up should take. In the absence of institutional policy on the matter, this leaves decisions about the type and duration of follow-up to the researchers and research ethics committees. And, as guidelines including the *ICH E6* and *Directive 2001/20/EC of the European Parliament and of the Council* (*2001/20/EC*) only name congenital anomalies or birth defects as serious adverse events in need of immediate reporting (see Table 6.5), researchers may choose not to follow-up after the pregnant woman has given birth to a live, and apparently physically normal, child.

Issues Not Addressed in the Guidelines

There are many issues, not specifically addressed in the nine major guidelines we reviewed, that must be considered when reviewing clinical trials pertaining to the management of medical conditions of pregnant women. We raise three such issues here (with no claim to be exhaustive).

First, involving pregnant women as research participants complicates aspects of statistics and study design. The relatively small number of potential participants, for example, means that the results may be less reliable than they would have been in a large subject pool, and research ethics committees may be more likely to reject the research on scientific, as opposed to strictly ethical, grounds (though the two are intertwined). The standard of care to manage medical conditions during pregnancy is controversial and imbued with uncertainty, owing to insufficient evidence upon which to guide care, particularly related to use of medications. Widespread off-label use of medications during pregnancy exists alongside widespread under-treatment, as when pregnant women and their health care professionals choose to forego medications they would otherwise use until they are no longer pregnant (see Baylis and MacQuarrie 2016). Unsettled views among experts on the standard of care in such cases can be expected to lead to disagreement about whether placebo control groups are acceptable and, if not, what the appropriate comparison product is. Research ethics committees may need additional guidance and expertise to address scientific matters such as these.

Second, financial inducements may be more problematic in research involving pregnant women. Proper prenatal care is expensive, and pregnant women with lower socioeconomic status might be disproportionately represented among research participants, as volunteering for research may provide them with access to prenatal care that they want, but cannot otherwise afford.[4] While the potential effects of financial inducements is acknowledged in most of the major guidelines, the wording in most cases is vague, and individual research ethics committees must determine when a financial incentive is large enough to be ethically problematic. Thus, there is likely to be considerable inconsistency in the way that this standard is applied in research ethics reviews.

Finally, research ethics committees are not encouraged to consider whether the exclusion of pregnant women from a particular clinical trial is, in fact, justified (see Kaposy 2016). Schonfeld (2013) points out that pregnant women tend to experience a *social vulnerability*: they are expected to conform to a certain set of social expectations, and in the context of research this results in an attitude of protectionism toward the foetus (200).[5] This attitude, combined with the particular way in which the major guidelines are worded, means that research ethics committees are unlikely to ask whether it is indeed appropriate to exclude pregnant women from participation in clinical trials that may, in fact, serve their health interests. With respect to the language of the guidelines, *ICH E8* is especially restrictive: "In general, pregnant women should be excluded from clinical trials where the drug is not intended for use in pregnancy" and advises that should a participant become pregnant while receiving a research drug that drug "should generally be discontinued if this can be done safely" (see Table 6.4). While more tempered, consider also the language in the *TCPS2:* "[w]omen shall not be inappropriately excluded from research on the basis of their reproductive capacity; or because they are pregnant or breastfeeding" (see Table 6.2). In a public consultation submission to the Canadian Interagency Panel on Research Ethics, Alana Cattapan (2014) noted that this way of wording the guidance does not put an onus on researchers to include pregnant women (3). By mentioning pregnancy only in terms of exclusion or the avoidance of pregnancy during clinical trials, researchers and research ethics committees may falsely assume that there is no legitimate means for pregnant participants to participate in clinical trials.

[4] We present this as a problem of justice, but note that disproportionately greater enrolment of participants with a lower socioeconomic status may also introduce a variable that biases or complicates the analysis (of investigational drugs on managing important medical conditions during pregnancy) and thereby diminish the generalizability of results.

[5] This is not a uniquely North American attitude: when asked whether the exclusion of pregnant women from German clinical trials was appropriate, leading reproductive toxicologists in the country responded that their inclusion was neither necessary nor desirable (Wild 2012, 85).

6.3 Research Ethics Review and Recommendations

In the previous section we summarised issues that are insufficiently covered in major national and international guidelines. In the section previous to that, we noted what a few research ethics committees have done to compensate for gaps in the guidance. In this section we make additional recommendations that individual research ethics committees and institutions should consider in their ethical review and oversight of clinical trials addressing clinically important questions in the management of medical conditions during pregnancy. In order to make our recommendations as practically applicable as possible, we frame them in terms of research ethics oversight throughout the lifecycle of the research.

Science Review of the Research Question(s) and Study Design

Research proposals must be scientifically as well as ethically sound. In light of the current situation where very few drugs are approved for use during pregnancy and the standard of care involves widespread off-label use of medications to manage women's medical conditions during pregnancy, justifying the clinical importance of the research question(s) is likely to be easily addressed. The greater challenge lies in the scientific assessment of the study design (and then the assessment of safety for pregnant women recruited into the trial). Contributing to this challenge is the complex nature of pregnancy superimposed on medical conditions, the absence of good empirical evidence to inform the study design, and the relatively smaller number of potential participants to recruit into clinical trials of this sort (though the potential participant pool could be higher in multijurisdictional studies with collaboration among research networks and patient groups).

The lack of evidence about how to manage medical conditions during pregnancy can complicate a clinical trial study design, and hence its scientific assessment by research ethics committees, in at least two ways. First, as noted above, absent an agreed upon and evidence-based standard of care, it can be difficult to determine the appropriate comparator: is it ethical to compare the drug under investigation to placebo, or would it be better to compare it to the standard of treatment in non-pregnant women? In her 2011 study on barriers to the inclusion of pregnant women in perinatal mental health research, Anna Brandon interviewed 14 perinatal mental health investigators, 4 (29 %) of whom had concerns about the use of a placebo arm in these clinical trials (8). Two of these investigators expressed the belief that placebo-controlled trials were not acceptable due to the adverse consequences posed by untreated mental illness, even though available treatments were untested in the pregnant population. Two more investigators cited explicit unwillingness on the part of their research ethics committees to approve placebo-controlled trials. While the exact reasons for this unwillingness are not specified, after years of being unable to meet research ethics committee demands one investigator gave up on the idea of placebo-controls and moved to observational trials (Brandon 2011, 8–9). One problem with a strong stance against placebo-controls, as other respondents in Brandon's study pointed out, is that genuine uncertainty continues to exist regarding the efficacy of medications, psychosocial interventions, and placebo responses in the treatment of mental illness in general. This is magnified in the perinatal population when the lack of information about effects on the foetus is taken into consideration (8–9). Second, given the lack of knowledge about the proper management of medical conditions in pregnancy, and the fact that pregnancy has well documented effects on pharmacokinetics, it can be a challenge to determine an appropriate dose or dosage regimen to test. Pre-clinical animal studies may be indicated to inform dose and safety, if appropriate animal disease models exist.

To assess these two specific challenges, as well as other factors in the scientific rationale and study design, specialised knowledge is needed. Given the relative lack of investigational drug research for managing medical conditions during pregnancy, many research ethics committees may lack specific expertise or knowledge to competently review such research proposals. In such cases, they should consult ad hoc advisors.[6] Depending on the clinical trial to be reviewed, research ethics committees may need, or need to solicit, the expertise of clinicians and clinician-researchers from the disease specialty in question, as well as obstetrics, gynaecology, or neonatology, and clinical trial design specialists, including statisticians. The specialised knowledge of relevant clinicians, for example, may be needed to offer insight as to whether a particular dose or dosage regimen is appropriate, given the physiological changes associated with pregnancy. Neonatologists may be consulted about foreseeable consequences (and how to minimise and manage them) of research participation for the foetus and the future newborn. Clinical trial design specialists and statisticians may be needed to help assess the adequacy of the study design, stopping rules, sample size and analysis strategy, given the particular medical complexities of the target population and lesser experience with factors that may affect retention of pregnant women as participants in research involving investigational drugs.

Research proposals must include a summary and discussion of relevant studies done to date, both on the investigational drug and the medical context within which the investigational drug is to be given. The science reviewers should pay particular attention to the quality and findings of prior studies, including serious adverse events reported. Where there seem to be gaps or 'red flags', it is appropriate to ask for more detail, ask for (or conduct) additional review of relevant literature, and assess the impact of available (and missing) knowledge about safety and study design. Likewise science reviewers should assess the proposed comparator and its justification.

The science review must consider the feasibility of recruiting sufficient numbers of pregnant women to such trials to obtain clinically significant results in a timely manner. Given the dearth of empirical evidence and the clinical importance of the

[6] For example Canada's *TCPS2* Article 6.5 and its Application provide guidance on parameters for involving ad hoc advisors.

research question(s), researchers may be overly optimistic about their ability to recruit sufficient participants and complete their trial. Being aware of this, science reviewers should assess whether the research is scientifically justified and whether the objectives, data collection, and analysis plan are achievable. Pilot studies may be needed to assess the feasibility of the study design and the potential to recruit sufficient participants. Particular features of pregnancy, the disease, and the investigational drug may need to be factored into how the research is powered. A number of more general factors may affect recruitment potential, including: a social bias against taking unproven drugs during pregnancy, the novelty of clinical trials to test drugs during pregnancy, the relatively short time-frame of pregnancy within which to recruit, the relatively few pregnant women with the medical condition in question who meet the inclusion criteria for the study, and the portion of eligible women who are likely to volunteer.

Given this, it is important that researchers carefully construct and frame the questions they are trying to answer. We anticipate that research in a pregnant population will take place only after the drug has been approved for the treatment of a particular condition, though not for the treatment of pregnant women with that condition.[7] For such studies, the research questions will focus on the efficacy of treatment in this specific pregnant population (not on whether the investigational drug works for the disease condition). Note that furthermore, and depending on the particular clinical trial, the target participants may already have been taking the drugs as indicated. It is only when they became pregnant that their use became off-label. Since the efficacy of such drugs will have been proven in other investigations, there are different questions that need to be answered in studies involving pregnant women: How does the physiology of pregnancy affect the way that this drug works? And, what effects does it have on the foetus? In other words, how does this drug work in a more medically complex population? Since researchers are not attempting to prove efficacy it may be ethical to have lower standards of evidence, at least for individual trials. The selection of data to be collected should anticipate its potential contribution to future meta-analyses so that conclusions from meta-analyses (as well as the proposed clinical trial) can help to guide future research and clinical practice.

It is also worth noting that, depending on the research questions being asked, there may be other options for study designs (see Healy and Mangin 2016; Farrell et al. 2016). While randomised-controlled trials are the 'gold standard' when it comes to demonstrating the efficacy or superiority of a medication, other methods may suffice for answering some research questions. For example, as Katherine L. Wisner and colleagues' meta-analysis on pharmacologic treatment of depression during pregnancy found, prospective case-controlled trials can yield clinically useful knowledge about the association between drug exposure and pregnancy outcome (Wisner et al. 1999, 1265), without requiring randomisation.

---

[7] Françoise Baylis and Scott A. Halperin (2012) have explored additional options within clinical trials methodologies for research involving pregnant women.

Exposure registries exist for many conditions and medications (Goldkind and Feibus 2010, 53), and regular analysis of this data could provide valuable knowledge about effects on the foetus and efficacy in pregnancy. The weakness of this approach, of course, is that it requires patients and their physicians to sign up and report the results themselves. Not all physicians will be knowledgeable about the existence of these registries, and some may be too busy – or may forget – to report on a regular basis. Reporting to exposure registries does not necessarily require physician involvement, and it is easy to imagine that participant-led initiatives could solve part of this problem by encouraging pregnant women to self-report. However, there is still a question of how to inform the maximum number of pregnant women about the option of contributing information to an exposure registry, given that they will be treated at different times, and in different locations.

This same weakness is not present in well-designed systematic database studies, which involve the examination of the health records of pregnant women who took a particular medication, and a comparison of their adverse pregnancy outcomes to the background rate (Goldkind and Feibus 2010, 53). These studies can be very informative when properly designed, though it may take a long time to get sufficient data, especially if few pregnant women have taken the drug under investigation.

Observational studies are another potential option, though they have weaknesses since the researchers do not control the dose or dosing regimen and are limited to investigating only those drugs that physicians already prescribe. With a large enough sample size the first problem could be overcome, to a certain degree, through a comparison of women who took different doses of the same medications. The second problem, however, is more difficult to solve. For those medications that are not widely used in pregnancy, pragmatic clinical trials[8] (i.e., trials designed to help choose between options of care) may be best suited to answering the research question. Because pragmatic clinical trials have fewer inclusion/exclusion criteria and offer more flexibility in terms of treatment options, they may be more attractive to both pregnant women and their physicians, allowing for greater participation and, therefore, more data.

To the extent that these and other alternative study designs can contribute to answering clinically important research questions, they should be considered in assessing the appropriateness and feasibility of the proposed study design. Some compromise may be unavoidable in achieving statistically significant and clinically important results. Yet alternative study designs and compromise do not obviate the need for clinical trials to answer some clinically important questions with respect to managing medical conditions during pregnancy. For example, clinical trials are better equipped to collect detailed and nuanced information about the pharmacokinetics and pharmacodynamics of drugs taken off-label during pregnancy.

[8] See Zwarenstein et. al. (2008) to distinguish between explanatory and pragmatic clinical trial study designs.

Pregnant Women as a Medically Complex Population

Mary C. Blehar et al. (2013) suggest that pregnant women should be "reclassified" as "a medically complex population, necessitating special scientific and ethical considerations" (e41). Shifting the common assumption that pregnant women are vulnerable to considering pregnant women as a medically complex population is advisable for a number of reasons. First, considering pregnant women in this way acknowledges the fact that they are able to protect their own interests, while at the same time recognising that the presence of the foetus adds both medical and ethical complexities. It also conveys the 'vulnerable situation' of the foetus without classifying it as a 'vulnerable participant' in need of special protection. The foetus, through its intimate connection to the pregnant woman, will be affected by any medication that she takes, but that same intimate connection makes it incorrect to think of the foetus as a participant distinct from the pregnant woman. Finally, this reclassification is necessary as the physiological changes that women undergo when pregnant have known effects on pharmacokinetics. An increase in blood flow through the kidneys causes some medications to be cleared at higher rates during pregnancy, while "increases in blood volume, decreases in gastric emptying time, changes to the concentration of sex hormones, [and] alterations in liver enzymes" can cause other changes in the way that medications work in the pregnant woman's body (Lyerly et al. 2008, 8).

Changing the classification may also help to mitigate the problems caused by the contested issue of 'minimal risk' in research ethics review. As Seema Shah et al. (2004) demonstrated, research ethics committees interpret this concept differently. For example, when asked about the level of risk posed to a child by an Magnetic Resonance Imaging Test (MRI) without sedation, 48 % of research ethics committee chairs indicated that this procedure involved minimal risk, 35 % said that it involved a minor increase over minimal risk, and 17 % rated it as a greater than minor increase over minimal risk (479). Because minimal risk is used in determining eligibility for research participation (with restrictions for those in vulnerable situations), if research ethics committees differ in their interpretation of minimal risk, then whether the research can proceed is contingent on the committee that reviews the protocol. As a 'medically complex' population, it may be ethical to expose participants to more than minimal risk (see Ballantyne and Rogers 2016), though risks should always be minimised, and should be justified by the potential benefits of the intervention to participants as well as the importance of the anticipated knowledge to be gained.

Research ethics committees should pay special attention to the proposed inclusion and exclusion criteria, as well as recruitment procedures, in order to determine whether they are sensitive to situations that may cause women, pregnant or not, to experience increased vulnerability in the research context. While in most jurisdictions (the United States is a notable exception), pregnant women are not considered to be in a vulnerable situation for the purposes of research ethics review, nevertheless, pregnant women may find themselves in other situations that confer vulnerability, and their pregnancy may, in fact, amplify the amount of vulnerability that they experience. For example, the high cost of prenatal care, especially in countries like the United States where health care coverage can be patchy, may make uninsured women more likely than insured or financially well-off women to participate in clinical trials because through their participation they would gain access to medical care they could not otherwise afford. Some pregnant women are in lower socioeconomic brackets – in an already vulnerable situation, which is amplified by their pregnancy – and research ethics review should be sensitive to this.

Balancing Risks and Potential Benefits

While it may be ethical to expose pregnant women and foetuses to risks that are 'greater than minimal,' the fact remains that drugs taken by a pregnant woman will have effects on the foetus. Foreseeable risks to the foetus and to the pregnant woman should be taken seriously, while recognising that excessive concern about foetal risk hinders research and health in this medically complex population.

The mere possibility of teratogenicity is not an acceptable reason to exclude pregnant women and women of childbearing potential from participation in studies from which they might benefit (see Kukla 2016). Research ethics committees should require scientific evidence or a strong justification to support protocols that stipulate that pregnant women are excluded from participation, or that the participation of women is contingent on the use of birth control. Research ethics committees should require all proposals for clinical trials to include a summary of research that has been done as to the teratogenicity/mutagenicity of the drug under investigation. If this information is not available for the investigational drug, then researchers may, at their discretion, appeal to research on similar medications. This ensures that estimations of risks to foetuses are based in fact rather than fear or speculation. Doing this will serve three purposes. First, the lack of evidence that the drug under investigation is harmful, or is extremely likely to be harmful, to the foetus may justify the inclusion of pregnant women in the study or, at the very least, justify the creation of a subsequent study with pregnant participants after the drug has shown promise with similar persons who are not pregnant. Second, the lack of evidence may inspire a move to studies in pregnant animals so that more data can be gathered as to the drug's teratogenic potential. Third, evidence of harm to the foetus, or high probability of harm, may justify the exclusion of pregnant women from research.[9]

Specific procedures such as imaging, biopsies, or blood draws may pose risks to pregnant women and/or foetuses. Research ethics committees should require and carefully assess a summary of the risks inherent to those procedures. Safe (or safer)

[9] Alternatively, it may justify a sub-analysis of data from pregnancy participants, or gathering additional data from these participants to contribute to understanding the effect of the investigational drug on the health of pregnant women and their foetuses (both during the study and in a follow up period).

alternatives should be noted or an explanation of why alternatives are not available should be provided. Doing so has a twofold purpose: evidence of danger to the woman or foetus may justify the exclusion of pregnant participants, and the absence of such evidence should encourage research ethics committees to advocate for research in this population.

While acknowledging that the foetus may be exposed to potential harms in research involving pregnant women, it must also be recognised that the foetus stands to benefit from the pregnant woman's good health. Consider, for example, depression. Anywhere from 8 to 20 % of pregnant women will experience perinatal depression, and a pregnant woman's mental illness has measurable effects on her future newborn. Newborns whose mothers suffered from clinical depression during pregnancy have higher cortisol levels and lower dopamine and serotonin levels, they exhibit less than optimal habituation, orientation, motor activity, and autonomic stability; they demonstrate greater arousal, less attentiveness, less physiological development, and increased irritability (Brandon 2011, 5). Perinatal depression is also associated with increased tobacco and alcohol use during pregnancy, increased risk of preeclampsia, and general illness during pregnancy, all of which can contribute to prematurity and lower birth weight: two of the biggest threats to infant health (Brandon 2011, 5). At least some of these health problems could be mitigated if women were provided with optimal care during pregnancy, and research ethics committees should keep this in mind when weighing the potential harms and benefits of participation in research on depression (see Healy and Mangin 2016).

Somewhat paradoxically, research involving pregnant women also helps to prevent future foetuses from being exposed to dangerous teratogens. While the thalidomide disaster is typically held up as an example of why pregnant women should not take drugs or be involved in research it actually demonstrates the opposite. Women who took thalidomide were not participating in clinical trials. They were simply taking an available drug (see Langston 2016). Careful and responsible research may have attenuated the magnitude of this disaster by providing information about the teratogenic potential of thalidomide, information that would have made physicians reluctant to prescribe it to pregnant women (Lyerly et al. 2008, 10–11). Likewise, the use of ACE inhibitors – widely prescribed for the treatment of hypertension – during the first trimester was recently linked to a small but statistically significant increase in the risk of foetal cardiovascular and neurological abnormalities (Lyerly et al. 2008, 10). Had researchers involved pregnant participants in relevant research, they could have prevented the congenital abnormalities that resulted from three decades of off-label use.

According to Brandon (2011), 64 % of pregnant women in the United States are prescribed one or more drugs for the management of chronic or acute illness during pregnancy (3) and, as noted previously, the teratogenic risk in pregnancy is still unknown for 91 % of drugs (Lo and Friedman 2002, 468). In the absence of robust evidence about the effectiveness of those drugs or their potential teratogenicity, taking off-label drugs exposes pregnant women and their foetuses to risks. The pregnant women are at risk because there is no evidence supporting the drug dosages being used in the context of a number of important physiological changes due to pregnancy. The foetuses, on the other hand, are at risk because their health outcome is intimately tied to that of the pregnant woman, and because a lack of knowledge about the teratogenic potential of drugs is not equivalent to knowledge about their safety. Understanding these and other risks involved and balancing them again potential benefits will be a challenge, but one that researchers, research ethics committees, trials sponsors and funding agencies must grapple with to optimise the medical management of pregnant women.

Evaluating and communicating about risks and potential benefits should occur at all steps of the research from design of the trial, recruitment, and consent, conducting the research interventions, and follow-up. Clear communication about goals, risks, and potential benefits is especially important during recruitment. The objective(s) of the clinical trial should be clearly presented to potential participants, be it to become more knowledgeable on the correct dosages for potentially useful drugs during pregnancy; how the physiological changes caused by pregnancy affect pharmacokinetics; whether a drug will have teratogenic effects on the foetus, etc.

The informed consent document should include statements that the drug(s) under investigation may expose pregnant women and foetuses to unforeseeable risks, that potential benefits are uncertain, and that the risks associated with a lack of information are not unique to research but are also present in off-label management of medical conditions during pregnancy. Potential participants should be made aware of how the risks (and potential benefits) of participation in a clinical trial compare to the risks (and potential benefits) associated with unproven and non-evidence based treatments, and their choice about research participation should be respected.

Follow-Up

As with all clinical trials, research ethics committees should require that research proposals involving pregnant women include a detailed plan for monitoring and responding to safety, efficacy (where feasible) and validity both during and after exposure to the investigational drug (and during and after pregnancy). An independent data and safety monitoring body (DSMB) is an important contributor to this process through its examination, at regular intervals, of data collected and research procedures undertaken. *ICH E8* (see Table 6.4) and CIOMS guidelines recommend follow-up after the birth of the child since some teratogenic effects may only become apparent with time. These guidelines are not specific as to the form or the duration of this follow-up, however involving a DSMB into the follow-up period seems apt.

The literature reveals that when research includes (or targets) pregnant participants, researchers are gathering at least some data on the outcomes of pregnancies. For example, in a trial measuring the efficacy of H1N1 vaccination in a pregnant population, the rate of stillbirths and spontaneous abortions (as compared to live births) was measured among women who had been trial participants, and compared to the rate of stillbirths and spontaneous abortions in the general population (Gorman 2010, 68). More generally, researchers report congenital abnormalities and birth defects as serious adverse events, in accordance with *ICH E6*.

While it is right to be concerned with whether drugs increase the rate of stillbirth, spontaneous abortion, congenital abnormalities, or birth defects, there are other potential consequences worth investigating.[10] For instance, what if taking a particular drug during pregnancy increases the risk that a child will develop asthma, a mental illness, or an intellectual disability? Or, what if the investigational drug affects a woman's ability to conceive in the future? The obvious way to generate answers to these important questions is for there to be long-term follow-up for several years after the end of a clinical trial, and for researchers to look for a broad range of adverse events including less than 'serious' adverse events.

Such follow-up need not be excessively burdensome or intrusive. Parents typically take their newborn children for regular check-ups during the first three years of life. With parental consent it should be possible for researchers to gain access to these records and track the child's development; researchers can note whether any trends, such as delayed physical or mental development, or a propensity toward illness, emerge from the data. It may also be possible to have a note placed on the child's health record so that future physicians are able to know that this child's mother participated in a clinical trial.

Follow-up on the health of the child raises ethical questions about consent. For example, when, if ever, should a child be made aware of her mother's participation in research while pregnant? Is maternal consent sufficient to justify follow-up with a child after birth, or should researchers also obtain consent from the legal father? Our position is that whether consent for follow-up is required depends on the nature of the proposed follow-up. If follow-up only involves access to a child's medical records for a specified amount of time, maternal consent is sufficient. However, if follow-up requires a child to undergo additional tests (beyond those that were part of routine childhood check-ups or clinically indicated), then consent should be obtained from all legally recognised parents.[11]

The trial design should also include appropriate follow-up with the women so that researchers can assess effects on their health. Online reporting platforms set up by trial sponsors could enable participants to report any adverse events that occur, perhaps even after the formal follow-up period. Information in these databases (with identifiers removed) should be made available for meta-analysis, allowing researchers to determine whether a particular outcome can be tied to participation in a particular trial. Reporting of a severe adverse event even after the formal follow-up period may give the trial sponsor grounds to contact other research participants, to warn them of what might happen. Notably, the usefulness of this follow-up method extends beyond trials involving pregnant participants.

Among the possible effects most difficult to assess are those affecting future pregnancies. Where potential effects on future pregnancies are foreseen, the follow-up period should extend throughout the women's years of child bearing potential. In any case, there should be a reporting mechanism available for a woman (or her doc-

[10] Anecdotally, it is the authors' experience serving on research ethics committees that much of the post-pregnancy follow-up involves only collecting information as to the outcome of the pregnancy (i.e., a live birth, the presence or not of any obvious congenital anomalies at birth).

[11] Applicable laws in the jurisdiction in question will take precedence.

tor on her behalf) to report a difficulty in a later pregnancy. This may justify researchers having access to her health records, along with those of other former participants, to determine whether there is an adverse effect worth reporting.

Given the dearth of evidence in the current standard of care for managing medical conditions during pregnancy, the follow-up should gather (and make available) as much clinically useful information as possible about the effects of the drug on both the woman and the foetus, regardless of whether those effects are 'serious'. This should be done to provide future patients with the best possible information, so that they can make informed decisions about whether to take drugs for medical conditions during pregnancy and, if they do, which drug(s) to take and at which dosage(s). As with any medical treatment, patients will have different opinions as to the potential consequences that they are willing to accept and the experienced and foreseeable burdens of their own medically complex situations will factor into that willingness. For example, some women may take a drug that is associated with a lower birth rate or an increased incidence of asthma. Other women may choose to forego the same drug.

6.4 Conclusion

In this chapter we have provided analysis and recommendations for research ethics review of clinical research involving pregnant women. See Table 6.6 for a summary of our recommendations.

Our focus has been on clinical trials of investigational drugs (including investigation of drugs used off-label during pregnancy) that address important clinical questions that pertain to medical conditions women face during pregnancy. It is important to encourage and facilitate such research to enable safe, effective, evidence-based management of medical conditions that are prevalent during pregnancy, such as lupus, high blood pressure, and mental illness. Our goals have been to improve research ethics review of research involving pregnant women by offering guidance where the major national and international guidelines are silent, and to help research ethics committees facilitate medically important clinical research involving pregnant participants.

The considerations and recommendations we have raised are not only for research ethics committees, however. Researchers, trial sponsors, and funding agencies are also important players in the clinical trial conceptualisation and approval process, and they too should facilitate drug research that is important to the well-being of women during pregnancy. While there are undoubtedly situations in which clinical research involving pregnant women are inappropriate due to the nature of foreseeable risks or frivolous research questions, the harm that occurs to pregnant women and their foetuses as a result of poor management of medical conditions during pregnancy in the absence of reliable evidence should be ameliorated. Current practices of blanket exclusion and research avoidance are not ethically justified.

Acknowledgements The authors gratefully acknowledge Angela Ballantyne, Françoise Baylis, Annette Rid, and Verina Wild for their helpful comments on a previous draft of this chapter.

Appendix

Table 6.1 Excerpt from US Department of Health and Human Services. 2009. *Code of Federal Regulations: Title 45, Part 46, Protection of Human Subjects*

Subpart B, §46.204 Research involving pregnant women or fetuses.
Pregnant women or fetuses may be involved in research if all of the following conditions are met:
(a) Where scientifically appropriate, preclinical studies, including studies on pregnant animals, and clinical studies, including studies on nonpregnant women, have been conducted and provide data for assessing potential risks to pregnant women and fetuses;
(b) The risk to the fetus is caused solely by interventions or procedures that hold out the prospect of direct benefit for the woman or the fetus; or, if there is no such prospect of benefit, the risk to the fetus is not greater than minimal and the purpose of research is the development of important biomedical knowledge which cannot be obtained by any other means;
(c) Any risk is the least possible for achieving the objectives of research;
(d) If the research holds out the prospect of direct benefit to the pregnant woman, the prospect of a direct benefit both to the pregnant woman and the fetus, or no prospect of benefit for the woman or the fetus when risk to the fetus is not greater than minimal and the purpose of the research is the development of important biomedical knowledge that cannot be obtained by any other means, her consent is obtained in accord with the informed consent provisions of subpart A of this part;
(e) If the research holds out the prospect of direct benefit solely to the fetus then the consent of the pregnant woman and the father is obtained in accord with the informed consent provisions of subpart A of this part, except that the father's consent need not be obtained if he is unable to consent because of unavailability, incompetence, or temporary incapacity or the pregnancy resulted from rape or incest;
(f) Each individual providing consent under paragraph (d) or (e) of this section is fully informed regarding the reasonably foreseeable impact of the research on the fetus or the neonate;
(g) For children, as defined in §46.402(a) who are pregnant, assent and permission are obtained in accord with the provisions of subpart D or this part;
(h) No inducements, monetary or otherwise, will be offered to terminate a pregnancy;
(i) Individuals engaged in the research will have no part in any decisions as to the timing, method, or procedures used to terminate a pregnancy; and
(j) Individuals engaged in the research will have no part in determining the viability of a neonate.

Table 6.2 Excerpt from Canadian Institutes of Health Research, Natural Sciences and Engineering Council of Canada, and Social Sciences and Humanities Research Council of Canada (2014). *Tri-Council Policy Statement: Ethical Conduct of Research Involving Humans (TCPS2)*

Article 4.3: Women shall not be inappropriately excluded from research solely on the basis of their reproductive capacity, or because they are pregnant or breastfeeding.
Application: Researchers should not exclude women from research on the basis of their reproductive capacity, or their pregnancy, or because they are breastfeeding, unless there is a valid reason for doing so.
Subjecting women of childbearing potential to inappropriate requirements precludes their participation in research. Exclusions should be made on the basis of clear criteria that reflect balanced attention to the potential benefits as well as the foreseeable risks of the research that may affect the welfare of women. For example, researchers should not require participants to use oral contraception, unless there is a valid reason for doing so.
In considering research on pregnant or breastfeeding women, researchers and REBs shall take into account foreseeable risks and potential benefits for the women and her embryo, fetus or infant, as well as the foreseeable risks and potential benefits of excluding pregnant or breastfeeding women from the research.

Table 6.3 Excerpt from Council for the International Organizations of Medical Sciences 2002. *International Ethical Guidelines for Biomedical Research Involving Human Subjects*

Guideline 17: Pregnant women as research participants
Pregnant women should be presumed to be eligible for participation in biomedical research. Investigators and ethical review committees should ensure that prospective subjects who are pregnant are adequately informed about the risks and benefits to themselves, their pregnancies, the fetus and their subsequent offspring, and to their fertility.
Research in this population should be performed only if it is relevant to the particular health needs of a pregnant woman or her fetus, or to the health needs of pregnant women in general, and, when appropriate, if it is supported by reliable evidence from animal experiments, particularly as to risks of teratogenicity and mutagenicity.

Table 6.4 Excerpt from International Council on Harmonisation (1997). *ICH Harmonised Tripartite Guideline: General Considerations for Clinical Trials E8*

3.1.4.3 Special Populations
(a) Investigations in pregnant women
In general, pregnant women should be excluded from clinical trials where the drug is not intended for use in pregnancy. If a patient becomes pregnant during administration of the drug, treatment should generally be discontinued if this can be done safely. Follow-up evaluation of the pregnancy, foetus, and child is very important. Similarly, for clinical trials that include pregnant women because the medicinal product is intended for use during pregnancy, follow-up of the pregnancy, foetus, and child is very important.

Table 6.5 Excerpt from International Council on Harmonisation (1996). *ICH Harmonised Tripartite Guideline: Guideline for Good Clinical Practice E6*

1.50 Serious Adverse Event (SAE) or Serious Adverse Drug Reaction (Serious ADR)

Any untoward medical occurrence that at any dose:

- results in death,
- is life-threatening,
- requires inpatient hospitalization or prolongation of existing hospitalization,
- results in persistent or significant disability/incapacity, or
- is a congenital anomaly/birth defect

Excerpt from the European Parliament and the Council of the European Union (2001). *Directive 2001/20/EC of the European Parliament and of the Council*

Article 2: Definitions

(o) 'serious adverse event or serious adverse reaction': any untoward medical occurrence or effect that at any dose results in death, is life-threatening, requires hospitalisation or prolongation of existing hospitalisation, results in persistent or significant disability or incapacity, or is a congenital abnormality or birth defect.

Table 6.6 Summary of recommendations for research ethics committees

1. Carefully consider the research question. Given the research question, is the involvement of pregnant women acceptable? Is the study design appropriate?

2. Consult experts—for example, specialists in the disease under investigation, obstetricians, neonatologists, or statisticians—as needed during science review.

3. Classify pregnant women as a 'medically complex' population (and avoid tendency to assume they are 'vulnerable'). Keep in mind, however, that pregnancy can amplify situations of vulnerability.

4. Understand that the interests of the woman and the interests of the foetus are not necessarily in conflict. The relationship between the two is intimate and interlinked. Benefits to women's health through evidence-based management of their medical conditions will also benefit foetuses to a certain degree.

5. Demand that the exclusion of pregnant participants be based on scientific evidence. When there is no evidence, and exclusion is not justified, ask whether there is a way to include pregnant women in research.

6. Remember that there are risks associated with not doing research, and take these into consideration when weighing risks and potential benefits.

7. Ensure that researchers have an adequate follow-up plan that spans several years, and considers a wider range of serious, and 'less serious', adverse events.

References

Ballantyne, A., and W. Rogers. 2016. Pregnancy, vulnerability, and the risk of exploitation in clinical research. In *Clinical research involving pregnant women*, eds. F. Baylis and A. Ballantyne, 139–159. Cham: Springer.

Baylis, F., and S.A. Halperin. 2012. Research involving pregnant women: Trials and tribulations. *Clinical Investigation* 2(2): 139–146.

Baylis, F., and R. MacQuarrie. 2016. Why physicians and women should want pregnant women included in clinical trials. In *Clinical research involving pregnant women*, eds. F. Baylis and A. Ballantyne, 17–31. Cham: Springer.

Blehar, M.C., C. Spong, C. Grady, S.F. Goldkind, L. Sahin, and J.A. Clayton. 2013. Enrolling pregnant women: Issues in clinical research. *Women's Health Issues: Official Publication of the Jacobs Institute of Women's Health* 23(1): e39–e45.

Brandon, A.R. 2011. Ethical barriers to perinatal mental health research and evidence-based treatment: An empirical study. *AJOB Primary Research* 2(1): 2–12.

Canada (Canadian Institutes of Health Research, Natural Sciences and Engineering Council of Canada, and Social Sciences and Humanities Research Council of Canada). 2014. *Tri-Council Policy Statement: Ethical conduct for research involving humans (TCPS2)*. http://www.pre.ethics.gc.ca/pdf/eng/tcps2-2014/TCPS_2_FINAL_Web.pdf. Accessed 3 Oct 2016.

Cattapan, A. 2014. *Public consultation submission to the Canadian Interagency Panel on Research Ethics*. http://www.pre.ethics.gc.ca/english/publicparticipation/comments-commentaires_2014/pdfs_comments/44.pdf. Accessed 27 June 2014.

Chervenak, F.A., and L.B. McCullough. 2011. An ethically justified framework for clinical investigation to benefit pregnant and fetal patients. *The American Journal of Bioethics* 11(5): 39–49.

CIOMS (Council for International Organizations of Medical Sciences). 2002. *International ethical guidelines for biomedical research involving human subjects*. Geneva: World Health Organization. http://www.cioms.ch/publications/layout_guide2002.pdf. Accessed 3 Oct 2016.

Council of Europe. 2005. *Additional protocol to the Convention on Human Rights and Biomedicine, concerning biomedical research*. http://conventions.coe.int/Treaty/en/Treaties/html/195.htm. Accessed 1 July 2014.

DHHS (US Department of Health and Human Services). 2009. *Code of Federal Regulations: Title 45, Part 46, Protection of Human Subjects*. http://www.hhs.gov/ohrp/regulations-and-policy/regulations/45-cfr-46/index.html. Accessed 3 Oct 2016.

Diav-Citrin, O., and G. Koren. 2011. Drug use during pregnancy. In *Therapeutic choices*, 6th ed, 1720–1729. Ottawa: Canadian Pharmacists Association.

Farrell, R.M., and R. Flyckt. 2016. Research on uterine transplantation: Ethical considerations. In *Clinical research involving pregnant women*, eds. F. Baylis and A. Ballantyne, 285–297. Cham: Springer.

Goldkind, S.F., and K.B. Feibus. 2010. Pregnant women in clinical trials: Scientific and ethical considerations. In *Enrolling pregnant women: Issues in clinical research*. Bethesda: Office of Research on Women's Health, National Institutes of Health. http://orwh.od.nih.gov/resources/policyreports/pdf/ORWH-EPW-Report-2010.pdf. Accessed 1 July 2014.

Gorman, R.L. 2010. The H1N1 trial. In *Enrolling pregnant women: Issues in clinical research*. Bethesda: Office of Research on Women's Health, National Institutes of Health. http://orwh.od.nih.gov/resources/policyreports/pdf/ORWH-EPW-Report-2010.pdf. Accessed 1 July 2014.

Healy, D., and D. Mangin. 2016. Does my bias look big in this? In *Clinical research involving pregnant women*, eds. F. Baylis and A. Ballantyne, 197–208. Cham: Springer.

ICH (International Council for Harmonisation). 1996. *ICH harmonised tripartite guideline: Guideline for good clinical practice E6*. http://www.ich.org/fileadmin/Public_Web_Site/ICH_Products/Guidelines/Efficacy/E6/E6_R1_Guideline.pdf. Accessed 1 Aug 2014.

ICH (International Council for Harmonisation). 1997. *ICH harmonised tripartite guideline: General considerations for clinical trials E8*. http://www.ich.org/fileadmin/Public_Web_Site/ICH_Products/Guidelines/Efficacy/E8/Step4/E8_Guideline.pdf. Accessed 1 Aug 2014.

IWK Health Centre. 2013. Research ethics board review criteria: Secondary reviewer. REB Version: 2013/09/25. In *Application materials & forms: REB reviewers' tools*. http://www.iwk.nshealth.ca/research/application-materials-forms. Accessed 13 May 2015.

Johnson, L.S.M. 2016. When hypothetical vulnerability becomes actual: Research participation and the autonomy of pregnant women. In *Clinical research involving pregnant women*, eds. F. Baylis and A. Ballantyne, 161–178. Cham: Springer.

Kaposy, C. 2016. Presumptive inclusion and legitimate exclusion criteria. In *Clinical research involving pregnant women*, eds. F. Baylis and A. Ballantyne, 51–62. Cham: Springer.

Kukla, R. 2016. Equipoise, uncertainty, and inductive risk in research involving pregnant women. In *Clinical research involving pregnant women*, eds. F. Baylis and A. Ballantyne, 179–196. Cham: Springer.

Langston, L. 2016. Better safe than sorry: Risk, stigma, and research during pregnancy. In *Clinical research involving pregnant women*, eds. F. Baylis and A. Ballantyne, 33–50. Cham: Springer.

Levine, R.J. 2010. IRB Perspective on inclusion of pregnant women in clinical research. In *Enrolling pregnant women: Issues in clinical research*. Bethesda: Office of Research on Women's Health, National Institutes of Health. http://orwh.od.nih.gov/resources/policyreports/pdf/ORWH-EPW-Report-2010.pdf. Accessed 1 July 2014.

Little, M.O. 2010. Treating important medical conditions during pregnancy. In *Enrolling pregnant women: Issues in clinical research*. Bethesda: Office of Research on Women's Health, National Institutes of Health. http://orwh.od.nih.gov/resources/policyreports/pdf/ORWH-EPW-Report-2010.pdf. Accessed 1 July 2014.

Lo, W.Y., and M.J. Friedman. 2002. Teratogenicity of recently introduced medications in human pregnancy. *Obstetrics & Gynecology* 100(3): 465–473.

Lyerly, A.D., M.O. Little, and R. Faden. 2008. The second wave: Toward responsible inclusion of pregnant women in research. *International Journal of Feminist Approaches to Bioethics* 1(2): 5–22.

Office of Sponsored Projects Administration. 2010. *Human subjects protection: A guide for researchers*. Southern Illinois University Carbondale. Updated May 7, 2010.

Schonfeld, T. 2013. The perils of protection: Vulnerability and women in clinical research. *Theoretical Medicine and Bioethics* 34(3): 189–206.

Shah, S., A. Whittle, B. Wilfond, G. Gensler, and D. Wendler. 2004. How do institutional review boards apply the federal risk and benefit standards for pediatric research? *JAMA* 291(4): 476–482.

The European Parliament and the Council of the European Union. 2001. Directive 2001/20/EC of the European Parliament and of the Council. *Official Journal of the European Communities*. http://www.eortc.be/Services/Doc/clinical-EU-directive-04-April-01.pdf. Accessed 1 July 2014.

WHO (World Health Organization). 2011. *Standards and operational guidance for ethics review of health-related research with human participants*. http://whqlibdoc.who.int/publications/2011/9789241502948_eng.pdf?ua=1. Accessed 1 July 2014.

Wild, V. 2012. How are pregnant women vulnerable research participants? *International Journal of Feminist Approaches to Bioethics* 5(2): 82–104.

Wild, V., and N. Biller-Andorno. 2016. Pregnant women's views about participation in clinical research. In *Clinical research involving pregnant women*, eds. F. Baylis and A. Ballantyne, 119–136. Cham: Springer.

Wisner, K.L., A.J. Gelenberg, H. Leonard, D. Zarin, and E. Frank. 1999. Pharmacologic treatment of depression during pregnancy. *JAMA* 282(13): 1264–1269.

World Medical Association. 2013. *WMA Declaration of Helsinki: Ethical Principles for Medical Research Involving Human Subjects*. http://www.wma.net/en/30publications/10policies/b3/. Accessed 3 Oct 2016.

Zwarenstein, M., S. Treweek, J.J. Gagnier, D.G. Altman, S. Tunis, B. Haynes, A.D. Oxman, D. Moher, and for the CONSORT and Pragmatic Trials in Healthcare (Practihc) groups. 2008. Improving the reporting of pragmatic trials: An extension of the CONSORT statement. *BMJ* 337: a2390.

Chapter 7
Pregnant Women's Views About Participation in Clinical Research

Verina Wild and Nikola Biller-Andorno

Abstract *In this chapter we first discuss the impact of thalidomide on the approach to drug use and clinical research with pregnant women in Germany. We then present results from a qualitative interview study conducted in Göttingen, Germany, in 2003. The interviews provide insights into women's thoughts and experiences in relation to decision-making during pregnancy, illness and treatment during pregnancy, their maternal-foetal relationship, and how they assess different types of clinical research scenarios. The results reveal the shortcomings of the current restrictive approach to drug therapy during pregnancy. They also help answer some conceptual questions about clinical research with pregnant women. The ultimate aim is to work towards a balanced approach that respects the autonomy and decision-making capacity of pregnant women, protects the foetus from preventable harm, and generates well-researched and officially approved drugs for use during pregnancy.*

Despite our knowledge concerning the underrepresentation of pregnant women in clinical research, and changes in regulations and laws over the last one or two decades towards a more inclusive policy, little about recruitment for clinical research involving pregnant women seems to have changed (Lyerly et al. 2008). Even for relatively common conditions, such as nausea, urinary tract infections or deep vein thrombosis, there is insufficient data from clinical research to recommend specific drug treatment in pregnancy (Che Yaakob et al. 2010; Matthews et al. 2010; Vazquez and Abalos 2011). This has potentially grave consequences for pregnant women's health. The question of *how* to include pregnant women in clinical research remains

V. Wild, MD, PhD (✉)
Institute of Biomedical Ethics and History of Medicine, University of Zurich, Zurich, Switzerland

Department of Philosophy, Ludwig-Maximilians-University Munich, Munich, Germany
e-mail: v.wild@lrz.uni-muenchen.de

N. Biller-Andorno, MD, PhD
Institute of Biomedical Ethics and History of Medicine, University of Zurich, Zurich, Switzerland

F. Baylis, A. Ballantyne (eds.), *Clinical Research Involving Pregnant Women*, Research Ethics Forum 3, DOI 10.1007/978-3-319-26512-4_7

to be answered. In relation to this, we consider one particular perspective: that of pregnant women. We are convinced that a discussion on how to include pregnant women in clinical research should not take place without giving pregnant women a voice. This chapter presents results from a qualitative interview study we conducted in Göttingen, Germany, in 2003. Before presenting the interview study we provide some contextual background.

7.1 Thalidomide as a Paradigm Shift in Drug Use

In 1954, the German company *Chemie Grünenthal GmbH* developed the drug Contergan® (thalidomide) and sold it to a total of 46 countries. The number of affected children with malformations related to the use of Contergan® during pregnancy is estimated to be 2,000 in Germany alone (Marquardt 1994). Even though teratogenic effects had been discussed occasionally among experts before these events, the experiences with Contergan® dramatically changed the way drug use during pregnancy was perceived:

> "In 1961, everything changed: Germany and the world understood in a revelation ["Heilsamer Schock"] that therapeutic drugs do not only characterize progress, but that they can hold an enormous, at first often unrecognized, potential for danger [...]. It was the fall of humankind after a naïve and careless use of drugs." (Müller-Oerlinghausen 2005, 33; authors' translation).

Thalidomide was also the catalyst of a fundamental paradigm shift for drug use during pregnancy: "Then came the Thalidomide catastrophe – and suddenly the world was aware that the human embryo was not sequestered in an impervious maternal body where it was shielded from all but genetic harm" (Wilson 1979, 205; see also Langston 2016). The pathophysiological reason for the teratogenic effect of thalidomide was discovered 50 years after its harmful use (Therapontos et al. 2009). No other drug has yet been identified that is considered as dangerous for foetal development as thalidomide (Schaefer et al. 2006, 3).

Justified Precaution or Irrational Panic?

Shortly after the experiences with Contergan®, drug companies started to introduce precautionary advice for drug use. Even when no damaging effects were found in preclinical research, the package inserts would recommend that the drug not be used during pregnancy, or if used during pregnancy then only under close medical supervision (Müller 1969, 1687). Meanwhile, medical experts and scientists noted and widely criticised the ongoing, dangerously high drug prescription rates during pregnancy, in the absence of clinical trial data and in contravention of manufacturers' recommendations (Scott 1974; Murphy 1984; Kasilo et al. 1988).

In subsequent years, scientists learned more about drug metabolism during pregnancy, particularly relating to changes in renal function, hormonal activity or increased volumes of body fluids, and yet the call for restrictive drug prescription practices intensified (Estler 1995; Paulus 1999; Garland 1998; Thürmann and Steioff 2001). In response to such calls, public concern about foetal harm from the use of therapeutic drugs during pregnancy grew, especially among pregnant women. This fear has since been criticised, however, as both inappropriate and alarmist. For example, a study in Canada showed that unfounded fears and worries about foetal harms tended to lead to abortions. The study concluded that if women were better informed about the known risks of therapeutic drugs, such abortions could be prevented (Koren et al. 1998). Another study, in Hungary, involving more than 60,000 pregnant women showed that only 1 % of malformations in children could be traced back to the harmful effects of drug use during pregnancy. In contrast, 16 % of abortions were performed in anticipation of foetal harm after drug intake (Czeizel 1999). The strong fear of foetal harm after drug intake has been characterised as 'irrational panic,' given that an 'indication for foetal abortion' on the basis of drug use during pregnancy is very rare (Paulus 1999).

The Situation in Germany

In Germany, no European or national law explicitly regulates clinical research on pregnant women,[1] and pregnant women remain widely excluded from clinical research. In 2003, German bioethicists started asking whether the current under-representation of pregnant women in clinical research was an appropriate protection, or whether it was a form of paternalistic and harmful overprotection (Biller-Andorno and Wild 2003a, b). Leading reproductive toxicologists criticised this bioethical perspective, and stipulated that the inclusion of pregnant women in clinical research was neither necessary nor desirable (Schaefer et al. 2004). From their point of view, collecting retrospective data was sufficient for the improvement of medical treatment during pregnancy and feminist claims could not justify research involving pregnant women: "It is not permissible to make a pregnant woman responsible for an experiment which she cannot judge sufficiently. The reference to the argument of equality [...] seeks to justify the potentially risky testing of questionably beneficial drugs with an emancipatory pretension" (Schaefer et al. 2004, A166; authors' translation). An expert opinion to the federal Enquête Commission, 'Ethics and law in modern medicine,' argued along similar lines. It referred to the autonomy and decision-making capacity of pregnant women and explicitly stated that pregnant women could not adequately judge the situation (Bobbert 2004; see also Kaposy 2016).

The categorical exclusion of pregnant women from clinical research for unspecified 'ethical reasons' seems to be accepted in Germany. For example, the author of

---

[1] The Additional Protocol to the European *Convention on Human Rights and Biomedicine* regulates clinical research on pregnant women, but Germany did not sign and ratify the Protocol so far (Council of Europe 2005).

a study on antibiotics in pregnancy claimed, "It goes without saying that clinical research on pregnant women are forbidden for ethical reasons" (Imhof 2005, 5; authors' translation). During the H1N1 epidemic in 2009, journalists wrote: "Vaccination [...] has not been tested on pregnant women at all. Clinical research on pregnant women is classified as unethical" (Winkelheide 2009; authors' translation).

7.2 The Idea for an Interview Study

In the early 1990s, lack of evidence to support therapeutic decisions during pregnancy increasingly led to the realisation that pregnant women needed to be included in clinical research in order to ensure high-quality care (Caschetta and Chavkin 1994; Merkatz et al. 1993). From a scholarly ethical perspective, the so-called 'vulnerability' of pregnant women as a justification of exclusion from clinical research was rejected, because there was no reason to believe that pregnant women could not make autonomous decisions, or that pregnant women were particularly prone to being exploited (Anderson 1994; Macklin 1994; Mastroianni et al. 1994; see also Ballantyne and Rogers 2016). As a significant result of this debate, since 2001 various regulations have been revised to encourage and regulate the inclusion of pregnant women in clinical research (CIOMS 2002; Council of Europe 2005; DHHS 2009). As noted above, however, this change in perspective has not yet been fully appreciated in Germany where there is a persistent reluctance to acknowledge that pregnant women are competent autonomous decision makers, and to call for equity in research participation. The dominant concern, rather, is to protect pregnant women (or, possibly more importantly, foetuses) from any kind of risk resulting from participation in clinical research.

In this context, we initiated an interview study to learn what German women thought about or experienced in relation to decision-making during pregnancy, illness and treatment during pregnancy, their maternal-foetal relationship, and how they assessed different clinical research scenarios. Our primary aim was to give voice to those who were at the centre of the debate about the inclusion of pregnant women in clinical research. Our interview study was conducted a decade ago, and published in German (see Wild 2010). With this English publication we broaden the scope of the discourse giving pregnant women in Germany a voice in the ethical debate about whether and how to perform clinical research that includes them.

Four assumptions grounded the development of our interview questions. We assumed:

(a) A restrictive approach to drug therapy during pregnancy could work to prevent foetal harm. Yet, it seemed to us that public and academic discourses overlooked the limitations of the restrictive approach, particularly in the case of more complex health care needs during pregnancy. We aimed to illustrate some of these limitations in the interviews.

(b) An aversion to the participation of pregnant women in clinical research. We wanted to know whether women themselves would wish to increase the participation of pregnant women in clinical research and how these women would react to different clinical research scenarios.
(c) An understanding of pregnant women's perceptions of the foetal-maternal relationship is important in determining how to include pregnant women in clinical research. We intended to explore this issue in the interviews.
(d) Decision-making capacity generally is not impaired during pregnancy and pregnant women are not 'vulnerable' per se. We aimed to find out more about situations that might increase situational vulnerabilities.

Method

In our study, 30 pregnant women or women who had recently given birth were interviewed from January to August, 2003. We conducted semi-structured interviews based on an interview guide. The study received ethics approval from the ethics committee of the University Medical Center Göttingen in 2002. Questions were grouped into six categories: (1) General questions; (2) Experience with medical care during pregnancy; (3) Participation in clinical research in general; (4) Participation in clinical research during pregnancy; (5) General issues on how to perform clinical research during pregnancy; and (6) The maternal-foetal relationship.[2]

The inclusion criteria were, 'pregnancy or recent birth' and 'willingness to participate in the interview'. In order to reach a sample of women with a diverse array of pregnancy and birth histories, we approached providers on the primary (private practice), secondary (catholic hospital), and tertiary levels (university hospital). In order to make contact with potential study participants, we approached the physicians in charge on random days and asked who among their patients might meet our two inclusion criteria. With the list of names they gave us, we approached each of those women individually, explained our study, and asked for permission to

[2] Questions per category: (1) General questions: (expected) date of birth, previous pregnancies, course of current pregnancy (disease, complications, etc.), professional background, and current job situation. (2) Experience with medical care during pregnancy: How did the clinical research participant experience drug intake? How well did she feel informed? Who or what helped to make decisions? How was the physician-patient-relationship? Did she change certain habits and if so, why? (3) Participation in clinical research in general: Did the interview partner ever participate in clinical research? Would the woman participate in clinical research that tests a new drug for a certain disease from which she suffers? Would the woman participate in clinical research as a healthy participant that tests a new drug, involves blood taking, and for which she receives a financial reward? (4) Participation in clinical research during pregnancy: What is the general opinion about clinical research involving pregnant women? Would the woman participate in certain types of clinical research? (5) General issues on how to perform clinical research: Involvement of the partner, decision-making procedure, role of the researcher. (6) The maternal-foetal relationship: How do women perceive the relationship?

interview them, to tape record the interview, and to use the data for future publication in anonymised form. After the women signed an informed consent form we began to record the interviews. Of the 31 women who were asked to participate in the interview study, one declined. The first 18 interviews were held at a catholic hospital. Ten more interviews were held at a university hospital. Two interviews were arranged with the help of a gynaecologist from a private practice and were held at the homes of the women. The interviews were transcribed verbatim. The qualitative analysis was performed through manual coding, and by categorising the coded sections in 16 different main categories (e.g., decision, birth, child, illness/disease, medication, physician, relation to the unborn, breastfeeding, etc.) with several sub-categories (Meuser and Nagel 1991; Mayring 2008).

Results

Drug Therapy During Pregnancy: The Restrictive Model

Many women claimed, without further explanation, that drug use during pregnancy should be avoided or that a restrictive approach to pregnancy care should be pursued: "Well, every woman knows that one should not take medication during pregnancy[3]" (Mrs. G[4]). Only one woman linked this explicitly to thalidomide:

> Mrs M: If you think about drugs during pregnancy, I think you will always come back to Contergan® and discuss it. The people affected by this are still alive. They are in our society and maybe it's also because it was the generation of my mother. I could have been affected by Contergan®. It has not disappeared from our minds, including pregnant women's minds.

Women showed a strong willingness to endure illness during pregnancy, without risking drug use or other possibly harmful medical intervention, such as an X-ray:

> Interviewer: What happened to your arm? (pointing to her arm which is in plaster).
> Mrs. V: I tripped a few days before giving birth and fell on my arm. That day I went to the hospital. But they did not want to do an X-ray because of my pregnancy and they put some cream on it [...]. I thought, maybe it's just a strain, but I couldn't move my arm properly.

After the woman gave birth, a fracture was diagnosed and properly treated.

There was, therefore, some support among study participants for the restrictive model of pregnancy care. Women chose to endure pain and other forms of discomfort, and to restrict drug therapy, to prevent harm to the foetus.

[3] All interview passages quoted here have been translated from German to English by the authors.

[4] The women received code-letters for the purpose of anonymisation.

Drug Therapy During Pregnancy: Deficiencies and Complications

The interviews also revealed a number of stories that challenged the restrictive model of pregnancy care. For example, one woman described how it was expected that pregnant women would endure symptoms without medication, and how she could hardly cope with the untreated illness:

> Mrs. M: I had strong reflux during pregnancy […]. There are these gels for the stomach, but they don't like to prescribe them. They start grumbling and ask: 'How bad is it for you?' before they prescribe anything. You really have to insist and plan before they give you a drug. Mostly they say: 'You have to endure it'.

Another woman (Mrs. Q) called the ambulance because she felt so weak from nausea and vomiting. The arriving health professionals [the woman said: "emergency physicians", but it might also have been paramedics] did not treat her or take her to a hospital. Instead, they said that she needed to see a specialised gynaecologist. She was told to drink tea with salt and sugar, even though, due to her severe nausea, drinking tea or water was not possible.

There were other women who wished to comply with the restrictive model of pregnancy care, but who then fell ill and experienced emotional or physical difficulties, resulting in their accepting the drug treatment. One woman, for example, who suffered from cervical insufficiency from week 22 onward and was treated with a cervical cerclage under general anaesthesia, said:

> Mrs. W: You know, you are not supposed to take drugs during pregnancy, and then they bang all this into you. You ask, for what is this and that, and it's just that 'you got to do it' and in it goes. And then they stood there again with their fucking needle and then all the drugs again […]. You are really happy when you get rid of one, and get rid of the next, and then they bang it back in again. Every woman loses her nerves here once. That's for sure.

The same woman explained her experience during the anaesthesia. According to her, one drug was not administered in order to minimise risks to her foetus. Mrs. W did not explicitly say what drug was administered, but in analysing her story it seems that a muscle relaxant was used and had an impact on her breathing, but that the effect of a sedative component to induce loss of consciousness was insufficient:

> Mrs. W: And then I could not breathe anymore. I wanted to tell them but I couldn't. I thought I would die, I couldn't breathe. I concluded my life was over. I tried to pant for air and move my hands but it was not possible anymore. Usually you get a drug for that.

Furthermore, women explained that they were worried, because they took medication during pregnancy and felt there was inadequate supervision and management of their treatment. One woman (Mrs. S), for example, was sent to a genetic consultation because she had taken Iberogast® in early pregnancy, an herbal drug used for stomach illness. The drug company writes on its homepage, "Although the drug is well tolerated, the herbal drug Iberogast® should be taken during pregnancy only after consulting a physician." (Iberogast 2013; authors' translation). Referral to a genetic counsellor made the woman fear the child would be genetically handicapped. During the consultation, the genetic history of the family was meticulously analysed and this perpetuated the fear. Finally, at the end of the consultation, the woman was relieved to be informed that Iberogast® includes liquorice root, which might induce pre-term contractions if taken in a high dosage (which she did not do).

Some women expressed distrust towards their physicians. One woman, for example, suffered from hay fever. The physician looked at the package insert and answered, rather imprecisely, that maybe, yes, she could take the drug. The woman was disappointed; she had hoped that the physician would offer better advice or search for a safer alternative.

Hence, problematic situations during pregnancy resulted from: untreated diseases despite an urgent need for treatment, fear of harming the future child by taking medication, inadequate management of drug use and its potential harms, overprotection by physicians, and deficient or missing information from physicians. These problematic situations could result in frustration, helplessness, insecurity, fear, pain, or physical burdens.

The Maternal-Foetal Relationship

We identified three central issues regarding the maternal-foetal relationship.

First, many women mentioned an unconditional love or affection towards the future child, and the ability to endure painful situations if this helped the foetus. The foetus was prioritised in many situations, and the pregnant woman's needs and wishes were described as secondary or even irrelevant: "For example, I do not eat fruit. But then I reminded myself all the time to eat enough milk-products and fruit. But I think this happens for many women automatically. You think about the baby, more than you do yourself." (Mrs. H)

Second, a single, generalised description of the relationship to the foetus is impossible. The descriptions varied from woman to woman, and sometimes the relationship also changed over the course of one pregnancy. Some spoke about a relationship, not to a different person but to a part of the 'self'; others had the impression that the foetus is a somewhat different person; others mentioned the gradual change in the relationship over the course of the pregnancy: "In the beginning you say 'I', 'I' feel bad. And later you say 'we'" (Mrs. R). Some women described the bond as very strong, while others struggled with bonding. Some women also explained that the relationship changed in certain situations, for example in the case of disease. Mrs. K, for example, suffered from severe nausea (hyperemesis with dehydration and convulsions) in her eighth week of gestation:

> Mrs. K: At the moment we are two.
> Interviewer: Why?
> Mrs. K: Because I was suffering so badly, and I thought this isn't yet as it should be.
> Interviewer: Did you already experience this feeling at the beginning of the pregnancy?
> Mrs K: I have only been thinking this since I felt so ill.

Third, the first perception of foetal movements was mentioned often as a crucial moment for the relationship between the pregnant woman and the foetus: "When something starts moving in the belly, you start thinking differently. That you are not alone anymore, but that you are two." (Mrs. G)

The answer to how women perceived their relationship to the foetus can best be captured by emphasising the *individual* and *dynamic* character of the unique maternal-foetal relationship.

Decision-Making During Pregnancy

In our interviews, more than half (18) of the women claimed not to have experienced a change in decision-making processes or capacity: "There is this saying that women become somewhat mentally incompetent during pregnancy, but I do not believe that." (Mrs. P). However, many factors influenced decisions made and the quality of the decision-making process in certain situations, for example: time pressure, lack of evidence for treatment decisions, or social and family context.

By contrast, nine study participants perceived a change in their decision-making processes. Seven of them explained that choices were more strongly deliberated and weighed, and that as a result decision-making sometimes took longer, or there was a new sensitivity regarding increased responsibility during pregnancy: "Suddenly you have to think of two. You don't carry responsibility only for yourself anymore. If you race with your car you could crash into a tree, but now you would crash two people into the tree" (Mrs. E). One woman mentioned that she felt her decision-making capacity was reduced during the process of giving birth, another woman said it was more difficult to make decisions during illness.

Two women expressed the view that their decision-making capacity improved during pregnancy: "If you know that you want to have the child, then you know pretty well what you want for yourself, in my experience." (Mrs. Q)

We identified two main groups among study participants. One group mentioned that their decision-making processes were unchanged, although they may have been influenced by different situational factors. This is the same as in non-pregnant individuals. The other group reported pregnancy-related changes in decision-making processes, such as needing more time to reach a decision due to the new responsibility of deciding for two, or increased clarity in decision-making. No woman claimed to be incompetent as a result of pregnancy.

Participation in Clinical Research

We first asked study participants whether, in the abstract, they would consider participating in clinical research of a new drug, if the research was for healthy, non-pregnant individuals and there was a small financial incentive for participation. None of the women in our study could imagine participating in such a trial. We then presented the study participants with other possible scenarios. For example, we asked if they would consider participating in clinical research in which a new drug was to be tested in non-pregnant individuals who were afflicted with a disease, and where there was no financial incentive for participation. Twelve women could imagine participating in such research, but some of these women introduced certain conditions for participation. Ten women could not imagine participating in such research.

We next asked our study participants whether they could imagine participating in clinical research during pregnancy. Eight women could not imagine participating in such research because of the risk of foetal harm. Nine women could imagine participating in such research, and they immediately described conditions under which they would participate: if the research offered a potential benefit for the future child, if it were a search for best treatment option, if no harm would come to the future child, or if there were no known effective treatment options. Eight women emphasised that the decision for participation should be made by each woman individually.

We then probed deeper with specific case scenarios:

(A) Potential benefit for pregnant women in general and potential individual benefit for research participant. Illness occurs during pregnancy and medication is necessary. There is a drug used in pregnancy that is not officially approved for such use. The clinical research investigates the effects of this drug.
(B) Potential benefit for pregnant women in general and potential individual benefit for research participant. Illness occurs during pregnancy and medication is necessary. A new drug is suggested to result in a better treatment option than alternative drugs currently in use. The clinical research investigates the effects of this new drug.
(C) Potential benefit for the foetus and potential harm for the pregnant woman. In early pregnancy the foetus is not developing properly. The administration of hormones is known to prevent preterm abortion or birth. The clinical research investigates the effectiveness of a newly administered higher dose of hormones that can result in severe nausea for the woman and might require sick leave from work.
(D) No potential individual benefit for the pregnant woman, some inconvenience for the pregnant woman, and minimal risk. The clinical research requires that more blood is taken at routine blood testing in order to analyse the level of a certain vitamin. The woman has to answer questions about her nutrition.
(E) The pregnant woman plans to have an abortion. Prior to the abortion, the clinical research investigates the effects of a drug that could have risks for the foetus (see Harris 2016).

The results are summarised in the following table (Table 7.1)[5]:

[5] Quantitative numbers that are presented in qualitative studies do not result in generalizable data, but we provide the numbers as additional, orienting information.

Table 7.1 Research during pregnancy

Scenario	Pro/Con and Preconditions formulated by the study participants
(A) Known but unapproved drug in case of illness	Pro: Knowledge gain
Yes: 27 women No: 3 women	Preconditions: Experienced clinician-researcher; drug use is necessary; low risk for the foetus
(B) New drug in case of illness	Pro: Potential benefit for the foetus; knowledge gain; prospect of improved therapy
Yes: 10 women No: 19 women	Preconditions: Low risk for the foetus; physician as supervisor of the clinical research; only in cases of severe disease; only in cases where there is no therapeutic alternative
	Con: Risk for the foetus/the woman
(C) Hormones for foetal growth	Pro: Potential benefit for the foetus
Yes: 10 women No: 19 women	Con: General disapproval of clinical research; interference with natural process of pregnancy
(D) Blood sample Yes: 24 women No: 1 woman	Pro: Knowledge gain
(E) Research before abortion	Pro: Knowledge gain
Yes: 7 women No: 15 women	Preconditions: Foetus is not viable; clinical research does not prolong the abortion process
	Con: No emotional space for thoughts about clinical research; bad conscience; the idea 'too cruel'; general disapproval of abortion

Additional Preconditions for Participation in Clinical Research

Physician We asked the women whether their decision to participate in clinical research would be affected if their own physician asked them to take part. With this question we wanted to find out whether a woman might be vulnerable to undue influence from her physician. About half of the women said it would increase the likelihood of participation if they were recruited by someone they knew and trusted; the other half said they would make their decision independently of the recruiter.

Time Most women claimed that they would need a certain amount of time to make the decision.

Partner 18 women claimed that a law requiring partners to support a potential decision to participate in clinical research would be inappropriate because the decision was theirs alone (discussion with a partner would be welcome, however). Nine women would prefer a law requiring joint decision-making.

Discussion

Drug Use During Pregnancy and Participation in Clinical Research

During the interviews, we had extensive conversations with study participants about their medical care and drug use during pregnancy (Wild 2010). Here, we can only provide a small portion of the stories told to us. The interviews shed light on the difficulties pregnant women encounter in the case of illness or drug use during pregnancy. While discussing these problems, none of the women proactively formulated the need for clinical research or a frustration that there were so few approved drugs on the market. It seemed to be accepted as given that drugs were not tested in pregnant women and that drug therapy is difficult during pregnancy. None of the women formulated arguments regarding injustice, disrespect for the autonomy of pregnant women or resulting harms from the systematic exclusion of pregnant women from clinical research.

Consistent with this, the women showed a general reluctance to imagine participation in clinical research, being pregnant or not. However, as soon as we described detailed scenarios of clinical research, we received different and more nuanced responses. For example, eight women would not consider participating in clinical research during pregnancy in general, and nine would consider participation under certain conditions, such as if there were no risk for the foetus. When we asked in more detail whether they would participate in clinical research that tested a drug already used during pregnancy, but not officially approved for such use, the majority of women (27 of 30) agreed they would participate in such research.

There seems to be an understandable, general reluctance to participate in clinical research, which is most probably connected to the common fear of drug use during pregnancy. But, conducting clinical research need not put pregnant women at a greater risk than usual. The therapeutic use of drugs that have not been officially approved for use in pregnancy is a risk that physicians and pregnant women currently take (see Baylis and MacQuarrie 2016). Given this disturbing status quo, it is understandable that the majority of women agreed to participation in a hypothetical clinical trial that would result in more detailed evidence on drugs used off-label. Such research would immensely benefit future pregnant women, their foetuses, and physicians. Much insecurity, fear and frustration, as well as physical discomfort, could be alleviated if more drugs were examined in detail and approved officially for use during pregnancy.

The Maternal-Foetal Relationship and Participation in Clinical Research

For the discussion of how to include pregnant women in clinical research, and especially how to assess possible risks, clarity on *who* is the research participant is crucial. This touches upon the understanding of the maternal-foetal relationship. In the ethical evaluation of a clinical research, for instance, one could focus on the autonomy of the pregnant woman, as only she is capable of making decisions. This might imply that the pregnant woman can – just as any non-pregnant individual can – decide to participate in high-risk clinical research or clinical research with high degrees of uncertainty. Alternatively, one could emphasise risks and potential benefits for the foetus who is not capable of decision-making. If this is the case, it might be justified for third parties to protect the foetus from possible harm, to insist on paternalistic measures to protect the foetus and therefore to be far more restrictive in relation to participation in clinical research.

How one understands the maternal-foetal relationship and foetal/women's rights, influences whether one puts the woman or the foetus at the centre of attention (see Ashcroft 2016). In this regard a parallel can be drawn to the feminist discourse on abortion. For example, Susan Sherwin writes:

> Feminists consider it self-evident that the pregnant woman is a subject of principal concern in abortion decisions. In most non-feminist accounts, however, not only is she not perceived as central, she is rendered virtually invisible. Non-feminist theorists, whether they support or oppose women's right to choose abortion, focus almost all their attention on the moral status of the developing embryo or the fetus. (2000, 375)

However, we believe that the question whether the woman or the foetus stands in the centre of attention creates a false dichotomy. In the case of research with pregnant women, the intention is (at least in most cases) for the foetus to be carried full-term, to be born, to become someone's child and be cared for. As Catriona Mackenzie claims, "in a context in which some one or more members of the moral community have decided to take parental responsibility for its [the foetus's] future well-being, it [the foetus] has moral significance by virtue of its relations with her or them" (Mackenzie 1992, 143). If such parental responsibility has been established, it is the researchers' duty to respect this and to care for the foetus' well-being too (Wild 2012).

Therefore, third parties have some responsibility towards the foetus, but only in conjunction with the needs of the pregnant woman. A fitting concept is to conceive of the pregnant woman and her foetus as a 'double-unit', consisting of two closely related parts (Wild 2012). Such a close relationship in which one part (the woman) provides the necessary condition for the coming into existence of the other (the foetus) does not exist in any other form of human relationship. Mackenzie puts the uniqueness of the double-unit into words:

> The experience of pregnancy, particularly in the early stages, is unique in the sense that it defies a sharp opposition between self and other [...]. The foetus, to the extent that it is experienced as part of the woman's body, is also experienced as part of her self, but as a part that is also other than herself [...]. It is a being, both inseparable and yet separate from her, both part of and yet soon to be independent from her. (1992, 148)

The perspective of the double-unit fits well with the results of our interviews. We did not find one prevailing model that favoured the foetus as an individual entity or that placed exclusive attention on the woman's position. Most women explained the maternal-foetal relationship as a dynamic relationship, without a clear distinction between self and other.

It seems appropriate that regulations on clinical research involving pregnant women should be based on this double-unit model that considers both the interests of the woman and those of the foetus. This chapter, however, does not provide details about what follows from this general claim. For example, we do not address how a research ethics committee should weigh a very likely, significant potential benefit to the woman against uncertain risk to the foetus, or whether research ethics committees should ban some risky research before even asking pregnant women whether they are willing to participate. Our general claim is that we must overcome an implausible dichotomy between the woman and the foetus, and adhere instead to the rule that a risk-benefit assessment of clinical research must always consider harm to the woman *in connection with the* foetus and vice versa, while considering the woman as the ultimate decision-maker over her participation in research with an overall acceptable risk-benefit ratio. If the foetus is harmed, for example because a preterm birth results from clinical research, this will affect the woman, too. And if a woman suffers from the side-effects of a drug, for example by inducing increased levels of mental stress, her foetus will also be affected. The double-unit model both respects the autonomy of the pregnant woman (respecting her ability to make responsible decisions for herself and her foetus/future child), while at the same time allowing researchers and the research ethics committee to also exercise responsibility in relation to the foetus/future child.

On this view, it would be ethical to conduct a randomised clinical trial comparing two antihypertensive drugs widely used off-label during pregnancy in pregnant women who suffer from gestational hypertension and who are already being treated off-label by one of these two drugs. Such a clinical trial would enhance therapeutic evidence for the use of those drugs during pregnancy and it could ultimately lead to an official drug approval for use in pregnancy. After being well informed about the research risks, it would be the pregnant women's decision to participate or not.

It is much more difficult, however, to justify research involving healthy pregnant women who want to bring their pregnancies to term where the research aims to test a new drug that has not yet been tested in healthy non-pregnant individuals. The researcher might knowingly create a potential preventable risk for the double-unit that might not be outweighed by potential benefits. An exception to this rule would be research that cannot generate meaningful results for the eventual care and treatment of pregnant women unless conducted in pregnant women. Another exception would be research in emergency situations such as a rapidly spreading epidemic.

In conclusion, researchers have a responsibility to protect the well-being of the foetus as soon as parental responsibility might be or has been established. But the foetus can only be considered a part of the pregnant woman, and so the woman is – after being properly informed – the one to decide whether she wants to participate in clinical research.

Risk-Benefit Assessment and Vulnerability

In our interviews, the women showed the usual abilities of adults to assess the risks and potential benefits of research and to articulate conditions under which research participation is imaginable. Our interview study showed no reason to doubt that pregnant women can make informed decisions about participation in clinical research.

However, we would like to point to a particular concern that occurs during pregnancy that might lead to a higher risk of exploitation. These are situations in which the pregnant woman feels moral pressure to do what is best for her future child. Findings from a British interview study (Mohanna 1997) as well as results from our interviews point to this concern. Scenario C involved the testing of a new hormone that might prevent preterm birth and would probably cause women to suffer from severe nausea. Whereas the women interviewed were rather reluctant about participation in most of the research scenarios presented that included a benefit for themselves, almost all of them readily agreed to participate in clinical research for the benefit of their foetus. In many other passages from the interviews, the women emphasised their will and wish to do the best for their future child, even if that interfered with their own needs. Hence, in risky clinical research with great potential benefit for the foetus, some women might overlook or downplay the risks to themselves.

Thus, our interviews indicate that research advertising a substantial benefit for the foetus might lead to potentially exploitative situations, thereby possibly necessitating special safeguards related to this specific kind of vulnerability. Clearly, more research is needed on this issue in order to better understand situations that increase the risk of vulnerability for pregnant women in order to adequately formulate the resulting responsibilities on the side of the researchers (see Ballantyne and Rogers 2016).

7.3 Conclusion

In this chapter we have given pregnant women a voice, and in our view more research of this kind is needed. The interviews we completed help deepen our understanding of pregnant women's views on clinical research participation, and show that women are capable of making careful and responsible decisions if adequately informed. The interviews also reveal possible situations of vulnerability that create certain obligations for clinician-researchers. In addition, the interviews show how challenging drug therapy during pregnancy can be because of the lack of clinical research data. In some cases, the restrictive model – administering no or as little medication as possible – worked well, but in many cases it did not. As we have shown, missing or inadequate drug therapy as well as drug therapy without adequate information can lead to significant emotional and physical distress.

Research capable of generating more detailed evidence on drugs that are widely used off-label during pregnancy is necessary and justified. Such research could immensely benefit future pregnant women, their foetuses, and health professionals, as much of the psychological and physical distress we described could be alleviated if more drugs were approved officially for use during pregnancy. More research and better information that takes the fears and worries of pregnant women concerning potential foetal harms into account is needed. The ultimate aim should be to work towards a balanced approach to research involving pregnant women that protects the foetus from preventable harm, that respects the autonomy and decision-making capacity of pregnant women, and that results in well researched and officially approved drugs for use in pregnancy.

References

Anderson, C. 1994. Panel backs pregnant women in trials. *Science* 263(5151): 1216.

Ashcroft, R. 2016. Ethical issues in a trial of maternal gene transfer to improve foetal growth. In *Clinical research involving pregnant women*, eds. F. Baylis and A. Ballantyne, 247–263. Cham: Springer.

Ballantyne, A., and W. Rogers. 2016. Pregnancy, vulnerability, and the risk of exploitation in clinical research. In *Clinical research involving pregnant women*, eds. F. Baylis and A. Ballantyne, 139–159. Cham: Springer.

Baylis, F., and R. MacQuarrie. 2016. Why physicians and women should want pregnant women included in clinical trials. In *Clinical research involving pregnant women*, eds. F. Baylis and A. Ballantyne, 17–31. Cham: Springer.

Biller-Andorno, N., and V. Wild. 2003a. Arzneimittelforschung an Schwangeren: Besonderer Schutz – aber kein Ausschluss aus der Forschung. *Deutsches Ärzteblatt* 100(15): A-970/B-814/C-760.

Biller-Andorno, N., and V. Wild. 2003b. Zwischen Protektion und Protektionismus – medizinethische Überlegungen zur Teilnahme schwangerer Frauen an klinischen Studien. In *Das Kind als Patient*, eds. C. Wiesemann, A. Dörries, G. Wolfslast, and A. Simon, 302–320. Frankfurt am Main: Campus Verlag.

Bobbert, M. 2004. Frauenheilkunde und Geburtshilfe. In *Gutachten "Probanden- und Patientenschutz in der medizinischen Forschung"*, eds. M. Bobbert, U.B. Brückner, and H. Lilie, 55–69. Erstellt im Auftrag der Enquête-Kommission ›Ethik und Recht der modernen Medizin‹ des Deutschen Bundestages. http://webarchiv.bundestag.de/archive/2005/0718/parlament/kommissionen/ethik_med/gutachten/gutachten_03_probandenschutz.pdf. Accessed 13 Nov 2014.

Caschetta, M.B., and W. Chavkin. 1994. FDA policy on women in drug trials. *The New England Journal of Medicine* 329(24): 1815–1816.

Che Yaakob, C.A., A.A. Dzarr, A.A. Ismail, N.A. Zuky Nik Lah, and J.J. Ho. 2010. Anticoagulant therapy for deep vein thrombosis (DVT) in pregnancy. *Cochrane Database of Systematic Reviews*. 6: CD007801.

CIOMS (Council for International Organizations of Medical Sciences). 2002. *International ethical guidelines for biomedical research involving human subjects*. Geneva: World Health Organization. http://www.cioms.ch/publications/layout_guide2002.pdf. Accessed 3 Oct 2016.

Council of Europe. 2005. *Additional protocol to the Convention on Human Rights and Biomedicine, concerning biomedical research. CETS No.: 195.* http://conventions.coe.int/Treaty/en/Treaties/html/195.htm. Accessed 6 Apr 2016.

Czeizel, A.E. 1999. The role of pharmacoepidemiology in pharmacovigilance: Rational drug use in pregnancy. *Pharmacoepidemiology and Drug Safety* 8(Suppl 1): S55–S61.

DHHS (US Department of Health and Human Services). 2009. *Code of Federal Regulations: Title 45, part 46, Protection of Human Subjects. See Subpart 46.111(b) (vulnerability) and Subpart 46.204 (research on pregnant women).* http://www.hhs.gov/ohrp/regulations-and-policy/regulations/45-cfr-46/index.html. Accessed 3 Oct 2016.

Estler, C.-J. 1995. Pharmakologische Grundlagen der Arzneimittelverordnung in der Schwangerschaft. *Zeitschrift für ärztliche Fortbildung (ZaeF)* 89: 743–748.

Garland, M. 1998. Pharmacology of drug transfer across the placenta. *Obstetrics and Gynecology Clinics of North America* 25(1): 21–42.

Harris, L.H. 2016. Clinical research involving pregnant women seeking abortion services: United States perspectives. In *Clinical research involving pregnant women*, eds. F. Baylis and A. Ballantyne, 265–284. Cham: Springer.

Iberogast. 2013. *Frequently asked questions*. http://www.iberogast.de/infothek/haeufige-fragen-faq/darf-iberogast-auch-waehrend-der-schwangerschaft-oder-in-der-stillzeit-eingenommen--werden.html#12. Accessed 6 Apr 2016.

Imhof, C. 2005. *Antibiotika-Therapie in der Schwangerschaft: Teratogene Risiken von Makroliden, Doxycyclin und Cotrimoxazol.* Dissertation zur Erlangung des Doktorgrades der Medizin. Universität Ulm. http://vts.uni-ulm.de/doc.asp?id=5696. Accessed 6 Apr 2016.

Kaposy, C. 2016. Presumptive inclusion and legitimate exclusion criteria. In *Clinical research involving pregnant women*, eds. F. Baylis and A. Ballantyne, 51–62. Cham: Springer.

Kasilo, O., M. Romero, M. Bonati, and G. Tognoni. 1988. Information on drug use in pregnancy from the viewpoint regional drug information centre. *European Journal of Clinical Pharmacology* 35(5): 447–453.

Koren, G., A. Pastuszak, and S. Ito. 1998. Drugs in pregnancy. *New England Journal of Medicine* 338(16): 1128–1137.

Langston, L. 2016. Better safe than sorry: Risk, stigma, and research during pregnancy. In *Clinical research involving pregnant women*, eds. F. Baylis and A. Ballantyne, 33–50. Cham: Springer.

Lyerly, A.D., M. Little, and R. Faden. 2008. The second wave: Toward responsible inclusion of pregnant women in research. *International Journal of Feminist Approaches to Bioethics* 1(2): 5–22.

Mackenzie, C. 1992. Abortion and embodiment. *Australasian Journal of Philosophy* 70(2): 136–155.

Macklin, R. 1994. Reversing the presumption: The IOM report on women in health research. *Journal of the American Medical Women's Association* 49(4): 13–16. 21.

Marquardt, E. 1994. Die Contergankatastrophe 1961 – Schock und erste Reaktionen. In *Contergan – 30 Jahre danach*, ed. Fritz U. Niethard, 79–84. Stuttgart: Enke.

Mastroianni, A.C., R. Faden, and D. Federman. 1994. *Women and health research: Ethical and legal issues of including women in clinical studies*. Washington, DC: National Academy Press.

Matthews, A., T. Dowswell, D.M. Haas, M. Doyle, and D.P. O'Mathúna. 2010. Interventions for nausea and vomiting in early pregnancy. *Cochrane Database of Systematic Reviews* 9: CD007575.

Mayring, P.A.E. 2008. *Qualitative Inhaltsanalyse: Grundlagen und Technik*, 10th ed. Weinheim/Basel: Beltz.

Merkatz, R.B., R. Temple, S. Sobel, K. Feiden, and D.A. Kessler. 1993. Women in clinical trials of new drugs – A change in Food and Drug Administration policy. *New England Journal of Medicine* 329(4): 292–296.

Meuser, M., and U. Nagel. 1991. ExpertInneninterviews – vielfach erprobt, wenig bedacht. In *Qualitativ-empirische Sozialforschung. Konzepte, Methoden, Analysen*, eds. D. Garz and K. Kraimer, 441–471. Opladen: Westdt. Verl.

Mohanna, K. 1997. Informed consent: Research in pregnancy brings special considerations. *British Medical Journal* 315(7102): 249–250.

Müller, W.C. 1969. Arzneimittelgaben während der Schwangerschaft. *Münchener Medizinische Wochenschrift* 111(34): 1687–1692.

Müller-Oerlinghausen, B. 2005. Die Rolle der Ärzteschaft bei der Aufklärung der Contergannebenwirkungen und die Auswirkung auf die deutsche Arzneimittelgesetzgebung. In *Die Contergankatastrophe – Eine Bilanz nach 40 Jahren*, eds. L. Zichner, M.A. Rauschmann, and K.-D. Thomann, 33–38. Darmstadt: Steinkopff Verlag.

Murphy, J.F. 1984. Drugs and pregnancy. *Irish Medical Journal* 77(2): 52–56.

Paulus, W.E. 1999. Pharmakotherapie in der Schwangerschaft. *Therapeutische Umschau. Revue thérapeutique* 56(10): 602–607.

Schaefer, C., H. Spielmann, and K. Vetter. 2004. Liberalisierung nicht erforderlich. *Deutsches Ärzteblatt* 101(4): A165–A166.

Schaefer, C., H. Spielmann, and K. Vetter. 2006. *Arzneiverordnung in Schwangerschaft und Stillzeit*. München/Jena: Elsevier (Urban & Fischer).

Scott, A. 1974. Drugs and pregnancy. *Journal of the Irish Medical Association* 67(5): 123–126.

Sherwin, S. 2000. Abortion through a feminist ethics lens. In *Ethical theory: A concise anthology*, eds. H. Geirsson and M.R. Holmgren, 374–390. Peterborough: Broadview Press.

Therapontos, C., L. Erskine, E.R. Gardner, W.D. Figg, and N. Vargesson. 2009. Thalidomide induces limb defects by preventing angiogenic outgrowth during early limb formation. *Proceedings of the National Academy of Sciences of the United States of America* 106(21): 8573–8578.

Thürmann, P.A., and A. Steioff. 2001. Drug treatment in pregnancy. *International Journal of Clinical Pharmacology and Therapeutics* 39(5): 185–191.

Vazquez, J.C., and E. Abalos. 2011. Treatments for symptomatic urinary tract infections during pregnancy. *Cochrane Database of Systematic Reviews* 1: CD002256.

Wild, V. 2010. *Arzneimittelforschung an schwangeren Frauen: Dilemma, Kontroversen und ethische Diskussion*. Frankfurt am Main: Campus Verlag.

Wild, V. 2012. How are pregnant women vulnerable research participants? *International Journal of Feminist Approaches to Bioethics* 5(2): 82–104.

Wilson, J.G. 1979. The evolution of teratological testing. *Teratology* 20(2): 205–211.

Winkelheide, M. 2009. *Schutz auf dem Prüfstand. Deutschlandfunk – Wissenschaft im Brennpunkt*. http://www.dradio.de/dlf/sendungen/wib/1025132/. Accessed 6 Apr 2016.

Part III

Chapter 8
Pregnancy, Vulnerability, and the Risk of Exploitation in Clinical Research

Angela Ballantyne and Wendy Rogers

Abstract *Pregnant women and their foetuses have long been regarded as vulnerable, where being vulnerable indicates a likelihood of suffering harm. This perception has led to the widespread exclusion of pregnant women from clinical research, in order to protect foetuses and the women who carry them from any dangers associated with exposure to experimental therapeutic products or interventions. This chapter explores the ways in which pregnant women are vulnerable, and the potential risk of exploitation if pregnant women are enrolled in clinical research. There are three overlapping sources of vulnerability: inherent, situational, and pathogenic, and each of these may be dispositional (i.e. potential) or occurrent (i.e. requiring immediate action to limit harm). We argue that while pregnant women may experience one or more forms of vulnerability, in general they are not at risk of exploitation during research because they do not provide researchers with the opportunity to conduct more efficient research. We conclude with policy suggestions for conducting research with pregnant women that responds to vulnerability, promotes autonomy, and supports fair access to research participation. We focus on pregnancy registries, parental consent, and minimal risk research limits.*

Pregnant women and their foetuses have long been regarded as vulnerable, where being vulnerable indicates the likelihood of suffering harm (Lange et al. 2013). This perception has led to the widespread exclusion of pregnant women from clinical research, in order to protect foetuses and the women who carry them from any dangers associated with exposure to experimental therapeutic products. In this chapter, we investigate the vulnerability of pregnant women and the ways in which exploitation may occur when pregnant women are involved in, or excluded from, clinical research. We argue that exclusion from clinical research creates its own

A. Ballantyne, BSc, PhD (✉)
Department of Primary Health Care and General Practice, University of Otago, Wellington, New Zealand
e-mail: angela.ballantyne@otago.ac.nz

W. Rogers, BMBS, PhD, FRACGP
Department of Philosophy and Department of Clinical Medicine, Macquarie University, Sydney, Australia

F. Baylis, A. Ballantyne (eds.), *Clinical Research Involving Pregnant Women*, Research Ethics Forum 3, DOI 10.1007/978-3-319-26512-4_8

vulnerabilities, and that it is possible to include pregnant women in clinical research without exploitation.

8.1 Background

Research ethics as a discipline rests on an assumption that research is a dangerous activity rather than a social good, fuelled in no small part by public outrage over unethical research[1] (United States 1979; Cartwright 1988; Marrus 1997; World Medical Association 2013). The potential dangers of research have been seen as especially relevant to those who are considered to be most vulnerable in the community, such as children, women of reproductive age, pregnant women, people in prison, people with cognitive impairment, and those highly dependent upon medical care, leading to their widespread exclusion from research. The last 40 years have seen a progressive effort to rebalance this perspective, and to recognise that research participation is of value to both the individual and the population or group of which they are a member. Recent policies have advocated for more medical research with women (NIH 1994), prisoners (Institute of Medicine 2007), and children (NIH 1998). Some have argued that pregnant women remain the last group to be routinely excluded from research (Lyerly et al. 2008).

Despite these initiatives, research ethics guidelines regularly stipulate that research with vulnerable populations or groups should be limited and/or subject to extra safeguards. The identification of those deemed vulnerable has changed over time; pregnant women are now no longer automatically included in all such lists, but their participation in research continues to attract special considerations in at least some jurisdictions. In the United States, pregnant women and their foetuses are categorised as a vulnerable research population and regulated under 45 CFR 46 Subpart B of the *Code of Federal Regulations* (DHHS 2009, 45 CFR 46). This categorisation is in tension with the views of the National Institutes of Health, who argue that pregnant women should be reconceptualised as 'complex' rather than 'vulnerable' (DHHS 2011). The Australian *National Statement on Ethical Conduct in Human Research* has a chapter devoted to "Ethical considerations specific to participants" that includes a section on pregnant women, along with other groups usually considered vulnerable such as children, persons with mental illness, prisoners, those in dependent relationships and people of Aboriginal and Torres Strait Islander descent. The *National Statement* requires that all research with pregnant women be scrutinised by full Human Research Ethics Committee (HREC) review,

[1] For example, in the United States, the Tuskegee study prompted the National Research Act 1974, which required the establishment of institutional review boards at institutions receiving federal grants and set up the National Commission for the Protection of Human Subjects of Biomedical and Behavioural Research that produced the *Belmont Report* in 1979. In New Zealand the research ethics framework was based on recommendations in the Cartwright Report that investigated unethical research on women conducted by Dr Green (Cartwright 1988).

rather than any expedited or alternative review pathways, thereby implying that all research with pregnant women is high-risk. Under these guidelines, interventional research with pregnant women must be limited to 'therapeutic research' (Australia 2007). In contrast, the Council for International Organizations of Medical Sciences (CIOMS) guidelines do not include pregnant women amongst those identified as potentially vulnerable. Rather, CIOMS Guideline 17 asserts, "Pregnant women should be presumed to be eligible for participation in biomedical research" (CIOMS 2002). This claim is qualified by linking the proposed research to the particular health needs of the woman or her foetus.

Thus, the international picture is mixed; while there are at least some efforts to remove a blanket categorisation of pregnant women as a vulnerable population, research ethics guidelines continue to require special protections for, or impose limits on, research participation by pregnant women. The reasons why pregnant women are considered vulnerable are rarely stated overtly. Clearly the kinds of considerations that obtain for other groups deemed vulnerable, such as lack of, or immature mental capacity, limited understanding, or risk of coercion, are applicable to few if any women by virtue of pregnancy alone. In the next section, we consider the relationship between vulnerability and pregnancy in order to clarify what kinds of vulnerability affect pregnant women, and how these may be relevant regarding participation in, or exclusion from, research.

8.2 Vulnerability

Vulnerability refers to the capacity to suffer or be harmed.[2] The risk of suffering may arise from an increased risk of harm (for example, because of a dangerous situation), or from a decreased capacity to protect oneself from harm (for example, because of impaired cognition), or from both. In clinical research, all participants are vulnerable to harms such as serious side-effects from the experimental intervention, or the wrongs of coercion or manipulation, invalid consent and so forth. This vulnerability lies at the heart of research ethics guidelines that aim to protect participants to the extent possible against potential harms and wrongs. Alongside this generic research-related vulnerability, there are certain populations or groups labelled as 'vulnerable,' where this denotes increased susceptibility to specific harms and wrongs and where this flags the need for extra protections over and above those offered to all research participants. Thus, there is ambivalence about vulnerability within research ethics: it is both a universal feature of all research participants, and a characteristic that arises in specific circumstances, affecting some research participants more than others (Rogers 2014).

[2] The concept of vulnerability is not limited to human persons, as many living creatures including non-human animals have the capacity to suffer harms, and are thereby vulnerable.

This ambivalence is reflected in philosophical approaches to conceptualising vulnerability.[3] One approach links the concept of vulnerability to the Latin word *vulnus*, meaning 'wound,' reflecting the universal capacity to suffer which is inherent to human embodiment. The second approach focuses on the contingent susceptibility of particular persons or groups to specific kinds of harm or threat by others.

On the universal view, to be vulnerable is to be fragile, susceptible to wounding and to suffering; and this fragility is an ontological condition of our humanity (Fineman 2008). Human vulnerability arises from our embodiment which makes us susceptible to affliction and injury, and ultimately death. In turn, our bodily vulnerability is linked to the social nature of human life which makes us both vulnerable to the actions of others and, to varying degrees at various points in our lives, dependent on the care and support of other people.

In contrast to this universal view of vulnerability, the particular view of vulnerability picks up on the notion that vulnerability is essentially relational and specific (Goodin 1985). That is, individuals or groups are vulnerable to particular others with respect to specific sorts of threats to their interests. Although everyone is potentially exposed to such threats, some persons or groups have little or no capacity to protect themselves, are more susceptible to exploitation, or at increased risk of being treated unjustly. On this view, vulnerable persons or groups are those with reduced capacity to safeguard their interests relative to other people; vulnerability is closely linked to relations of power, where the vulnerable lack power and are dependent upon others for meeting their needs. Whereas the universal approach to conceptualising vulnerability stresses our common embodied humanity and equal susceptibility to suffering, the particular approach stresses the ways in which various inequalities, for example in resources or power, make some especially susceptible to harm or exploitation by others.

The particular view of vulnerability dominates research ethics. It has led to the creation of lists of 'the vulnerable' that, as Hurst notes, may be so extensive as to include almost everyone, and thereby lack any utility in alerting researchers and ethics review bodies to particular increased risks affecting specific individuals or groups (Hurst 2008). As well as being overly broad, the lists of those presumed vulnerable fails to recognise the complex ways in which vulnerability may arise, and can lead to stereotyping and paternalism (Luna 2009). Consider, for example, the Australian *National Statement* requirement regarding research with pregnant women:

> Research to which this chapter applies must be reviewed and approved by a Human Research Ethics Committee (HREC) rather than by one of the other processes of ethical review described in paragraphs 5.1.7 and 5.1.8, except where that research uses collections of non-identifiable data and involves negligible risk, and may therefore be exempted from ethical review. (Australia 2007, 47)

---

[3] The following discussion of vulnerability draws upon existing accounts of a typology of vulnerability developed by one of the authors (Rogers) in collaboration with others (i.e., Rogers et al. 2012; Lange et al. 2013; Mackenzie et al. 2014).

Full ethics committee review of all research involving pregnant women is required, unless the research 'uses collections of non-identifiable data and involves negligible risk', just because the research participants are pregnant. This approach fails to discriminate between low and high risk research and treats pregnant women paternalistically by implying that they need a higher level of research protection regardless of whether the research collects qualitative data about dietary habits or involves administration of novel chemical agents.

8.3 Typology of Vulnerability

We believe that using 'vulnerability' to signify more than ordinary susceptibility to harms or wrongs can be useful in research ethics. However, in order to preserve the notion of universal susceptibility to harm, and to avoid the problems of stereotyping individuals on the basis of their membership in a specific population or specific groups, we draw upon a typology of vulnerability (Rogers et al. 2012; Mackenzie et al. 2014). The typology identifies three overlapping sources of vulnerability: inherent, situational, and pathogenic. Each of these may be dispositional (i.e. potential) or occurrent (i.e. requiring immediate action to limit harm). The typology brings together the morally salient features of both universal and particular vulnerability in order to achieve a more nuanced understanding of the nature and sources of vulnerability. As the typology identifies different sources of vulnerability, it can avoid stereotyping by suggesting responses that are sensitive to the individual and her context (Lange et al. 2013).

Inherent Vulnerability

Inherent vulnerability refers to sources of vulnerability that are fundamental to the human condition-to the fact that we are finite creatures with unavoidable needs. It captures the notion of the shared capacity to suffer that underpins the universal understanding of vulnerability. Inherent vulnerability arises from our embodied and social natures and our dependence on others; we all feel the pangs of hunger, and are all liable to harm when, as babies and at times of frailty, we depend upon the care of others for our survival. Regarding pregnant women, pregnancy creates inherent physical vulnerabilities such as increased metabolic requirements and decreased physiological reserves as pregnancy progresses; pregnancy may limit physical prowess; existing health problems may be exacerbated; or pregnancy-related disorders, such as pre-eclampsia, may emerge. Other inherent vulnerabilities related to pregnancy include the potential effect of pregnancy on the woman's sense of identity (for example, change in body shape, altered food preferences and increased fatigue), and the increased dependence on care that occurs, especially if the pregnancy is complicated. In most cases, pregnant women are deeply invested in the health and well-being of their foetuses. This concern for the welfare of the foetus is a further source of inherent vulnerability, as foetal interests may trigger new sources of vulnerability for pregnant women. For example, antenatal interventions for the health of the foetus may expose pregnant women to various physical or psychological harms (see Ashcroft 2016). These harms may include exploitation where the woman's depth of concern leads her to participate in risky research to potentially benefit the foetus. Inherent vulnerability cannot be eliminated, but appropriate safeguards and responses can mitigate, at least to some extent, potential harms. Thus, ensuring adequate nutrition, healthcare, housing, and supports for pregnant women can decrease the likelihood of harms relating to the inherent vulnerability of pregnancy.

Situational Vulnerability

Situational vulnerability is context-specific, and is caused or exacerbated by personal, socio-economic, political, or environmental factors. As situational vulnerability captures the notion that some individuals or groups are more vulnerable than others, it encompasses the particular approach to understanding vulnerability. Pregnant women may be situationally vulnerable when they cannot easily afford the extra costs of living associated with pregnancy, when the pregnancy adversely affects their interpersonal relationships, or creates a need for health care in contexts where such care is unavailable. Other examples of pregnancy-related situational vulnerability include blame, disapproval, or worse if the circumstances of the pregnancy attract criticism from others, and possibly even interference (such as unsolicited advice to do or refrain from doing certain activities while pregnant). Unlike inherent vulnerability, situational vulnerability is not an inevitable part of what it is to be a human being. Rather, situational vulnerability can be significantly reduced or prevented because it results from human interactions; it is a consequence of the situation of *this* pregnant woman in *this* family, *this* society, and *this* culture.

Inherent and situational vulnerability may be causally interconnected. Situational vulnerability can give rise to inherent vulnerability – for example, family rejection on account of an unplanned pregnancy can cause or exacerbate ill health. Some kinds of inherent vulnerability will render people more liable to situational vulnerability: thus, pregnancy-related health care may incur costs that tip a woman into poverty, while a woman with an intellectual disability may be at greater risk of an unplanned pregnancy that will lead to her loss of independence.

Pathogenic Vulnerability

The third type of vulnerability, pathogenic vulnerability, is a subset of situational vulnerability. The notion of pathogenic vulnerability draws attention to the way that some situational vulnerabilities are caused or exacerbated by morally dysfunctional relationships, or social structures characterised by injustices such as domination, oppression, or abuse. People of colour are pathogenically vulnerable in societies that tolerate or encourage racism, while women are at increased risk of harm in societies that ignore or fail to support victims of gender-based violence. The notion of pathogenic vulnerability also highlights the way that some interventions designed to protect against vulnerability can have the paradoxical effect of increasing it. Pregnant women are pathogenically vulnerable when workplaces have unjust policies regarding employment in pregnancy; when pregnancy triggers oppressive or violent responses in men who do not wish to take responsibility for their role in the pregnancy; or when cultural/religious traditions advocate punishing women whose pregnancies occur outside of sanctioned relationships. Pathogenic vulnerability is morally wrong: the harms and wrongs that occur as a result of pathogenic vulnerability are the result of unjust human actions.

8.4 Vulnerability in Pregnancy

Using this typology, we propose that the vulnerability faced by pregnant women as the result of historic exclusion from research is a type of pathogenic vulnerability. Pregnant women are vulnerable when their need for health care during pregnancy is hampered by a lack of evidence about which interventions are safe and effective for conditions that occur commonly in pregnancy, such as influenza or depression. This lack of evidence is the result of policies aimed at protecting pregnant women and their foetuses from any risks associated with research, but such policies exacerbate rather than ameliorate vulnerability, as women are left with few or no treatment options with known effects in pregnancy (Cragan 2014). While the intention of protective policies may have been motivated by concerns for the well-being of pregnant women and their foetuses, the effect is unjust as the women thereby lack information about treatment options – information that would ordinarily be available to non-pregnant patients.

Foetal Vulnerability

Thus far we have not considered the foetus as a vulnerable subject distinct from the pregnant woman. Foetuses have a limited range of interests. They cannot, for example, be oppressed, have their autonomy infringed, or suffer psychological harms.

This means that there are fewer ways in which foetuses can be vulnerable compared to pregnant women. Nonetheless, foetuses have a unique inherent vulnerability related to their life stage characterised by rapid development. This rapid growth creates the potential for even transient exposures to foreign materials to cause enduring harm. Substances ingested by pregnant women, such as therapeutic or recreational drugs and alcohol, may lead to harms including spontaneous abortion, stillbirth, preterm delivery, structural malformations, dysmorphic features and neurodevelopmental disorders (Cragan 2014). Likewise foetuses may be harmed by exposure to infections that may be more or less serious for the women (for example rubella, toxoplasmosis, or syphilis). While foetal health is intrinsically linked to the health and behaviours of pregnant women, the two are not directly correlated. Behaviour that is debilitating for the pregnant women, such as severe nausea and vomiting, may have no effect on foetuses, whereas drugs, such as thalidomide, that have limited effect on the pregnant women, may be catastrophic for the foetuses.

Foetuses may be situationally vulnerable where their well-being depends upon antenatal interventions (such as, intra-uterine surgery for spina bifida or intra-uterine blood transfusion for haemolytic anaemia) that are lacking for financial or resource reasons, or where the women lack access to adequate nutrition or healthcare. And, insofar as women lack treatments of known safety and efficacy during pregnancy, foetuses may be vulnerable to harm from exposure to interventions untested in pregnant women, or to the effects of women's untreated or under-treated ill health.

Previous and existing protectionist policies limiting the participation of pregnant women in research seem to be largely motivated by concern for the well-being of the foetus, as such restrictions are not routinely imposed upon non-pregnant women. The foetus is regarded as especially vulnerable to the effects of experimental agents, a legacy of the disorders of limb development that occurred after the prescription of thalidomide for morning sickness (see Langston 2016). Although foetal vulnerability has driven regulation around research with pregnant women, we do not consider this to be the sole or overriding moral concern. The rest of the chapter proceeds on the following assumptions: foetuses have interests and associated vulnerabilities; they have some moral status and are due consideration; and, in general, pregnant women are the best stewards of foetal interests.

8.5 Exploitation

Vulnerability is intimately linked to exploitation: to exploit is to take unfair advantage of another's vulnerability, and vulnerability is concerning because it exposes one to the risk of exploitation and other harms. The presence of vulnerability, however, does not automatically entail exploitation. Other factors determine whether the vulnerabilities of pregnant women and their foetuses are likely to be systemically exploited by researchers. Here, we explore the relationship between vulnerability and exploitation and ask whether pregnant women are especially at risk of exploitation in clinical research.

In ordinary speech, exploitation can be used in a morally neutral or a morally pejorative sense.[4] 'The gymnast exploits her natural flexibility' is a neutral example of exploitation, while 'Slavery involves the use and exploitation of one person for another's gain' is a morally pejorative example of exploitation. The latter is sometimes referred to as 'wrongful exploitation' (Valdman 2009).

Here we focus on the morally pejorative uses of exploitation as they apply to vulnerable people participating in clinical research. We compare two accounts of wrongful exploitation in the research ethics literature – unfair benefits accounts and Kantian instrumental use accounts – in order to further elucidate the potential relationship between vulnerability and exploitation.

Unfair Benefits Accounts of Exploitation – Consent, Choice and Fairness

Exploitation as unfair benefits is the dominant view of exploitation in research ethics (Phillips 2011; Emanuel et al. 2002). The most popular version of the unfair benefits model of exploitation is that described by Alan Wertheimer (Wertheimer 1996). Wertheimer argues that exploitation may be mutually beneficial, mutually consensual, and non-coercive. The moral wrong, he argues, lies in the unfair distribution of benefits between the transacting parties. He refers to this as mutually advantageous exploitation.

The unfair benefits accounts of exploitation are useful in research ethics because they clearly delineate consent and unfairness. Informed consent is perhaps the dominant ethical construct in research ethics. Yet cases that seem prima facie to satisfy the requirements of informed consent may still be criticised on the grounds of exploitation, as occurred with the AZT trials in Africa (Crouch and Arras 1998) and the cancelled Surfaxin trials in Bolivia (Hawkins and Emanuel 2008).

In the mid-1990s a series of WHO-endorsed clinical trials, commonly referred to as the 'AZT trials', were initiated in developing countries to test the efficacy of short-course doses of zidovudine (AZT) in preventing perinatal transmission of HIV. The '076' regime of AZT had been adopted as the standard of care for HIV-positive pregnant women in the United States and other developed countries in 1994. But the 076 regime was considered too expensive for use in developing countries (at US$800 per pregnant woman and child pair) and inappropriate given prevailing conditions in these countries. Internationally sponsored trials of shorter, cheaper doses of AZT were therefore initiated in developing countries (mostly in Africa). Critics called these trials, which involved more than 17,000 pregnant

[4] The term exploitation is also often used in a third way, as a moral amplifier, to signify that the author vehemently objects to a practice. Such examples are excluded from our analysis because they are not intended to refer to exploitation in a technical sense.

women, exploitative. Placebo-controlled trials could not be approved in the US, and there were no plans in place to ensure that the trial communities in developing countries would receive post-trial access to the short-course regime even if it proved successful.

Similarly, in 2000 a private American drug company, Discovery Labs, planned a multi-centre, double-blinded, randomised, two-arm, placebo-controlled Phase III trial, involving 650 premature infants in Bolivia, to test the efficacy of a new synthetic surfactant called Surfaxin. The Bolivian hospitals were selected because while they could not routinely provide surfactant treatment for respiratory distress syndrome (RDS) (and therefore there was no current best available treatment against which to test Surfaxin), they could support and run the sophisticated ICU facilities promised by the sponsor in return for hospitals' participation. The neonates enrolled in the study would be intubated and either given air suffused with Surfaxin (treatment arm) or air without any drug (placebo arm). The study was halted before recruitment due to international accusations of exploitation (Hawkins and Emanuel 2008).

In both of these cases, researchers sought and obtained informed consent from the research participants or their guardians. As such, the ethical criticisms of these trials are not focused on consent, but rather on exploitation (because the participants and their communities were not treated fairly). They did not receive sufficient or fair benefit in exchange for their participation. People are subject to exploitation when their unmet needs create a power differential in the relationship, making them vulnerable to the more powerful party.

Why might someone consent to an arrangement or relationship where she does not receive sufficient or fair benefits? Consent here refers to the fact that prospective research participants have mental capacity and are thereby capable of weighing their options, considering the pros and cons and providing informed consent to participate in a trial. Wertheimer has argued that it is rational to consent to an unfair transaction when the transaction represents the greatest benefit available, even if by some external standard this benefit is not deemed fair. From this it follows, that when an exploitee's circumstances are such that a transaction or offer is more valuable to her than to the exploiter, she may agree to unfair terms in an effort to secure the transaction. Thus a defect in consent is neither a necessary nor sufficient condition for mutually advantageous exploitation.

However, some argue that seemingly rational and otherwise un-coerced choices are nonetheless constrained in morally worrying ways if made under conditions of desperation or from a position of unequal bargaining power (Wilkinson 2003). On this view, even notionally advantageous exploitation in research is ethically concerning because it involves constrained or coerced consent rather than genuine consent, where the latter is understood as the meaningful exercise of autonomy. In general, research ethicists have been reluctant to define consent in this thicker way, because it seems to obscure the distinction between objections to research on the grounds of invalid consent and those of exploitation. Instrumental use accounts of exploitation take this approach and are discussed in more detail below.

Thus even if there is valid consent, a transaction may be criticised on the separate grounds of unfair benefit. Mutually advantageous exploitation does not harm the exploitee outright; rather, it harms her indirectly by denying her a good or a benefit to which she is fairly entitled. The exploitee gains an initial benefit in that her position is potentially improved (for example, she gains access to potential treatment via research participation), but she does not benefit sufficiently (as when access to the intervention is limited to the duration of the trial; or when the trial intervention does not match her most urgent health needs). Most ethics guidelines warn against exploiting research participants by denying them fair benefits, yet there is no agreement about what is unfair (Ballantyne 2010). Extreme examples such as the AZT and the Surfaxin trials spark many (but not all) commentators' intuitions about unfairness, generating sufficient momentum to prevent, halt, or adjust the terms of the trial. While there is some consensus that exploitation involves unfairness, there is no consensus about what is unfair.

In summary, the unfair benefits accounts of exploitation defines exploitation as a consensual, mutually beneficial transaction where the exploiter takes advantage of an inequality in bargaining potential arising from the exploitee's vulnerable circumstances, and uses this to extract an unfair proportion of the benefits from their cooperative relationship.

Instrumental Use Accounts of Exploitation

Instrumental use accounts of exploitation are characterised by Kantian concepts such as wrongful use, taking advantage and treating another as a mere means (Buchanan 1985; Seigel 2008). Proponents argue that exploitation is wrong because it is objectionable to treat others as mere means, but they acknowledge that the objective measure of whether someone was treated as a mere means will often be the unfair terms of the transaction in question (Wendler 2000).

Advocates of instrumental use accounts deny that one can consent to exploitation as they tend to invoke a more substantive conception of consent than that of the unfair benefits accounts (Zion et al. 2000). While the exploitee may agree to the transaction in the narrow sense of providing informed consent, this agreement does not constitute genuine autonomous consent, as a fully autonomous agent would not freely consent to being used as a mere means. In such cases, the consent is the result of vulnerability arising from an impoverished set of opportunities or options; there is no scope for the full exercise of the individual's autonomy and, thus, it may seem rational to agree to an exploitative offer.

Instrumental use accounts of exploitation link vulnerability with exploitation, in that situational vulnerability, for example due to a lack of resources, creates opportunities for exploitation. As Oyewale Tomoori asks: "In an environment where the majority can neither read nor write and is wallowing in poverty and sickness, hunger and homelessness, and where the educated, the powerful, the rich, or the expatriot is a semi-god, how can you talk of informed consent?" (Zion et al. 2000, 16).

According to instrumental use accounts of exploitation, exploitation occurs because the exploitee lacks viable alternatives, and thus may agree to a transaction where she is treated as a mere means to an end, and for this reason consent is not valid.

In this instance, exploitation is seen to be wrong because it harms the dignity of the person to be treated as a mere means. Instrumental use has implications for the person's sense of agency, self-worth, empowerment, and confidence, as illustrated by Wood: "Proper respect for others is violated when we treat their vulnerabilities as opportunities to advance our own interests or projects. It is degrading to have your weakness taken advantage of and dishonourable to use the weakness of others for your own ends" (1997, 15). This kind of exploitation reflects pathogenic vulnerability.

In summary, instrumental use accounts of exploitation claim that all exploitation is non-consensual, uses the exploitee as a mere means, and is thereby harmful. In contrast, unfair benefit accounts of exploitation differentiate between two distinct categories of objection to research – failure to obtain informed consent, and failure to provide fair benefits to research participants – where only the latter is exploitation. This distinction is considered important within research ethics, with its strong historical focus on the importance of informed consent. Research ethicists have sought an account of exploitation that identifies a distinct moral wrong, and this is perhaps the primary reason for the dominance of the unfair benefits accounts of exploitation in the research ethics literature (Participants 2002).

8.6 Vulnerable But Not Exploited

As described above, pregnant women are vulnerable in a variety of ways. However, the majority of these vulnerabilities do not create opportunities for exploitation with regard to research. For example, a pregnant patient with breast cancer who has reached the end of conventional medical therapy is inherently vulnerable due to her medical condition. This may lead her to want to participate in a clinical trial of a new drug as a last attempt at prolonging her life long enough to deliver her baby. This type of research participation is not exploitative, despite the pregnant woman's vulnerability and desperation; there is no suggestion that she will receive an unfair share of benefits or be coerced into the trial, or that her vulnerability and desperation necessarily differs from, or is more significant than, that of any person with a terminal illness who wishes to try to prolong her life for a significant event.

Even in cases of pathogenic vulnerability, there may be no exploitation. For example, an HIV-positive pregnant woman in Tanzania with no access to state-funded HIV drugs for herself or her future child, might join a clinical trial of AZT in the hope that this will prevent maternal-foetal transmission of HIV, thereby saving her baby from contracting HIV. Vulnerability in this case is pathogenic, because global justice requires the provision of life saving treatment in this circumstance given the availability and cost of AZT. Even given these unjust background conditions, however, it is not necessarily the case that research participation in this instance is exploitative. It may be unfair that the pregnant woman should have to join a clinical trial to access anti-retrovirals, but the trial itself may be structured to provide fair benefits to participants and their community (including for example, post-trial access to the drug, treatment for family members, ante-natal check-ups and auxiliary care for the newborn) (see Little et al. 2016).

Populations that are characteristically exploited in research offer particular advantages to researchers, which make the research cheaper or easier to perform. For example, prisoners were used extensively in drug research because, being a captive population, their movements, diet, exercise and access to drugs (smoking and alcohol) were open to greater control and closer monitoring than those of participants recruited in the community. This feature reduced variables and led to potentially more reliable results (IOM 2007). Patients in low and middle income countries are now targeted for drug research because there are higher numbers of treatment naïve patients in these populations. At least some of this research exploits the situational vulnerability of populations where limited local healthcare may lead to patients being eager to join trials, which in turn makes recruitment quicker and the trials cheaper (Ballantyne 2010).

In contrast, the inclusion of pregnant women in research does not make research easier, quicker, or cheaper. In fact the physiological state of pregnancy may confound research results. Despite the fact that pregnant women are labelled vulnerable, there is no incentive for researchers to systematically recruit, let alone exploit pregnant patients. So while vulnerability and exploitation are intimately linked in the research ethics literature, they should be disentangled in any analysis of the ethics of research with pregnant women. The focus should be on delineating different sources of vulnerability and addressing these, so as to better facilitate research involving pregnant women.

There is, however, one vulnerability of pregnant women that may create opportunities for exploitation, and this is vulnerability arising from women's concern for their foetuses. Clinical trials that aim to test interventions for the benefit of the foetus may be exploitative in terms of both fair benefits and instrumental use. There is significant social pressure on pregnant women to modify their behaviour for the sake of their pregnancy, to undergo, for example, significant changes in diet and exercise, avoid potential toxins, manage stress, and use dietary supplements. This pressure may become overwhelming in cases where the health of the foetus could potentially benefit from an experimental intervention. On fair benefits accounts of exploitation, a pregnant woman entering a clinical trial for an intervention to potentially benefit the foetus is exploited only if she does not receive a fair share of the benefits. This requires both that the researchers/sponsors stand to gain substantial benefit from the intervention (through patents or enhanced reputation), and that the pregnant woman benefits too little. Determining how much the pregnant woman benefits in these cases is difficult, as at least some of the woman's interests may be tied up with the welfare of the foetus, such that she may see any potential medical benefit to the foetus as a benefit to herself. It is worth noting that, given the controversial nature of foetal research, including surgery, pregnant women are likely to be given an extensive opportunity to consider personal and foetal interests during the consent process.

On instrumental use accounts of exploitation, enrolling pregnant women in research to benefit the foetus may exploit pathogenic vulnerability arising in situations where women's control over their bodies is reduced by pregnancy – for example where partners or other family members take a proprietorial interest in the foetus and insist that the woman participate in foetal research.

8.7 Framework for Addressing Vulnerability in Research with Pregnant Women

Here we develop a framework for including pregnant women in research that addresses the vulnerabilities we have identified, avoids exploitation, and promotes/supports autonomy to the extent possible. Just research requires a balance between fair access to the benefits of research, facilitating autonomy, and appropriate mechanisms to address vulnerabilities in the research population.

As we have argued, pregnant women are subject to inherent, situational, and pathogenic vulnerabilities. Pregnancy is a state of inherent vulnerability due to the physiological changes associated with even a healthy pregnancy, and the impact this has on the women's sense of self, such as the increased need for care and assistance during pregnancy. Pregnant women may be situationally vulnerable if pregnancy tips them into unemployment, or triggers ill health for which the treatments are unaffordable. Finally, vulnerability faced by pregnant women as the result of historic exclusion from research is a type of pathogenic vulnerability.

Vulnerability is a prerequisite of exploitation but does not necessarily entail exploitation. At certain points in history, researchers have systematically exploited vulnerable populations or groups, where these populations or groups were both vulnerable *and* valuable to researchers (valuable in terms of reducing complexity and/or costs). Pregnancy, on the other hand, complicates rather than simplifies research. We have argued that pregnant women are not especially vulnerable to exploitation in research because they do not present researchers with any increased efficiency over and above that which is available with a non-pregnant research population. As such, although pregnant women might be vulnerable, they are not targets of research exploitation.

While in the research ethics literature vulnerability and exploitation are closely intertwined, our analysis disentangles vulnerably and exploitation. Researchers and regulators should focus on addressing the relevant vulnerabilities. They should not conflate vulnerability with exploitation, nor use the latter as a reason to limit all research with pregnant women.

Addressing Vulnerability in Research

Here we provide some options for addressing issues of vulnerability within pregnancy-related research; these include systematic collection of clinical data relating to treatment during pregnancy, attention to informed consent, and minimal risk research limits.

Two key principles relating to ethical research with vulnerable populations or groups are (1) to limit research to that which cannot be conducted in non-vulnerable populations or groups, and (2) to be especially cognisant of risk and inconvenience for research that is conducted with vulnerable groups.[5] One research strategy that potentially meets these requirements is the collection of data on the clinical treatment of pregnant women in registries. Pregnant women are a unique population, as there are no other humans who can stand in for them in clinical research. They already take many medications off-label, thus registry-based research would not expose them to extra risk over and above that incurred by their clinical treatment. Yet despite this extensive 'natural experiment', data about the outcomes of off-label treatment in pregnant women are not systematically collected and analysed. We believe pregnant women represent an excellent population within which to expand the routine analysis of clinical data for research purposes. This proposal is consistent with what the US Institute of Medicine defines as a learning health care system "in which knowledge generation is so embedded into the core of the practice of medicine that it is a natural outgrowth and product of the healthcare delivery process and leads to continual improvement in care" (2007, 6).

Faden and colleagues (2013) have recently argued for a learning health care system that bridges the entrenched split between clinical and research ethics, by fostering systematic research on clinical data. Weakening the distinction between research ethics and clinical ethics is controversial, because it potentially creates research participants out of all patients seeking clinical care. Nonetheless, there is an important feature about this kind of research, which is that the clinical care of pregnant women would not be dictated or even altered by research aims. The duty to provide patients with optimal clinical care based on their values and the professional judgement of their clinicians remains inviolable. Rather, a learning health system requires that pregnant women receiving care have their data routinely collected for analysis.[6]

In such a system, collection of data for research would be routine, but pregnant women would have the right to opt-out of research uses of that data. Pregnant women would need to be informed that their data (clinical notes, laboratory samples, DNA) were being collected for future unspecified research relating to

[5] Some research ethics guidelines require minimising risk with vulnerable populations. We argue that risk should be minimised to the greatest extent possible with all research populations and we prefer an ethical model that focuses on risk/potential benefit balance rather than risk minimisation.

[6] See, for example, the combined clinical care and research registry described in Bentley et al. (2007).

pregnancy, and offered the option of withdrawing their data from the research pool. This approach would balance the need for research involving pregnant women, minimise risk, optimise clinical care, and promote patient autonomy.

Maintaining public research registries of this nature is costly. In our view, justice supports the dedicated use of public funds to redress the lack of data about treatments during pregnancy. A just health care system should provide clinical care based on a strong evidence base. Currently, however, clinical care for pregnant women is significantly less evidence-based than clinical care for other populations. This creates a specific and ameliorable vulnerability.

In our view, industry should contribute to the cost of pregnancy-related research registries. At present, the pharmaceutical industry excludes pregnant women from the vast majority of trials, and requires the use of contraception (and tests of compliance) in female participants of reproductive age. Women who become accidentally pregnant are typically removed from trials, although their pregnancy, and the health of the foetus/infant, may be tracked. Pharmaceutical companies are aware that their products may be used off-label to treat pregnant women, but they avoid liability and responsibility to the health of pregnant women and their foetuses by not marketing their products for use during pregnancy. We believe that industry has a social responsibility to ensure that its products are safe for those that need and use them, and that it cannot avoid this responsibility via selective labelling. Yet we understand that in many cases it may not be suitable, responsible, or safe to expose pregnant women to new chemical agents or other interventions at the time these are first being tested in the general population (Baylis and Halperin 2012). We suggest, therefore, that pharmaceutical companies have the option of either: (a) submitting Phase I clinical trial data for the use of their products in pregnant women as part of the standard drug approval process where this is appropriate; or (b) purchasing a buy-out of this obligation by making a lump-sum payment to the regulatory agency. This approach leaves industry free to make the complicated and nuanced decision regarding the ethics, feasibility, and liability issues associated with the use of each intervention within a pregnant population. Funds raised from buy-out fees could be used to support public pregnancy registries.

Pregnancy registries are consistent with notions of fair research with vulnerable populations. This kind of research minimises risk, involves only interventions administered with the intention of benefiting the individual, and does not create additional burdens over and above those entailed in clinical care (see Healy and Mangin 2016). Part funding by industry would alleviate some of the costs, and create pressure to take seriously the health-research needs of pregnant women. However, while registries based on existing clinical care will fill many of the current gaps, the need for additional kinds of research with pregnant women remains.

Whether research is based on registries of routine clinical care, or on more formal clinical trials, pregnant women may be situationally vulnerable with regards to consent. This is especially the case where the research is intended for the health needs of the foetus, rather than the woman, because there are already considerable pressures on pregnant women to prioritise the welfare of foetuses above their own. Pregnant women are expected to alter a wide range of their behaviours in order to minimise any risk of mishap to the pregnancy. These pressures may well align with women's own desires for the best possible pregnancy outcomes, but as noted previously, this emotional bond creates a vulnerability that may be exploited when pregnant women are offered participation in research that is risky for them while offering unknown benefits for the foetus. This claim about vulnerability is supported by research that found women were considerably more willing to participate in a clinical trial for the potential benefit of the foetus than they were to participate in a clinical trial for their own health benefit (Rodger et al. 2003).

We support the position that pregnant women who identify their foetuses as patients[7] should be offered the opportunity to consider participation in research relevant for the well-being of the foetus (see Ashcroft 2016). However, not all pregnant women may identify their foetuses as patients, and this is an issue that must be explored sensitively with women on a case-by-case basis, in situations that allow the pregnant woman to arrive at her own decision. This may require one-on-one consultation with the woman, absent the presence of family members or others who may tend to favour the welfare of the foetus over that of the woman, or whose presence, intentionally or otherwise, may prevent the woman from expressing her own views.

In research aimed at foetal well-being, we do not support the requirement for consent from the father of the foetus, as mandated in some jurisdictions. For example, in the United States, if the research holds out the prospect of direct benefit solely to the foetus, then the consent of both the pregnant woman and the (putative) father must be obtained (DHHS 2009, 45 CFR 46.204(e)). This requirement is problematic, as it presumes that the woman and the putative father have the same attitude towards the foetus, and that they both see it as a potential patient warranting an experimental intervention. This kind of requirement creates pathogenic vulnerability as the woman loses the capacity to control interventions in her own body, and may be prevented from engaging in research in which she would like to participate. Either way, given that any intervention for the well-being of the foetus must be mediated in and through her body, we maintain that she is the sole person who should be able to grant or deny such intervention.

Where research intended to benefit the woman may pose a threat to the welfare of the foetus, researchers must take into consideration potential bias towards foetal well-being, and strive to contextualise the risks in a nuanced manner. This includes both explaining the risks of the experimental intervention compared to no intervention, and to existing treatments that in themselves may be untested in pregnant women. It is possible that the risks may be lesser in a clinical trial than in using interventions such as medications off-label, and this point should be clearly explained (see Kukla 2016).

Some jurisdictions classify all pregnancy-related research as high risk (Australia 2007), or apply minimal risk requirements to some research with pregnant women. For example, in the United States, research with the foetus or pregnant women is

---

[7] For an account of the ethical concept of the foetus as patient, and associated beneficence-based obligations, see McCullough et al 2005.

limited to that which poses minimal risk to the foetus, if there is no prospect of benefit to either the pregnant women or foetus (DHHS 2009, 45 CFR 46.204 (b)). Minimal risk "means that the probability and magnitude of harm or discomfort anticipated in the research are not greater in and of themselves than those ordinarily encountered in daily life or during the performance of routine physical or psychological examinations or tests" (DHHS 2009, 45 CFR 46.102). Given the complex nature of vulnerability and the complex interwoven relationship between the pregnant woman's interests and foetal interests, and the general lack of research with pregnant women, we argue that pre-emptory limits of pregnancy-related research are counter-productive. Researchers, IRBs, and pregnant women should be free to determine when the risks of research are outweighed by the potential benefits. The *Code of Federal Regulations* does in fact allow for this sort of exception to the minimal risk requirement, but requires review by a panel of experts, a public meeting, and sound ethical principles to guide the research (DHHS 2009, 45 CFR 46.207). This cumbersome process inappropriately limits research. We believe it would be more valuable to use available funds upstream to support public engagement: for example, through deliberative forums[8] to allow pregnant women to be part of setting the research agenda.

8.8 Conclusion

Vulnerability is not a simplistic concept that can be used to label whole populations or groups of research participants. Here we have described how inherent, situational, and pathogenic vulnerability apply to pregnant women. This nuanced perspective requires flexibility on the part of researchers and research ethics review committees to tailor research to the needs of different groups of pregnant women, and to ensure that pregnant women are supported to make informed choices about the risks and potential benefits of research. We have differentiated vulnerability and exploitation and argued that while pregnant women may be vulnerable in various ways, they are not especially vulnerable to exploitation in the research context, because they do not provide researchers with the opportunity to conduct cheaper and more efficient research. However, pregnant women are highly vulnerable due to the lack of proven safe and effective treatments for health problems that occur during pregnancy and, to this end, we suggest strategies to broaden the relevant evidence base.

[8]Deliberative forums, such as citizens' juries, provide opportunities for members of the public to receive expert information and articulate, share and deliberate about relevant issues. Their informed views are taken to be reflective of community values (Murphy 2005).

References

Ashcroft, R. 2016. Ethical issues in a trial of maternal gene transfer to improve foetal growth. In *Clinical research involving pregnant women*, eds. F. Baylis and A. Ballantyne, 247–263. Cham: Springer.

Australia (The National Health and Medical Research Council, the Australian Research Council and the Australian Vice-Chancellors' Committee). 2007. *National statement on ethical conduct in human research* (updated March 2014). Canberra: Commonwealth of Australia.

Ballantyne, A.J. 2010. How to do research fairly in an unjust world. *American Journal of Bioethics* 10(6): 26–35.

Baylis, F., and S. Halperin. 2012. Research involving pregnant women: trials and tribulations. *Clinical Investigation* 2(2): 139–146.

Bentley, S.M., J.L. Melville, B.D. Berry, and W.J. Katon. 2007. Implementing a clinical and research registry in obstetrics: Overcoming the barriers. *General Hospital Psychiatry* 29(3): 192–198.

Buchanan, Allen. 1985. *Ethics, efficiency, and the market*. Totowa: Rowman and Allanheld.

Cartwright, S. 1988. *The report of the committee of inquiry into allegations concerning the treatment of cervical cancer at national women's hospital and into other related matters 1988*. Auckland: GPO.

CIOMS (Council for International Organizations of Medical Sciences). 2002. *International ethical guidelines for biomedical research involving human subjects*. Geneva: World Health Organization. http://www.cioms.ch/publications/layout_guide2002.pdf. Accessed 3 Oct 2016.

Cragan, J. 2014. Medication use during pregnancy. *BMJ* 349: g5252.

Crouch, R.A., and J.D. Arras. 1998. AZT trials and tribulations. *Hastings Centre Report* 28(6): 26–34.

DHHS (US Department of Health and Human Services). 2009. *Code of Federal Regulations: Title 45, Part 46, Protection of Human Subjects. See Subpart B – Additional Protections for Pregnant Women, Human Fetuses and Neonates Involved in Research*. http://www.hhs.gov/ohrp/regulations-and-policy/regulations/45-cfr-46/index.html#subpartb. Accessed 3 Oct 2016.

DHHS (Department of Health and Human Services, Public Health Service, National Institutes of Health, Office of Research on Women's Health). 2011. *Enrolling pregnant women: Issues in clinical research*. Bethesda: National Institutes of Health.

Emanuel, E., D. Wendler, and C. Grady. 2002. What makes clinical research ethical? *JAMA* 283(20): 2701–2711.

Faden, R., N. Kass, S. Goodman, P. Pronovost, S. Tunis, and T. Beauchamp. 2013. An ethics framework for a learning health care system: A departure from traditional research ethics and clinical ethics. *Ethical Oversight of Learning Health Care Systems, Hastings Center Report Special Report* 43(1): S16–S27.

Fineman, M. 2008. The vulnerable subject: Anchoring equality in the human condition. *Yale Journal of Law and Feminism* 20(1): 1–23.

Goodin, R. 1985. *Protecting the vulnerable: A reanalysis of our social responsibilities*. Chicago: The University of Chicago Press.

Hawkins, J., and E. Emanuel (eds.). 2008. *Exploitation and developing countries: The ethics of clinical research*. Princeton: Princeton University Press.

Healy, D., and D. Mangin. 2016. Does my bias look big in this? In *Clinical research involving pregnant women*, eds. F. Baylis and A. Ballantyne, 197–208. Cham: Springer.

Hurst, S.A. 2008. Vulnerability in research and health care: Describing the elephant in the room? *Bioethics* 22: 191–202.

Institute of Medicine. 2007. *Ethical considerations for research involving prisoners*. Washington, DC: The National Academies Press.

Kukla, R. 2016. Equipoise, uncertainty, and inductive risk in research involving pregnant women. In *Clinical research involving pregnant women*, eds. F. Baylis and A. Ballantyne, 179–196. Cham: Springer.

Lange, M.M., W. Rogers, and S. Dodds. 2013. Vulnerability in research ethics: A way forward. *Bioethics* 27(6): 333–340.

Langston, L. 2016. Better safe than sorry: Risk, stigma, and research during pregnancy. In *Clinical research involving pregnant women*, eds. F. Baylis and A. Ballantyne, 33–50. Cham: Springer.

Little, M.O., A.D. Lyerly, A.C. Mastroianni, and R.R. Faden. 2016. Ethics and research with pregnant women: Lessons from HIV/AIDS. In *Clinical research involving pregnant women*, eds. F. Baylis and A. Ballantyne, 227–246. Cham: Springer.

Luna, F. 2009. Elucidating the concept of vulnerability: Layers not labels. *International Journal of Feminist Approaches to Bioethics* 2: 121–139.

Lyerly, A.D., M. Little, and R. Faden. 2008. The second wave: Toward ethical inclusion of pregnant women in clinical research. *International Journal of Feminist Approaches to Bioethics* 1(2): 5–22.

Mackenzie, C., W.A. Rogers, and S. Dodds. 2014. What is vulnerability and why should it matter for moral theory? Introduction. In *Vulnerability: New essays in ethics and feminist philosophy*, eds. C. Mackenzie, W.A. Rogers, and S. Dodds, 1–29. New York: Oxford University Press.

Marrus, M.R. 1997. *The Nuremberg war crimes trial 1945–46: A documentary history*. Boston, Massachusetts: Bedford Books; for primary source data go to Harvard Law School Library. 2003. Nuremberg trials documents: a digital collection. Harvard University on-line. http://nuremberg.law.harvard.edu/php/docs_swi.php?DI=1&text=overview. Accessed 7 July 2014.

McCullough, L.B., J.H. Coverdale, and F.A. Chervenak. 2005. A comprehensive ethical framework for responsibly designing and conducting pharmacologic research that involves pregnant women. *American Journal of Obstetrics and Gynecology* 193(3 SUPPL.): 901–907.

Murphy, N.J. 2005. Citizen deliberation in setting health-care priorities. *Health Expectations* 8: 172–181.

NIH (National Institutes of Health). 1994. *NIH guidelines for the inclusion of women and minorities as subjects in clinical research*. http://grants.nih.gov/grants/guide/notice-files/not94-100.html. Accessed 29 June 2014.

NIH (National Institutes of Health). 1998. *NIH policy and guidelines on the inclusion of children as participants in research involving human subjects*. http://grants.nih.gov/grants/guide/notice-files/not98-024.html. Accessed 29 June 2014.

Participants (Participants in the 2001 Conference on Ethical Aspects of Research in Developing Countries). 2002. Ethics: Fair benefits for research in developing countries. *Science*. 298: 2133–2134.

Phillips, T. 2011. From the ideal market to the ideal clinic: Constructing a normative standard of fairness for human subjects research. *Journal of Medicine and Philosophy* 36(1): 79–106.

Rodger, M.A., D. Makropoulos, M. Walker, E. Keely, A. Karovitch, and P.S. Wells. 2003. Participation of pregnant women in clinical trials: Will they participate and why? *American Journal of Perinatology* 20(2): 69–76.

Rogers, W. 2014. Vulnerability and bioethics. In *Vulnerability: New essays in ethics and feminist philosophy*, eds. C. Mackenzie, W. Rogers, and S. Dodds, 60–87. New York: Oxford University Press.

Rogers, W., C. Mackenzie, and S. Dodds. 2012. Why bioethics needs a concept of vulnerability. *International Journal of Feminist Approaches to Bioethics* 5(2): 11–38.

Seigel, A. 2008. Kantian ethics, exploitation, and multinational clinical trials. In *Exploitation and developing countries: The ethics of clinical research*, eds. J.S. Hawkins and E.J. Emanuel, 175–205. Princeton: Princeton University Press.

United States (National Commission for the Protection of Human Subjects of Biomedical and Behavioral Research). 1979. *The Belmont Report: Ethical principles and guidelines for the protection of human subjects of research*. http://www.hhs.gov/ohrp/humansubjects/guidance/belmont.htm. Accessed 26 Mar 2015.

Valdman, M. 2009. A theory of wrongful exploitation. *Philosophers' Imprint* 9(6): 1–14.

Wendler, D. 2000. Informed consent, exploitation and whether it is possible to conduct human subjects research without either one. *Bioethics* 14(4): 310–339.

Wertheimer, A. 1996. *Exploitation*. Princeton: Princeton University Press.

Wilkinson, S. 2003. *Bodies for sale: Ethics and exploitation in the human body trade*. London: Routledge.

Wood, A. 1997. Exploitation. In *Exploitation: Key concepts in critical theory*, eds. K. Nielsen and R. Ware, 2–26. Atlantic Highlands: Humanities Press International.

World Medical Association. 2013. *WMA Declaration of Helsinki: Ethical Principles for Medical Research Involving Human Subjects*. http://www.wma.net/en/30publications/10policies/b3/. Accessed 3 Oct 2016.

Zion, D., L. Gillam, and B. Loff. 2000. The Declaration of Helsinki, CIOMS and the ethics of research on vulnerable populations. *Nature Medicine* 6(6): 615–617.

Chapter 9
When Hypothetical Vulnerability Becomes Actual: Research Participation and the Autonomy of Pregnant Women

L. Syd M. Johnson

Abstract *Various national and international research guidelines and regulations limit the inclusion of pregnant women in clinical research by classifying them as vulnerable. This exclusion has widely acknowledged negative consequences for the health of women, foetuses, and future children. Another negative consequence is the threat to a pregnant woman's autonomy and agency when she is treated as a 'vulnerable' person without cause. Research guidelines and regulations around the world continue to be overly protectionist. The limitations on autonomy they imply (and create) infantilise pregnant women, treating them 'as if' they are vulnerable – as if they, in fact, lack autonomy and the capacity to make informed choices about their own research participation. The hypothetical 'as if' becomes actual as research guidelines and regulations effectively reduce the autonomy and agency of pregnant women, making them unable to protect their own interests, including their interests in protecting their future children.*

Globally, research regulations and guidelines (hereafter jointly referred to as guidelines) concerning the inclusion and exclusion of pregnant women wrestle with the question of how to balance the acknowledged need for research during pregnancy with the need to protect pregnant women and their foetuses from possible harm. Pregnant women are frequently conceptualised as a 'vulnerable population' or 'special group' in need of protection, yet this categorisation of pregnant women requires interrogation. Of particular concern is the possibility that the guidelines themselves foster the perception that pregnant women are vulnerable. In so doing, they wrong pregnant women by failing to respect their autonomy and agency. It would be ironic if guidelines that have as foundational principles respect for persons and their autonomy effectively diminished or constrained the autonomy of pregnant

L.S.M. Johnson, PhD (✉)
Humanities, Kinesiology and Integrative Physiology, Michigan Technological University, Houghton, MI, USA
e-mail: lsjohnso@mtu.edu

F. Baylis, A. Ballantyne (eds.), *Clinical Research Involving Pregnant Women*, Research Ethics Forum 3, DOI 10.1007/978-3-319-26512-4_9

women in a misguided and paternalistic effort to protect them unnecessarily. It would be doubly ironic if pregnant women and their future offspring were in fact harmed by overly protectionist and exclusionary policies.

Pregnant women are frequently included among 'vulnerable' or 'special' groups in research guidelines, several of which are considered here. The *Common Rule* of the United States Department of Health and Human Services is regulatory and has the force of law; its companion, the *Belmont Report*, provides the ethical framework for the protection of human participants.[1] Four documents, representing a range of national and international guidelines, are also examined: The WHO Council for International Organizations of Medical Sciences *International Guidelines for Biomedical Research Involving Human Subjects* (CIOMS 2002); Canada's *Tri-Council Policy Statement: Ethical Conduct for Research Involving Humans* (TCPS2) (Canada 2014); the South African Medical Research Council *Guidelines on Ethics for Medical Research: General Principles* (South Africa 2000); and The Indian Council of Medical Research *Ethical Guidelines for Biomedical Research on Human Participants* (India 2006).

There are more than a 1,000 national and international laws, regulations, and guidelines governing research with humans (DHHS 2012). For obvious practical reasons, they will not all be considered in this chapter. The guidelines highlighted here meet two criteria: they state that the principles of respect for persons and autonomy are foundational, and they contain explicit guidance concerning research with pregnant women. For example, the *Declaration of Helsinki* was excluded because, while it states that medical research should "ensure respect for all human subjects," it does not explicitly refer to autonomy, nor to pregnant women (World Medical Association 2013, 7). An exhaustive survey of national and international guidelines would be beyond the scope of this chapter, the goal of which is to explore the tension between respect for autonomy, and the exclusion of pregnant women as research participants. All of the guidelines considered here allow for the inclusion of pregnant women in research, but not without restrictions that call into question the commitment to recognising and respecting the autonomy and decision-making of pregnant women.

9.1 Regulations and Guidelines Regarding Research During Pregnancy

For decades, several federal agencies of the United States government charged with regulating medicines, medical devices, and clinical research have called for greater inclusion of women in research. At the same time, the US *Common Rule*, the code

[1] The use of the term 'subject' has been widely replaced by 'participant' in recent years. Both the *Common Rule* and *the Belmont Report* use 'subject,' while *CIOMS guidelines*, *TCPS2*, and the guidelines of South Africa and India usually refer to 'participant.' 'Participant' will be the preferred terminology used in this chapter.

of regulations for the protection of human research participants, calls for "additional protections for pregnant women, human fetuses, and neonates involved in research" (DHHS 2009, 45 CFR 46.204). The 'additional protections' include limiting both therapeutic and non-therapeutic research to that with the same threshold of risk permissible for research on children: minimal risk. The *Common Rule* states that pregnant women or foetuses may be involved in research if all of the following conditions are met: (a) preclinical studies, including studies on pregnant animals and nonpregnant women have been conducted previously and provide data for assessing the risks to pregnant women and foetuses; (b) the risk to the foetus is caused solely by interventions that have the prospect of direct benefit for the woman or foetus, or, the risk to the foetus is not greater than minimal and the purpose of the research is to obtain information that cannot be obtained by other means; (c) any risk is the least possible for achieving the research objectives; (d) the research may directly benefit the pregnant woman, the pregnant woman and her foetus, or the foetus alone (DHHS 2009, 45 CFR 46.204).

The *Common Rule* identifies three groups as in need of special regulation and protection: Prisoners, children, and a group comprising pregnant women, human foetuses, and neonates. What is notable about the restrictions on research involving pregnant women is how similar they are to regulations concerning research on children. On the one hand, it might seem appropriate to provide the same protections to foetuses and children: neither are autonomous, neither can consent to, nor voluntarily withdraw from, research participation, and there is a long and tragic history of nonvoluntary, exploitative research on vulnerable and institutionalised children, such as those subjected to hepatitis research at Willowbrook (Robinson and Unruh 2008), and the Kennedy-Krieger lead abatement study (Mastroianni and Kahn 2002). There are no comparable reasons, however, to presume that adult, pregnant women in full possession of decision-making capacity and autonomy should not be free to choose to participate in research. Yet the restrictions in the *Common Rule* are widely interpreted as requiring the exclusion of pregnant women from clinical research. For example, the US Food and Drug Administration webpage for pregnancy registries, where women who take drugs during pregnancy are encouraged to report on their outcomes, includes the following statement:

> Since drug companies can't test medicine on pregnant women, they may have little or no information about how these medicines could effect a woman or her fetus. Pregnancy registries are the best way to learn and to help women decide about taking medicines. These studies can also help improve the information for pregnant women that is provided on drug labels. (FDA 2011)

While it is clearly false that 'drug companies can't test medicine on pregnant women,' and likewise false that uncontrolled, voluntary, ad hoc registries are the 'best way' to learn about the effects of drugs on pregnant women and foetuses, the *Common Rule* is interpreted – even by US federal agencies – as saying just that. In a paper reporting on the data obtained from one pregnancy registry, for the Eli Lilly antipsychotic drug Olanzapine, the authors note that "Due to ethical constraints restricting inclusion of pregnant or breastfeeding women in clinical trials, there is a paucity of data available on the use of antipsychotic drugs in this population" (Brunner et al. 2013). The registries hardly remedy that paucity of data: during the 24 year study period, from 1986 to 2010, only 610 reports of Olanzapine use during pregnancy or breastfeeding were recorded in the worldwide registry.

CIOMS International Guidelines for Biomedical Research Involving Human Subjects states in Guideline 16 that:

> A general policy of excluding from such clinical trials women biologically capable of becoming pregnant is unjust in that it deprives women as a class of persons of the benefits of the new knowledge derived from the trials. Further, it is an affront to their right of self-determination. (CIOMS 2002, Guideline 16)

At first blush, the statement is broadly supportive of including consenting pregnant women in clinical research. Guideline 17 adds that pregnant women should be "presumed to be eligible for participation in biomedical research." Yet this guideline contains an apparent qualification: "Research in this population should be performed only if it is relevant to the particular health needs of a pregnant woman or her fetus, or to the health of pregnant women in general." This statement might be interpreted to mean that only therapeutic research is permitted, or it might mean that pregnant women should be included only in research specific to pregnancy, pregnancy-related conditions, or foetal health.

The CIOMS guidelines commendably attempt to be inclusive, and to guard against automatic and unjustified exclusion of women who are pregnant, or who could become pregnant. Assuming the intent of these guidelines is to be inclusive of individual pregnant women who choose to enroll in potentially therapeutic research relevant to health needs that are *not* related to their pregnancies, the language of the guidelines could be amended to reflect that.[2]

Regarding research involving women, Canada's *TCPS2* is clear about the need for inclusion:

> Women have historically been inappropriately excluded from participating in some research… The inclusion of women in research advances the commitment to Justice, improves the generalizability of research findings to women where that is the goal of the research, and is essential to ensure that women and men benefit equally from research. (Canada 2014, 50)

The guidelines are less clear, however, about when exclusion is permissible, stating that "Women shall not be *inappropriately* excluded from research solely on the basis of their reproductive capacity, or because they are pregnant or breastfeeding" (Canada 2014, 4.3) [emphasis added]. The guidelines do not define what would constitute inappropriate exclusion, but advise research ethics boards to "take into account foreseeable risks and potential benefits for the woman and her embryo, fetus or infant…" (Canada 2014, 4.3). Like the CIOMS guidelines, the *TCPS2*

[2] The CIOMS guidelines are currently under revision, and the guidelines concerning women and pregnant women are expected to be substantially changed. This may result in greater clarity on this point.

professes to support the inclusion of pregnant women, but is silent on how far that inclusion should extend.

South Africa's research guidelines are sparse and relatively nonspecific regarding research with pregnant women and women of childbearing potential. Pregnant women are among those needing "special consideration," a category that includes children, prisoners, students, and "people with mental disabilities" (South Africa 2000, 7.1.3). Like CIOMS and *TCPS2*, South Africa's guidelines caution against the exclusion of pregnant women without justification:

> The exclusion of pregnant women from research should be adequately justified, both in terms of protecting the health of the fetus and from the perspective that such exclusion is scientifically supportable. (South Africa 2000, 7.1.3.1)

India's guidelines apply to both pregnant and nursing women, and are among the most restrictive, permitting research only if related to pregnancy, lactation, or foetal health:

> Pregnant or nursing women should in no circumstances be the participant of any research unless the research carries no more than minimal risk to the fetus or nursing infant and the object of the research is to obtain new knowledge about the foetus, pregnancy and lactation [sic]. (India 2006, IV.i.)

There are many conditions that can affect the health of pregnant women and their foetuses, and many of them are not specifically pregnancy-related. Like other women, pregnant women have chronic illnesses such as asthma, diabetes, hypertension, auto-immune disorders, and depression that require treatment. Many chronic health conditions of women are known to adversely affect the health and development of the foetus if not adequately managed during pregnancy. But with a paucity of data demonstrating the safety and effectiveness of most drugs for pregnant women or foetuses, pregnant women and their healthcare providers must make choices about treatment options without knowing whether treatments are safe and effective for pregnant women, or what the effects on their foetuses might be. Even those guidelines that attempt to be inclusive rather than exclusive of pregnant women fail to acknowledge that the health needs of pregnant women and their foetuses extend beyond specifically pregnancy-related health concerns.

9.2 Autonomy and Decision-Making in Clinical Research Regulations and Guidelines

Autonomy, understood literally and most basically, is self-rule, and autonomous actions are those that are voluntary and self-legislated. Respect for autonomy, as an element of respect for persons, is a foundational bioethical principle, and widely regarded as an inviolable 'first principle' of clinical research ethics (United States 1979; CIOMS 2002; Lupton and Williams 2004). It protects research participants from wrongful exploitation by making the informed consent of persons with decisional capacity the cornerstone of ethically conducted research.

The centrality of autonomy is evident in the guidelines for research on humans, although precisely what is meant by autonomy is not always completely clear. The definition and meaning of autonomy remains a matter of debate in the bioethics literature as well (Kukla 2005; Mackenzie 2010). For the purposes of this chapter, the working definitions of autonomy as they are found in the guidelines will be used in examining whether the guidelines themselves may impact the autonomy of pregnant women.

The *Belmont Report*, defines an autonomous person as "an individual capable of deliberation about personal goals and of acting under the direction of such deliberation" (United States 1979, B.1). *CIOMS* defines autonomous persons as those who have the capacity for self-determination, and who are capable of deliberation about their personal choices (CIOMS 2002). In a similar vein, the *TCPS2* states that "Autonomy includes the ability to deliberate about a decision and to act based on that deliberation. Respecting autonomy means giving due deference to a person's judgment and ensuring that the person is free to choose without interference" (Canada 2014, 1.1). Autonomy, so operationalised, requires capacities for deliberation and self-determination, and the freedom to act on one's deliberate choices. These descriptions of autonomy focus both on traits of autonomous persons (capacities for deliberation and self-determination), and on autonomous actions (actions that are free from interference, and that spring from the deliberate choices of individuals).

South Africa's guidelines stress that respect for the autonomy of the participant requires that "freedom of choice must be safeguarded" (South Africa 2000, 3.1.3.i), and that autonomy is "a participant's fundamental right" to consent, or refuse consent, to participate in research (South Africa 2000, 5.3.2.2.). Similarly, the Indian Council's guidelines cite respect for individual autonomy as one of four universal ethical principles, but do not define autonomy beyond operationalising it in the context of the informed consent process, as the freedom to voluntarily choose whether to participate in research (India 2006, III.I.1).

Although respect for autonomy is considered a fundamental ethical principle guiding clinical research, the concept of autonomy remains vaguely defined in the research guidelines, and is primarily operationalised as the functional requirement that the informed consent of research participants or their proxies (in the case of children or others with diminished autonomy) be obtained. Historically, guidelines governing research on humans have responded to the abuse of human participants, and in particular those who were involuntarily or non-voluntarily subjected to research without knowledge or consent. The regulatory solution to the exploitation of involuntary/non-voluntary research subjects was the foundational emphasis on respect for autonomy as a recognition of individual freedom of choice and self-determination. That is, where respect for autonomy is the safeguard against the use of research participants without their consent or against their will, research guidelines define respect for autonomy as a negative right, a freedom from coercion or constraints on the exercise of voluntary choice.

It is generally presumed that adults who possess decision-making capacity are competent to provide voluntary, informed consent for research participation.

Moreover, it is presumed that adults are capable of making decisions that are in their own best interests, and that adult individuals are in the best position to judge what is in their own interests, and to weigh the risks and potential benefits of their choices and activities. This presumption extends beyond consent in medical or research contexts, to encompass most aspects of adult life, including decisions concerning the conduct of daily life. It includes, as well, important decisions regarding procreation, such as when and how to procreate, and when and how to avoid procreating.

There are exceptional cases where it is reasonable to suspect that the capacity and competence required for voluntary consent are absent, impaired, or diminished. Persons with cognitive or intellectual disabilities as a result of acquired brain injuries, illness, or congenital conditions in some cases may not be capable of providing informed consent for research participation. Additionally, there are other groups who are categorised as especially vulnerable to coercion or exploitation, and for whom special consideration and protection is required. Such groups include prisoners, about whom there are valid concerns about coercion and their freedom to make voluntary, autonomous choices about research participation. The inherent constraints on freedom and autonomy experienced by prisoners, as well as their long history of exploitation and involuntary research participation, justifies special precautions and protections for this group.

Finally, some others might have their autonomy or decision-making capacity diminished by illness, including psychiatric illness, or by social, political, or economic circumstances that make them especially vulnerable to coercion, inducement, or exploitation. In all of these cases, there are good reasons to be concerned about the autonomy of individuals within these groups, and their ability to provide uncoerced, voluntary and informed consent for research participation. All research guidelines agree that individuals or groups who already have diminished or threatened autonomy, either as a result of social, economic, or political circumstances, age, or diminished decision-making capacity, and so forth, must be afforded special protections against abuse and exploitation. That is, those who, for a variety of reasons, lack autonomy, or the ability to exercise their autonomy and make free, voluntary choices regarding research participation must be protected.

9.3 The Vulnerability of Pregnant Women

Infants and children, prisoners and institutionalised persons, persons with significant cognitive disabilities, and pregnant women are frequently included among the vulnerable or special groups entitled to protections that take the form of more restrictive regulations on the research that can be conducted with persons in these groups. Yet the conditions that result in vulnerability to exploitation, or concerns about autonomy and the capacity to provide voluntary, informed consent to research participation do not apply to all pregnant adult women as a group. Pregnant adult women, like all adult women, can and should be presumed to be competent to make decisions about their participation in research. Being pregnant, by itself, is not the kind of special circumstance that requires challenging that presumption (see Wild and Biller-Andorno 2016; Ballantyne and Rogers 2016).

The US *Common Rule*, and the guidelines of India and South Africa all include pregnant women among the vulnerable or special groups in need of greater protection. As pregnant women are frequently included among vulnerable groups, it is necessary to consider what is meant by vulnerable, just in case that term does describe or apply to pregnant women. It is not obvious what pregnant women as a group have in common with other vulnerable or special groups, namely children, prisoners, persons with intellectual disabilities, or, more generally, persons with diminished autonomy or external constraints on their autonomy. This is not to deny that in some circumstances, some individual pregnant women are vulnerable, but being pregnant by itself is not of necessity a condition resulting in vulnerability.[3]

The *Common Rule* notably does not define vulnerability, but merely lists vulnerable populations, i.e. "children, prisoners, pregnant women, mentally disabled persons, or economically or educationally disadvantaged persons" (DHHS 2009, 45 CFR 46.111). South Africa's guidelines, on the other hand, outline six characteristics of vulnerable communities: limited economic development; inadequate protection of human rights; discrimination on the basis of health status; inadequate understanding of scientific research; limited availability of health care and treatment options; and limited ability of individuals in the community to provide informed consent (South Africa 2000, 7.1.3.8). It is not clear, however, how these characteristics specifically align with the classes of persons whose welfare requires special protection – pregnant women, children, prisoners, people with mental disabilities, the elderly, students, and persons in dependent relationships (South Africa 2000, 7.1.3). Elsewhere, the South African guidelines state that pregnant women are "usually competent to consent," but they stop short of saying that pregnant women should be presumed to be competent:

> Pregnant women are *usually competent to consent* to health research, but the circumstances may sometimes compromise their decision. Where possible, the father of the unborn child should be included in making the decision. (South Africa 2000, 5.3.1.1.3) [emphasis added]

By setting aside the general presumption that adult pregnant women are competent and capable of consenting to research participation, the South African guidelines may create the very vulnerability that seemingly justifies the restrictions that protect the welfare of pregnant women, and may also effectively diminish their autonomy and ability to consent voluntarily.

CIOMS notably provides a specific definition of vulnerability as "a substantial incapacity to protect one's own interests" (CIOMS 2002, Guideline 12). "Vulnerable persons are those who are relatively (or absolutely) incapable of protecting their own interests. More formally, they may have insufficient power, intelligence, educa-

[3] For a detailed discussion of types of vulnerability that can be experienced by pregnant women, see Ballantyne and Rogers (2016). Notably, Ballantyne and Rogers argue that while pregnant women *can* be vulnerable in a number of ways, that vulnerability does not create incentives or opportunities for them to be exploited in research, because involving pregnant women in research does not make research easier, quicker, or cheaper.

tion, resources, strength, or other needed attributes to protect their own interests" (CIOMS 2002, Guideline 13). In the CIOMS guidelines, then, we see a concrete definition of vulnerability as the relative or absolute incapacity or inability to protect one's own interests. That is, vulnerability can be defined as either lacking the traits of an autonomous person (the incapacity to protect one's own interests), or being unable to act autonomously owing to constraints on one's freedom to act (the inability to protect one's own interests). CIOMS has an expansive list of persons who might be vulnerable including: subordinate members of hierarchical groups, such as employees, medical students, and members of the armed forces; the elderly; residents of nursing homes; people receiving welfare benefits or social assistance; the poor and unemployed; emergency room patients; some ethnic and racial minority groups; homeless persons, nomads, refugees; prisoners; politically powerless individuals; and members of communities unfamiliar with modern medical concepts (CIOMS 2002, Guideline 13).

TCPS2 defines vulnerability similarly as "A diminished ability to fully safeguard one's own interests in the context of a specific research project. This may be caused by limited capacity or limited access to social goods, such as rights, opportunities, and power. Individuals or groups may experience vulnerability to different degrees and at different times, depending on their circumstances" (Canada 2014, 218). Historically vulnerable groups and individuals include "children, the elderly, women, prisoners, those with mental health issues, and those with diminished capacity for self-determination" (Canada 2014, 8).

Individual women and pregnant women can be or can become vulnerable in all of the ways noted above. For example, women who are members of historically marginalised racial or ethnic groups, women with intellectual disabilities, refugees, and unemployed and economically disadvantaged women might all be vulnerable as potential research participants. But not all individual members of historically or currently oppressed, marginalised, or vulnerable groups are equally oppressed, marginalised, or vulnerable (Denny and Grady 2008, 412). Some individual women might be multiply disadvantaged by virtue of race, age, economic or educational status, language, and so on. An affluent, white, English-speaking woman with access to private health insurance in the US would be far less vulnerable – if vulnerable at all – than a poor, Spanish-speaking, Latina woman without medical insurance in the US. A pregnant woman with special needs, whether economic, health-related, or other, might be more vulnerable than a more favourably situated pregnant woman, but her vulnerability is not by virtue of being pregnant, but rather results from other needs or circumstances (Denny and Grady 2008). Those other needs and circumstances can affect all women, or indeed, all persons. Vulnerability, then, is contextual, and it is not clear what pregnancy adds to the equation, in terms of the vulnerability of the pregnant woman.

9.4 Vulnerable Groups and the Maternal-Foetal Dyad

If all pregnant women constitute a single class, that class could be considered vulnerable owing to the vulnerability of some individual members. Such a classification, if itself justifiable, might then justify paternalistically protectionist policies under what Miller and Wertheimer characterise as "group soft paternalism," which protects groups as a whole when some members of the group are "not capable of acting autonomously" (Miller and Wertheimer 2007, 28). All individuals in the group would be subject to the same protections, whether individually needed or not. Miller and Wertheimer claim that group soft paternalism is the ethical justification for research regulations generally, and not just those that refer to special protections for certain vulnerable groups. Research guidelines restrict what investigators and researchers can do, but they also importantly (and paternalistically) restrict the freedom of research participants because they create situations in which some research protocols – including some that might benefit individuals within the group – will never be proposed or instituted given certain protectionist policies (Miller and Wertheimer 2007, 28). This type of group paternalism can explain the special protections afforded to pregnant women as a group, but it also could explain and would justify similar restrictions on clinical research with all human participants. After all, nearly any group can be constituted so as to include members who are vulnerable, thus invoking protections for the group as a whole.

Pregnant women have but two things in common: they are women, and they are pregnant. As an entire class, they have a multitude of differences, and their classification as a single vulnerable group is arbitrary. Many other groups, with at least as much or more in common, might have many more characteristics of vulnerability when taken as a whole. The class of "all human males," for example, would include millions of individuals with a wide range of vulnerabilities that would warrant special consideration or protection under the descriptions of vulnerability offered by the guidelines. Thus, the justification for classifying pregnant women as a vulnerable group cannot be found in the actual vulnerability of certain individuals within the group. Nothing about pregnancy itself diminishes the autonomy and the decision-making capacity of pregnant women, nor does it necessarily result in disadvantage or dependence for the pregnant woman. Women do not lose their capacities when they become pregnant, and are not vulnerable *per se* (Lupton and Williams 2004, 1308). They remain capable of protecting their own interests.

There is another way to constitute pregnant women as a vulnerable group, and that is by combining the interests of pregnant women with the hypothetical and contested interests of their foetuses. If foetuses have interests of their own (a controversial proposition not addressed here), it is obvious that they would not be capable of protecting their own interests, any more than infants or young children would be. Foetuses lack autonomy and decision-making capacity, and are completely dependent. Thus, if foetuses are considered persons under research guidelines, they would qualify as vulnerable persons. Foetuses are not explicitly defined as persons in research guidelines, but the maternal-foetal dyad might nonetheless be considered a vulnerable group, owing to the inherent vulnerability of one of its members, even if that vulnerable member is human, but not uncontroversially a person. One does not have to be a person to be morally significant. As such, the indeterminate moral status of the foetus qua person is not an obstacle to considering foetuses within a moral framework. If nothing else, they are potential future persons, and are morally significant entities to the pregnant women who bear them.[4]

There is a unique intertwinement between the pregnant woman and her foetus (Little 1999). It is an intertwinement of two biologically distinct and socially different organisms. That intertwinement, and the dependence of the foetus upon the pregnant woman does not obliterate the independence or identity of the pregnant woman. Yet conflating pregnant women and the foetuses they carry does appear to be the rationale behind classifying pregnant women as a vulnerable group. It's a conflation that reconstitutes pregnant women as containers of vulnerable foetal persons/patients, and fails to recognise them as distinct and non-vulnerable individuals. The presumption that pregnant women require greater protection in research rests primarily, if not entirely, on the presumed need of the foetus for protection. The interests of the pregnant woman are subsumed for purposes of research and assessing the risks of research. For example, guidelines generally refer to foetal risks rather than risks to the pregnant woman. The *Common Rule* literally classifies "Pregnant Women, Human Fetuses and Neonates," as a group, and explicitly regulates "Research on pregnant women or fetuses" together. The disjunction "or" suggests a possible separation of research on pregnant women and foetuses, but it is not at all clear how research on foetuses might be accomplished as an "or," without simultaneously involving pregnant women, since "access to the fetus can only be obtained through its mother" (ACOG 2005).

Given the language and content of the *Common Rule*, pregnant women and foetuses are clearly considered as a grouped entity, but it is the vulnerability of foetuses, rather than the vulnerability of pregnant women, that underlies the classification as "vulnerable" and provides the justification for restrictions on allowable research. The *Common Rule* does differentiate between research that might benefit a pregnant woman alone, or a woman and foetus, and that which might

---

[4] Existing research guidelines do not take a stand on the alleged personhood or moral status of the foetus. For example, CIOMS refers to "the person the fetus is destined to become" (CIOMS 2002, Guideline 17), while explicitly noting that discussion of the moral status of foetuses and embryos proved a sticking point in considering research on the products of conception (CIOMS 2002, Introduction). *TCPS2* defines the foetus as "a human organism during the period of its development beginning on the 57th day following fertilization or creation, excluding any time during which its development has been suspended, and ending at birth" (Canada 2014, 183). The definition of a "participant" in *TCPS2* plausibly encompasses foetuses, but without defining participants as necessarily being persons: "An individual whose data, or responses to interventions, stimuli, or questions by a researcher are relevant to answering a research question; also referred to as "human participant," and in other policies/guidance as "subject" or "research subject" (Canada 2014, 215). The *Common Rule* defines a foetus as "the product of conception from implantation until delivery" (DHHS 2009, 45 CFR 46 Subpart B 46.202) without making reference to its personhood. India and South Africa do not define persons or foetuses in their guidelines, and are silent on the matter of foetal personhood.

benefit the foetus alone. In the latter case, the *Common Rule* requires consent from both the pregnant woman and the father of the foetus (DHHS 2009, 45 CFR 46.204). Research on neonates, by contrast, requires consent from "either parent," but not both (DHHS 2009, 45 CFR 46.205). As such, it is strange that research that potentially benefits a foetus alone, and which can only be performed through the body of a pregnant woman, requires the consent of the putative father. Such a requirement is consistent with the way the regulations otherwise treat pregnant women and their foetuses as a grouped entity consisting not of individuals with different vulnerabilities, but as having the vulnerability of foetuses. Were such consent requirements placed on women in general, they would be clearly and objectionably paternalistic. They are likewise paternalistic with respect to pregnant women because they fail to acknowledge their bodily autonomy, as well as their capacity to protect their foetuses, their own health, and their interests in both.

CIOMS restricts research to that which is "relevant to the particular health needs of a pregnant woman or her fetus, or to the health needs of pregnant women in general…" (CIOMS 2002, Guideline 17). The CIOMS guidelines seemingly permit research relevant only to the health needs of a pregnant woman, rather than the woman and her foetus, thus effecting a separation of risk to the foetus and the pregnant woman, so long as the pregnant woman is adequately informed about the risks to both. CIOMS guidelines are generally more permissive than other documents regarding research during pregnancy, but there is a question about how research would be interpreted under these guidelines if it is relevant to all women, and not just pregnant women. *TCPS2* states that "researchers and REBs shall take into account foreseeable risks and potential benefits for the woman *and* her embryo, fetus or infant" (Canada 2014, 4.3) [emphasis added]. South Africa's guidelines situate the risk with the foetus rather than the pregnant woman (South Africa 2000, 7.1.3.1.). India's guidelines are among the most restrictive, and also explicitly situate the risk with the fetus: "Pregnant or nursing women should in no circumstances be the participant of any research unless the research carries no more than minimal risk to the foetus or nursing infant…" (India 2006, IV.i.). India's guidelines apparently prohibit research that might benefit the pregnant woman alone, even if it presents only minimal risk to the foetus.

It should be obvious that the risk of harm to a gestating foetus does not render a pregnant woman herself vulnerable in the sense that she is relatively or absolutely incapable of protecting her own interests. While the actual health-related interests of a pregnant woman can be separated from the possible health of her foetus, it is not at all obvious that the foetus' health can be separated from the health of the pregnant woman. The health of the foetus is in many important ways dependent on the health of the pregnant woman, but not, generally speaking, the other way around.[5] This one-way dependence of foetal health on the health of the pregnant woman does not support the notion that pregnant women themselves are vulnerable

[5] There are rare situations in which a woman's health or life might be endangered by pregnancy or pregnancy-related conditions, such as pre-eclampsia, HELLP Syndrome, and amniotic fluid embolism. In such cases, however, the pregnant woman's health is not dependent on the foetus' health,

and cannot protect their own interests, nor does it support their need for greater protection from the risks of research. Rather, it supports the need for research on pregnant women that is responsive to all the health needs and concerns of pregnant women, separate from their specifically pregnancy-related health needs and concerns. This is both because in the maternal-foetal dyad, the health of the pregnant woman affects the health of her foetus, and because pregnant women have an interest in what happens to their foetuses and future offspring. To the extent that research guidelines restrict the access of pregnant women to potentially beneficial clinical research, those guidelines wrong them by diminishing their ability to protect their own interests, including their interests in the health of their foetuses and future children.

9.5 Vulnerability to Exclusion

Neither CIOMS nor *TCPS2* categorise pregnant women as vulnerable, but rather single out pregnant women in an effort to caution against their automatic exclusion, thus acknowledging that pregnant women are actually vulnerable to unjust exclusion. Other groups cited as historically vulnerable are not "vulnerable" in the same way. That is, the other groups are not vulnerable to unjust exclusion, but rather are vulnerable to exploitation and over-use, owing to their easy accessibility or diminished autonomy. It is the vulnerability to exploitation, abuse and over-use that research guidelines seek to mitigate through restrictions on the use of vulnerable participants (see Ballantyne and Rogers 2016).

The vulnerability-to-exclusion of pregnant women is quite different from the vulnerability to exploitation, abuse, and over-use, and thus needs some explication. If vulnerable persons are those who, as described by CIOMS, are relatively or absolutely incapable of protecting their own interests, then restrictive research guidelines render pregnant women vulnerable in the following way. All pregnant women, like all persons, have an interest in participating in research so that evidence exists for the safety and effectiveness of the treatments they might need (see Healy and Mangin 2016). Pregnant women not only have this interest with respect to their own health needs, but also with respect to the safety and effectiveness of treatments that might affect the health or development of their foetuses. Pregnant women are excluded from research because of social, ethical, and regulatory attitudes about protecting foetuses absolutely from risk. These attitudes, as they inform and are enforced by research guidelines, present a barrier to participating in research – a barrier that pregnant women are unable to overcome. Thus, pregnant women are vulnerable specifically when they are unable to protect their own interests in participating in research. That is, they are vulnerable to exclusion that thwarts their inter-

---

per se. Rather, being pregnant adversely affects the woman's health, regardless of the condition of the foetus.

ests in potentially beneficial research participation, and the availability of safe and effective treatments for both themselves and their foetuses.

To sum up, pregnant women are part of a maternal-foetal dyad in which the foetus is vulnerable as a result of its incapacity, dependence, lack of autonomy, and absolute inability to protect its own interests. Membership in the maternal-foetal dyad doesn't make pregnant women themselves vulnerable. They are more accurately described as "complex" (NIH-ORWH 2011, 25). Indeed, they are complex through and through, from their mercurial, and intertwined biology, to the medical, scientific, and moral considerations relevant to their inclusion in clinical research. But pregnant women are importantly capable of making informed decisions regarding research participation, and protecting their own interests, including their interests in protecting their foetuses. Guidelines that effectively treat pregnant women as if they have the capacities of children or foetuses do violence to the ideals of autonomy and self-determination and cause a moral injury to women by failing to respect them as persons. While it is a social good that research guidelines protect the welfare of both pregnant women and their foetuses – just as they do for all research participants – it is also important that they do so in a way that lives up to their own ideals and principles, by respecting the autonomy of participants with decisional capacity, and avoiding paternalistic protectionism. In the case of pregnant women and their foetuses, paternalistic protectionism fails to protect pregnant women's interests by putting up barriers to potentially beneficial research, and it also fails to acknowledge the capacity of pregnant women to protect both their own interests, and the health of their foetuses, thus creating a vulnerability-to-exclusion.

9.6 Manufacturing Maternal Foetal Conflict

All research guidelines surveyed here recognise that autonomy should be respected and protected in those who have it. In cases where individuals lack autonomy, as in the case of infants and children, the appropriate and guiding principle of proxy consent for research involving children and other dependents is the protection of their health and welfare. Foetuses can be protected in the same way that children are protected by their parents – through voluntary, informed consent for interventions from capable, competent, autonomous persons acting on behalf of the dependent foetus. The maternal-foetal dyad is composed of a vulnerable, dependent, incapable future person, and a presumptively autonomous person capable of making decisions and protecting her own and her foetus' interests (see Wild and Biller-Andorno 2016). The welfare of a future child is a voluntary constraint on the choices of a pregnant woman, and she can generally be presumed to have that welfare in mind when making decisions that affect both members of the dyad.

This is not to deny that there are cases where foetal harm occurs as a result of a pregnant woman's actions, or her failure or inability to prevent harm, just as it is true that children can come to harm when their parents cause or fail to prevent harm. But the same presumptions that favour parental autonomy in the case of existing children favour the autonomy of pregnant women in the case of gestating foetuses. Parents have wide discretion to make choices that affect the health and welfare of their children, and that discretion is justified both by parental autonomy interests, and by the belief that, in most cases, parents are in the best position to safeguard the welfare of their children and act in their best interests because of the unique and intimate filial bond between them. Where social or legal interference is warranted, it is because children are endangered, for example, by abuse or neglect. That is, paternalistic state intervention is justified when it serves and protects the interests of vulnerable children, but not as a default position, or when it does not protect the interests of the vulnerable. Paternalist research guidelines that promote the exclusion of pregnant women might be justifiable if they protected vulnerable foetuses or pregnant women from harm, but they do not. Restrictions that effectively exclude pregnant women from research have resulted in a state of affairs in which there is a paucity of evidence to inform treatment decisions. The exclusion of pregnant women from research merely relocates the risks of research to the risks of inadequate or inappropriate treatment for pregnant women and their foetuses.

The person best situated to weigh the risks and potential benefits of research that involves the maternal-foetal dyad is the one member of that dyad who is not vulnerable: the autonomous woman, whose interests are uncontroversially affected – because her health is affected, and because the foetus can only be accessed through her body – and who is capable of making informed decisions that protect her own interests. Given relevant information, pregnant women are also capable of making decisions on behalf of their foetuses, no less so than mothers are capable of making similar decisions for their existing children.

Failure to recognise and acknowledge the interests pregnant women have in both their own health and the health of their foetuses creates a contrived maternal-foetal conflict that assumes that the foetus needs protection not only from external risks, but also from the woman who is gestating the foetus (see Ashcroft 2016). It ignores the fact that “in the majority of cases, the interests of the pregnant woman and her foetus converge rather than diverge” (ACOG 2005, 9). Disregarding the convergence of interests in the maternal-foetal dyad itself harms pregnant women, by failing to recognise that the lack of clinical research and access to relevant knowledge actually diminishes the ability of pregnant women to make choices that could be in the interests of their foetuses and future children.

9.7 From Hypothetical to Actual Vulnerability

The inclusion of pregnant women among vulnerable research populations is inappropriate. The risks that appear to justify excluding pregnant women on the basis of vulnerability are not risks to the women themselves, but rather risks to their vulnerable foetuses. Being pregnant does not create any special or unique vulnerability to exploitation or abuse in research (see Ballantyne and Rogers 2016). The research-related vulnerability of pregnant women, as acknowledged by both CIOMS and

TCPS2, is unjust exclusion from research, rather than exploitation, abuse, or overuse. If pregnant women themselves are especially vulnerable to exploitation or abuse because they cannot protect their own interests, it is only a hypothetical rather than an actual vulnerability. That hypothetical vulnerability is based on a confused hypothesis that conflates pregnant women and foetuses, and also conflates risks to pregnant women and risks to foetuses.

At the present time, many research guidelines steeply discount the risks to pregnant women resulting from their exclusion from research participation, and greatly magnify the risks to foetuses from research participation. By privileging unknown, hypothetical foetal risks, research guidelines disregard the known risks to the health and care of pregnant women that result from a lack of research and lack of adequate, empirically informed medical care (see Kukla 2016). Vulnerability-to-exclusion for both women and foetuses, results from a research oversight system that defaults to the position that hypothetical risks to a foetus are almost always excessive, and to be avoided. The hypothetical risks of research are informed by an abundance of caution and a paucity of data, while the risks of exclusion, such as inadequate or uninformed medical treatment for conditions that affect the health of pregnant women and foetuses, are quite real. Thus, hypothetical risks become actual risks, and hypothetical vulnerability becomes actual vulnerability when pregnant women as a class are excluded from clinical research that could benefit them, their foetuses, and future pregnant women by providing evidence for the safety, effectiveness, and true relative risks of medical interventions.

To restate the obvious, pregnancy does not by itself diminish the decision-making capacity of pregnant women. Nor does pregnancy by itself render pregnant women incapable of exercising their autonomy, or providing voluntary, informed consent for research participation. The exclusion of pregnant women from clinical research has widely acknowledged negative consequences for the health of women, their foetuses, and their possible children. Additionally, there is a threat to pregnant women's autonomy and agency when research guidelines treat them as vulnerable, and reduce them to mere containers for non-autonomous, vulnerable research subjects – their foetuses. Paternalistic and unwarranted restrictions on research participation constrain the ability of pregnant women to act on and protect their own interests.

Research regulations and guidelines around the world thus add insult to injury: the limitations on autonomy they imply, create, and reify have the effect of infantilising pregnant women, treating them as if they are vulnerable, as if they lack autonomy and the capacity to protect their own interests and make informed choices about their own research participation. The hypothetical "as if" then becomes actual as research guidelines effectively diminish the autonomy and agency of pregnant women, rendering them unable to protect both their own interests, and the interests of their future children.

References

ACOG (American College of Obstetricians and Gynecologists). 2005. Maternal decision making, ethics, and the law. ACOG committee opinion no. 321. *Obstetrics and Gynecology* 106(5 Pt 1): 1127–1137.

Ashcroft, R. 2016. Ethical issues in a trial of maternal gene transfer to improve foetal growth. In *Clinical research involving pregnant women*, eds. F. Baylis and A. Ballantyne, 247–263. Cham: Springer.

Ballantyne, A., and W. Rogers. 2016. Pregnancy, vulnerability, and the risk of exploitation in clinical research. In *Clinical research involving pregnant women*, eds. F. Baylis and A. Ballantyne, 139–159. Cham: Springer.

Brunner, E., D.M. Falk, M. Jones, D.K. Dey, and C.C. Shatapathy. 2013. Olanzapine in pregnancy and breastfeeding: A review of data from global safety surveillance. *BMC Pharmacology and Toxicology* 14(38). doi:10.1186/2050-6511-14-38.

Canada (Canadian Institutes of Health Research, Natural Sciences and Engineering Council of Canada, and Social Sciences and Humanities Research Council of Canada). 2014. *Tri-Council Policy Statement: Ethical conduct for research involving humans (TCPS2)*. http://www.pre.ethics.gc.ca/pdf/eng/tcps2-2014/TCPS_2_FINAL_Web.pdf. Accessed 3 Oct 2016.

CIOMS (Council for International Organizations of Medical Sciences). 2002. *International ethical guidelines for biomedical research involving human subjects*. Geneva: World Health Organization. http://www.cioms.ch/publications/layout_guide2002.pdf. Accessed 3 Oct 2016.

Denny, C., and C. Grady. 2008. Research involving women. In *The Oxford textbook of clinical research ethics*, eds. E.J. Emanuel, C. Grady, R.A. Crouch, R.L. Kie, F.G. Miller, and D. Wendler, 407–422. Oxford: Oxford University Press.

DHHS (US Department of Health and Human Services). 2009. *Code of Federal Regulations: Title 45, Part 46, Protection of Human Subjects*. http://www.hhs.gov/ohrp/regulations-and-policy/regulations/45-cfr-46/index.html. Accessed 3 Oct 2016.

DHHS (US Department of Health and Human Services, Office for Human Research Protections). 2012. *International compilation of human research standards*. http://www.hhs.gov/ohrp/international/intlcompilation/intlcompilation.html. Accessed 3 Apr 2015.

FDA (US Food and Drug Administration). 2011. *Science and research: Pregnancy registries* (last updated 22 July 2011). http://www.fda.gov/ScienceResearch/SpecialTopics/WomensHealthResearch/ucm251314.htm. Accessed 27 Feb 2015.

Healy, D., and D. Mangin. 2016. Does my bias look big in this? In *Clinical research involving pregnant women*, eds. F. Baylis and A. Ballantyne, 197–208. Cham: Springer.

India (Indian Council of Medical Research, Director-General). 2006. *Ethical guidelines for biomedical research on human participants*. New Delhi. http://icmr.nic.in/ethical_guidelines.pdf. Accessed 8 Apr 2015.

Kukla, R. 2005. Conscientious autonomy: Displacing decisions in health care. *Hastings Center Report* 35(2): 34–44.

Kukla, R. 2016. Equipoise, uncertainty, and inductive risk in research involving pregnant women. In *Clinical research involving pregnant women*, eds. F. Baylis and A. Ballantyne, 179–196. Cham: Springer.

Little, M.O. 1999. Abortion, intimacy, and the duty to gestate. *Ethical Theory and Moral Practice* 2(3): 295–312.

Lupton, M.G.F., and D.J. Williams. 2004. The ethics of research on pregnant women: Is maternal consent sufficient? *International Journal of Obstetrics and Gynaecology* 111(12): 1307–1312.

Mackenzie, C. 2010. Conceptions of autonomy and conceptions of the body in bioethics. In *Feminist bioethics: At the center, on the margins*, eds. Jackie L. Scully, Laurel E. Baldwin-Ragaven, and Petya Fitzpatrick, 71–90. Baltimore: Johns Hopkins University Press.

Mastroianni, A.C., and J.P. Kahn. 2002. Risk and responsibility: Ethics, *Grimes v Kennedy Krieger*, and public health research involving children. *American Journal of Public Health* 92(7): 1073–1076.

Miller, F.G., and A. Wertheimer. 2007. Facing up to paternalism in research ethics. *Hastings Center Report* 37(3): 24–34.

NIH-ORWH (United States. Department of Health and Human Services, Public Health Service, National Institutes of Health, Office of Research on Women's Health). 2011. *Enrolling pregnant women: Issues in clinical research*. Bethesda: National Institutes of Health. http://orwh.od.nih.gov/resources/policyreports/pdf/ORWH-EPW-Report-2010.pdf. Accessed 8 Apr 2015.

Robinson, W.M., and B.T. Unruh. 2008. The Hepatitis experiments at the Willowbrook State School. In *The Oxford textbook of clinical research ethics*, eds. E.J. Emanuel, C. Grady, R.A. Crouch, R.L. Kie, F.G. Miller, and D. Wendler, 80–85. Oxford: Oxford University Press.

South Africa (South African Medical Research Council). 2000. *Guidelines on ethics for medical research: General principles, including research on vulnerable groups, international collaboration and epidemiology*. http://www.kznhealth.gov.za/research/ethics1.pdf. Accessed 26 Mar 2015.

United States (National Commission for the Protection of Human Subjects of Biomedical and Behavioral Research). 1979. *The Belmont Report: Ethical principles and guidelines for the protection of human subjects of research*. http://www.hhs.gov/ohrp/humansubjects/guidance/belmont.html. Accessed 26 Mar 2015.

Wild, V., and N. Biller-Andorno. 2016. Pregnant women's views about participation in clinical research. In *Clinical research involving pregnant women*, eds. F. Baylis and A. Ballantyne, 119–136. Cham: Springer.

World Medical Association. 2013. *WMA Declaration of Helsinki: Ethical Principles for Medical Research Involving Human Subjects*. http://www.wma.net/en/30publications/10policies/b3/. Accessed 3 Oct 2016.

Chapter 10
Equipoise, Uncertainty, and Inductive Risk in Research Involving Pregnant Women

Rebecca Kukla

Abstract *I examine how equipoise and uncertainty ought to be managed in ethically sound and scientifically valid clinical research involving pregnant women. Drawing on recent work in philosophy of science, I argue that it is built into the internal nature of practical reason that equipoise and uncertainty are always relative to a set of values and interests. In brief, this is because a higher evidence bar will raise the risk of false negatives exactly as much as it will lower the risk of false positives, and typically both kinds of errors have costs. Furthermore, interventions during pregnancy are likely to engage sets of values and interests that are deeply held, particularly prone to intense ideological and cultural pressures, and highly variable. Pregnant women have interests and agency of their own, and are caretakers of their foetuses' well-being. As such, they have an especially important role in determining what counts as legitimate equipoise for the purposes of clinical research in which they may participate. I conclude that for both epistemological and ethical reasons, pregnant women should be given the epistemic tools to make informed, value-relative determinations of scientific uncertainty, and they should be included in the initial process of determining research questions and designing trials.*

The participation of pregnant women in clinical research remains so rare that almost all drugs approved since 1980 are of 'unknown risk' to foetuses. The US Food and Drug Administration has approved only a dozen medications for use during pregnancy, and all are medications for conditions directly related to pregnancy (Lyerly et al. 2011). For over 90 % of drugs, we simply have no data concerning their use during pregnancy, and hence little to no knowledge of the risks or the potential benefits to either pregnant women or foetuses.[1]

[1] 90 % of medications fall under what the FDA, up until June 30, 2015, classified as 'Category C.' This means that either studies in animals have revealed adverse effects on foetuses (teratogenic or embryocidal or other) and there are no controlled studies in women, or studies in women and animals are not available.

R. Kukla, PhD (✉)
Department of Philosophy and Kennedy Institute of Ethics,
Georgetown University, Washington, DC, USA
e-mail: rkukla@gmail.com

F. Baylis, A. Ballantyne (eds.), *Clinical Research Involving Pregnant Women*,
Research Ethics Forum 3, DOI 10.1007/978-3-319-26512-4_10

As a result, we have disgracefully little understanding of how to care for women safely and effectively during pregnancy, or of how to protect foetuses when the women who bear them face medical problems. But pregnant women do, of course, get sick and need medical care; indeed, two thirds of pregnant women end up taking at least four medications during pregnancy (Lyerly et al. 2008), with over 50 % taking at least one during their first trimester (Mitchell et al. 2011). When they have medical needs, their care becomes in effect an uncontrolled post-market experiment; this is so whether they receive or abstain from unapproved treatments.[2] As Lyerly, Little, and Faden powerfully put the point, "Pregnancy, it turns out, is an 'off-label' condition" (2008, 6).

Not only is such care suboptimal, but it leaves pregnant women and their doctors in an epistemological and ethical quandary (see Baylis and MacQuarrie 2016). Pregnant women are routinely advised to 'weigh the risks and potential benefits' before deciding whether to take a medication during pregnancy. But this is frustrating advice at best: given little to no information about risks and potential benefits, one cannot meaningfully weigh them. Asking this of pregnant women charges them with an unfeasible and inappropriate task; it amounts to making them responsible for gaps in knowledge that are best laid at the doorstep of our scientific policies and practices. Since women generally care deeply about protecting the health of their future children, the task is not only hopeless but also stressful, with lots of opportunity for anxiety and guilt.

In the face of such unknowns, there is a general social sentiment that it is best for women to 'err on the side of safety' and refrain from treatment during pregnancy if possible. But this is misleading, and part of a larger ideology of pregnancy within which we see the foetus as fragile and in need of being kept pure, and see the risks of intervening as more salient than the risks of not intervening (Kukla 2005; Lyerly et al. 2007, 2009). In fact, there is often *no reason whatsoever* to think that abstaining counts as the safe choice. Inhabiting a healthy body that is in a position to eat well and exercise is generally better for foetuses than inhabiting a sick body. Absent any good quality research on the effects of drugs on pregnant women and foetuses, we simply have no idea, much of the time, whether treating or abstaining from treatment counts as erring on the side of safety. Ideology aside, until we conduct research, no-one knows how to make the safer choice, no matter how diligent they may be.

Many chapters in this book develop the case for increased participation of pregnant women in research, and I hope I have already underscored the importance of this inclusion. My focus will be on the place where epistemological concerns and ethical constraints on such research intersect. In pointing out that we often simply don't know whether treating a pregnant woman with a medication will be safer or riskier than leaving her untreated, we are – in the parlance of research ethics – noting that we are in a 'state of equipoise' with respect to these two options. That is, we are genuinely uncertain which option has a better risk profile. If we are in such a state

[2] Although the so-called 'thalidomide disaster' is often cited as a reason to be wary of doing research involving pregnant women, it seems better interpreted as a lesson in the risks of this sort of post-market experimentation (see Langston 2016).

of equipoise, then, ceteris paribus, there is no known problematic added risk from running a trial comparing a treatment and a control group. Our unwillingness to run such trials draws upon an unstated pregnancy exceptionalism, in which normal epistemic and ethical standards for the acceptability of a trial are irrationally discarded.

But the concept of equipoise is a vexed one that is both epistemologically and ethically contested and complex. One of my goals is to analyse what equipoise means in the context of research involving pregnant women, and in turn what sort of equipoise might be ethically required as a condition on proceeding with such research. In doing so, I will draw on active debates in the philosophy of science concerning so-called 'inductive risk' and the role of values in scientific inference – debates that have yet to be mined for their importance in the context of clinical research ethics.

10.1 Equipoise and Uncertainty in Clinical Research

The 'principle of equipoise,' for the purposes of research ethics, states that it is an ethical condition upon the initiation and continuation of a trial that we can maintain an honest null hypothesis concerning whether the treatment arm will do better than the control arm. There has been a great deal of debate among bioethicists as to what this principle demands exactly, and whether and why it is in fact an ethical constraint on trials. The general motivating idea is that if we already know, in advance, that some participants will fare worse than others, we ought to provide them with the better option from the outset; hence genuine uncertainty about the outcome of the trial is a precondition for its ethical acceptability. Each part of this constraint is vexed, however. In particular, it is neither clear who the 'we' is that is supposed to be able to maintain an honest null hypothesis, nor what it means to be in a 'genuine' or 'honest' state of uncertainty, nor just how uncertain one has to be to count as uncertain.

The question of *whose* equipoise matters is a particularly thorny one. When Charles Fried first formulated the principle of equipoise in 1974, he required that individual physician-researchers be in a cognitive state of uncertainty. Later writers have mostly agreed that such equipoise is both too fragile, since all sorts of bits of evidence and other pressures can sway an individual, and not especially epistemologically or ethically interesting, since this kind of equipoise is indexed to a contingent psychological state and not to actual quality of evidence or to social uptake. Benjamin Freedman (1987), Eugene Passamani (1991), and others have argued instead for the importance of what is typically dubbed 'clinical equipoise,' or uncertainty within the expert community of clinicians. Passamani's version of clinical equipoise requires "a community of competent physicians who would be content to have their patients pursue any of the treatment strategies being tested in a randomized trial, since none of them has been clearly established as preferable" (1991, 1590).

Others have broadened this to include not just consensus that all the strategies are equal, but legitimate expert dissent over which is preferable.

Although something like clinical equipoise is the most popular version in the literature, some authors have widened the scope of the relevant community yet further. Robert Veatch (2007) has argued that 'participant equipoise' should be the most relevant ethical condition on a trial. That is, what we need is for the potential participant herself to be epistemically indifferent to which trial arm she ends up in. Karlawish and Lantos (1997) have pointed out that community values and cultural traditions might make one intervention preferable to another in a particular local context, regardless of expert clinical opinion. This means that we cannot infer directly from clinician equipoise to patient equipoise, nor assume that clinical opinion will translate directly into practice.[3] Ubel and Silbergleit (2011) remind us that different sorts of experts may not agree as to whether an issue is settled, and they suggest taking practitioners' and not just researchers' uncertainty into account. For reasons that will be become clear in subsequent sections, I think a wide range of stakeholders' equipoise might turn out to be relevant for any given trial. This means that a determination that there exists the right sort of uncertainty to count as equipoise will require the integration of multiple types of stakeholder perspectives. This is a complicated exercise in social epistemology and ethics.

Not only is it non-obvious whose equipoise matters, but it is also not clear what we mean by 'genuine' uncertainty or an 'honest' null hypothesis. Freedman and others have pointed out that 'theoretical equipoise,' in which the evidence favouring two options is *exactly* balanced, is exceptionally fragile and useless as any kind of practical standard. Any data point at all would destroy such equipoise, by changing the balance of evidence, however slightly. Moreover, even if we could sustain theoretical equipoise long enough to start a trial, it is implausible that anyone would fund or find value in a clinical trial if there were *no reason whatsoever* to think that one arm represents any improvement at all over the other. In order to make a convincing case for the need for a clinical trial, researchers have to give at least *some* reason for thinking that they are onto something plausibly promising, thereby destroying theoretical equipoise from the start. Thus how much uncertainty we need in order to count our null hypothesis as an honest one involves a complicated judgement. Indeed, I will argue in later sections that no universal standard for how much uncertainty is enough could be invoked, even in principle.

Finally, it is not obvious why equipoise should be an ethical constraint on trials at all. Some bioethicists (for example, Miller and Joffe 2011) have argued that the principle of equipoise ought to be rejected altogether. For instance, it is unclear why we want to prohibit a suspected differential between trial arms. If we were to design

[3] An excellent concrete example of their point is discussed by Ballantyne et al. (2016). A placebo trial on the effect of probiotics during pregnancy had trouble recruiting participants, because women who found out about the trial wanted to use the product rather than risk being assigned to the placebo arm. It appears that the cultural meanings surrounding 'natural' supplements made the intervention preferable despite a lack of evidence concerning safety or efficacy. One can assume that the reaction to a trial involving a prescription medication would have been quite different (see Ballantyne et al. 2016).

a trial in which one group received the standard of care, and another received a new intervention that we strongly suspected would be an improvement over the standard of care (but that was not yet available for general use), it does not seem like we would be doing an injustice or a problematic harm to anyone in the clinical trial. And such a trial might be helpful because, for instance, it might be unclear whether the new intervention is enough of an improvement to be worth its higher cost.

I suggest that equipoise is epistemically and ethically important because we don't want to knowingly disadvantage participants by enrolling them in a clinical trial. Furthermore, we want the trial to address a legitimately open question of some sort – otherwise we are using resources and imposing burdens on participants with no possibility of real scientific payoff, which is epistemologically and ethically unsound (see also Borgerson 2014). I have argued elsewhere (Kukla 2007) that we do not necessarily need to have equipoise about which trial arm will do better – this is a question internal to the trial design, and it fails to situate the trial in its social and practical context. Rather, we need to have equipoise with respect to whether the intervention being tested is legitimately one that should be made available, all things considered (including cost, efficacy, ease of distribution, and so forth), to the target population, given the alternatives. Thus whether a trial meets the equipoise requirement is properly not just an internal methodological question, but one that looks to implementation in the real world.[4]

It seems, on first blush, that the principle of equipoise has been irrationally ignored in the domain of research involving pregnant women. While no one has ever claimed that equipoise on its own is a sufficient ethical condition for the acceptability of a trial, it does seem that our defeasible attitude should be something like the following. If we have no good reason to think that we are raising anyone's harm-benefit ratio by enrolling them in a clinical trial, and if there is potentially welfare-enhancing scientific knowledge to be gained by conducting the trial, then the presumption is in favour of the trial being ethical and worthwhile. But in the case of pregnancy, nearly uniquely, we seem to jettison this common-sense principle.[5] We overwhelmingly opt against conducting clinical research involving pregnant women, even when the participants are at least as likely to benefit as they are to be harmed. This seems prima facie irrational.

When it comes to research involving pregnant women, determining whether we are in a state of equipoise before beginning a trial may be complicated by the fact that the impact of the trial on pregnant women and on foetuses may be quite different. If, for instance, we suspect that pregnant women, on balance, will experience slightly elevated risk from a research intervention, while foetuses will experience

[4] As I have argued (2007), this reading of the principle of equipoise requires that we also add on a separate 'Minimum Standard Requirement,' which is that nobody in a clinical trial receive care that we have good reason to think is inferior to what they would be entitled to receive outside of the trial.

[5] Historically, other groups, most notably women in general, were routinely excluded from research out of a similar protectionist, anti-interventionist impulse. However, this tendency was explicitly counteracted by changes in the National Institutes of Health (NIH) guidelines in 1992. By 2001, women's participation in research already outpaced men's (Prout and Fish 2001).

slightly elevated potential benefits, does this 'balance out' and count as a state of equipoise? Do we need to be in a state of uncertainty with respect to both parties, or do we aggregate their risks and potential benefits?

This puzzle seems acute if we cast the principle of equipoise as one specifically intended to guarantee that neither the pregnant women nor their foetuses receive suboptimal care. However, I have framed the principle of equipoise as requiring uncertainty about whether a treatment is worth making available, all things considered, rather than as a question about patient care in particular. In this context, the fact that research participation potentially affects two people seems to raise no qualitatively distinctive complications.[6]

I hope to have shown in this section both that the principle of equipoise is ethically important and epistemologically vexed. There is no agreement in the literature as to what counts as the relevant sort of uncertainty, whose uncertainty it is, or how to detect it. In the following section, I draw on a recent debate in philosophy of science to show that the problem of settling what counts as uncertainty and determining when we have it is even more complex than we have thought. In particular, I will argue that there is, even in principle, no answer to the question of when we have genuine uncertainty – or when we can maintain an 'honest null hypothesis' – that isn't indexed essentially to the values and interests of particular stakeholders.[7] Furthermore, as I will show in the section after that, in the case of research involving pregnant women, the values and interests of different stakeholders will diverge greatly, and will be shaped by all sorts of ideological and cultural forces. Thus, the application of the principle of equipoise in the case of such research will require us to think carefully, both in general and on a case-by-case basis, about whose equipoise matters, what values are shaping judgements of uncertainty, and how to integrate different perspectives.

10.2 Inductive Risk, Interests, and Uncertainty

For decades now, scholars have worried about whose equipoise should matter when it comes to the ethical acceptability of a trial. But a natural answer would seem to suggest itself, at least in theory. Whether there is enough evidence to reject a null hypothesis is ultimately an objective scientific question, and different people disagree on this only because *either* they have different information from one another, *or* not everyone is reasoning well about the evidence. So, one might think, the

---

[6] It is an incontestable fact that interventions during pregnancy can affect two beings. This is so completely independently of whether we think the foetus is a person, or more generally what its moral status is. At least potentially, the foetus will become a person eventually, who will have been affected by what was done to the body in which it gestated.

[7] Ubel and Silbergleit (2011) have defended this claim as well, although they do not draw on the philosophy of science toolbox to make this argument, nor do they consider how this applies to the case of research on pregnant women in particular.

equipoise that really matters is that of an ideal reasoner, who competently assesses all the available evidence, without any bias. Of course, in practice this would be hard to implement, since we disagree on such matters. Our information *is* incomplete and our reasoning *is* biased. But we should care about the judgements of experts more than those of laypeople, and do our best to identify and eliminate bias and to dismiss biased opinions.

In this section, I argue that this response is a non-starter, and indeed that there is, even in principle, no such thing as an objective and unbiased assessment of when a null hypothesis can be honestly maintained or rejected. I argue this with two goals in mind. First, I want to show that the epistemological problems surrounding the principle of equipoise are (even) more difficult and interesting than they have been portrayed as being, and hence that any serious research ethics will have to engage with some equally serious issues in social epistemology. Second, I want to demonstrate that when it comes to pregnancy and research involving pregnant women, the kinds of subjective factors and biases that I will be arguing are ineliminable are especially rich and problematic.

We all acknowledge that values and interests ineliminably shape science, in that they shape what we study, how we study it, and what we do with the results. But scientists and philosophers alike have long held onto an ideal of a pure, value-free core of scientific method and reasoning, in between conception and uptake, that can be truly unbiased and objective.[8] One might think it is possible, using proper scientific precepts, to objectively assess the resulting evidence and determine whether it is strong enough to call for acceptance or rejection of a hypothesis.

In 1953, however, Richard Rudner argued that hypothesis acceptance and rejection necessarily and ineliminably involves value judgements.[9] All scientific evidence is inductive, he pointed out, and hence no scientific hypothesis is ever verified with certainty. Whenever we accept a hypothesis *or* reject it *or* decide that we cannot yet accept it and are still in a state of equipoise, our decision always involves epistemic risk. In setting a bar for how much evidence we need in order to accept or reject a hypothesis – to end a state of equipoise, that is – we cannot look to any independent objective answer. Rather, as Rudner put it, "how sure we need to be before we accept a hypothesis will depend on how serious a mistake would be" (1953, 2). To use his examples:

> If the hypothesis under consideration were to the effect that a toxic ingredient of a drug was not present in lethal quantity, we would require a relatively high degree of confirmation or confidence before accepting the hypothesis – for the consequences of making a mistake here are exceedingly grave by our moral standards. On the other hand, if say, our hypothesis stated that, on the basis of a sample, a certain lot of machine-stamped belt buckles was not

[8] The idea that science is governed by a 'value-free ideal' has been standard for over a century, and has been explicitly debated and upheld since Weber's (original) publication of "'Objectivity'in social science and social policy" in 1904 (Weber 1949). For a good overview of the picture of science as having a value-free core even while its entry and exit conditions, as it were, are clearly value driven, see Kitcher (2011). See also the *Stanford Encyclopedia of Philosophy* entry on "Scientific Objectivity."

[9] A similar argument was foreshadowed in Churchman (1948).

defective, the degree of confidence we should require would be relatively not so high. (Rudner 1953, 2)

Thus, the decision whether to accept a hypothesis in light of the data depends on a necessarily value-laden judgement about how to balance the inevitable *inductive risks* – that is, the risk of a false positive (accepting the hypothesis when it is in fact false), and the risk of a false negative (rejecting the hypothesis when it is in fact true).[10] If our hypothesis is that a toxic ingredient in a drug is not present in lethal quantities, we will likely be highly intolerant of false positives; this means upping our standards for hypothesis acceptance, which means tolerating a higher risk of false negatives. If, on the other hand, our hypothesis is that a lot of belt buckles are not defective, we will likely tolerate a far higher risk of false positives, thereby lowering our risk of false negatives. No interest-independent standards can govern such inductive risk judgements.

Furthermore, Rudner's references to what 'we' would require in order to accept a hypothesis are crucially misleading. There is no unified 'we.' Different stakeholders will have different values and interests that will, quite rationally, lead them to balance inductive risks differently. Consider a clinical trial designed to show that a drug is more effective than the standard of care for some condition. The drug company executive's main stake is in getting the drug to market if possible, so she will be especially concerned to avoid false negatives. She has good reasons to reject the null hypothesis on relatively weak evidence. From a patient's point of view, however, effectiveness is the primary concern, and taking a less effective drug would be the greater loss, so the patient will be more invested in avoiding false positives, and would want a higher standard for hypothesis acceptance.

A crucial point, here, is that there is no correct standard written into the universe – any judgement involves uncertainty and any judgement involves a balancing of the two kinds of epistemic risk. So it makes no sense to say that we should just go with the 'safer' evidence bar. A higher evidence bar will raise the risk of false negatives exactly as much as it will lower the risk of false positives, and typically both kinds of errors have costs. For instance, the drug company executive stands to take a reputational hit if her company markets a drug that eventually turns out to be ineffective, while the patient misses out on a superior treatment if the drug turns out to be effective. There is no truth of the matter as to where the safest bar is located, except relative to some particular set of values, interests, and stakes.

To return to the language of equipoise: there is *no truth of the matter* as to whether a state of equipoise is appropriate, even given full access to the best available information and perfect probabilistic reasoning, *except relative to a set of interests* that guides how one balances inductive risks. Even if two people have the same information, assign the very same probabilities to a hypothesis given that information, and are equally and maximally rational, one of the two might remain in a state of equipoise while the other does not.

[10] Hempel (1965) first coined the term 'inductive risk' and framed the problem in this precise way.

Early discussions of inductive risk balancing presumed that this was a conscious step that occurred at the end of a clinical trial, once all the data was in. However, in recent years, discussions of inductive risk have become much more vigorous and interesting because of a series of arguments demonstrating that inductive risk judgements are woven throughout the entire process of scientific inquiry and are often implicit, or even built into institutional practices and policies.

In her influential article, "Inductive Risk and Values in Science," Heather Douglas (2000) argued that there are multiple points during the course of research at which we must make inductive risk judgements in the face of uncertainty. For instance, in clinical trials on the carcinogenic effects of dioxin, expert pathologists differed in their classifications of the same slides of rat liver tumours as benign or malignant. Confronted with borderline cases, interests will help determine how we classify and code what we see, even given baseline competence and honesty. Such *in medias res* inductive risk judgements are interesting because they cannot possibly be codified and made transparent as part of the protocol; expert judgement is required in moving from perception to codification.

Another example comes from Torsten Wilholt (2009). In testing the toxicity of Bisphenol A in rats, it turns out that 90 % of government-funded clinical trials report significant effects from low dose exposure, while 0 % of industry-funded trials report such effects. It also turns out that different strains of rats are differentially sensitive to oestrogen, a compound similar to Bisphenol A, and that the industry-funded trials were more likely to use the less sensitive rats. Here the relevant interested inductive risk judgement is, as it were, embodied in the choice of rat population. But there is, of course, no correct rat population with the correct level of sensitivity to oestrogen built into the structure of the universe.

For that matter, as Douglas points out, our establishment and institutionalisation of $P = .05$ as the bar for statistical significance is an institutionalised inductive risk judgement. It settles what counts as a 'significant' difference between trial arms, but by its very nature it gives no guarantee that the difference did not show up by chance. A difference with a higher P-value might still reflect a real effect, and a difference with a lower P-value might still show up by coincidence. Whether a single trial comes up with a statistically significant result is typically far from determinative when it comes to ending equipoise, but the practice of setting a bar for statistical significance amounts to institutionalising one particular type of inductive risk balancing judgement – one that may be inappropriate to the epistemic needs of some stakeholders by defining significance either too stringently or too liberally, given their values and interests.

Thus, questions about whose equipoise matters are crucial, not just because different people will often have different subjective levels of uncertainty about the same questions, but because these differences cannot be cancelled out by combining good information with objective or unbiased assessments of the evidence, even in principle. In order to decide whether the equipoise condition on running a trial has been satisfied, we need to figure out whose values should be relevant, and we need to examine and critically assess which implicit or explicit values and interests are in fact shaping our inductive judgements. For instance, as I reviewed earlier, most discussions of equipoise have presumed that some version of experts' equipoise is what we care about. But Miller and Joffe (2011) remind us that experts not only make mistakes, but are sometimes systematically and persistently wrong. As they point out, experts stood by hormone replacement therapy (HRT) for women despite what now seems to be a deeply inadequate evidence base, and we now know that HRT did no clear good and substantial harm.[11] One possible explanation of this sort of systematic error is that it results from an idiosyncratic inductive risk balancing based on a set of values and interests not shared with other stakeholders – for example, a deep professional and personal investment in a research programme or an approach to treatment.

10.3 Equipoise and Inductive Risk in Research Involving Pregnant Women

I have argued that values and interests ineliminably shape judgements of equipoise. When it comes to pregnancy, the values that have shaped our inductive risk balancing have been particularly recalcitrant, and typically tacit. Idiosyncratic ideological forces that may not stand up to scrutiny have governed our judgements about which trials are acceptable. Uncovering these forces will help clarify the epistemological and ethical situation faced by those of us who think we need more high-quality information about how to treat pregnant women and foetuses. At the same time, questions about whose equipoise matters are especially interesting and pressing in the case of this research.

The topic of equipoise has received relatively little attention in the literature on research involving pregnant women, and to the best of my knowledge that attention has been exclusively focused on experimental maternal-foetal surgery.[12] Maternal-foetal surgery is a dramatic and extremely invasive intervention that offers no prospect of medical benefit to pregnant women, and hence an analysis of its ethics will not generalise easily to other sorts of research involving pregnant women, particularly to clinical trials.

As we saw at the start, researchers have generally been exceptionally unwilling to pursue trials involving pregnant women, in the name of protecting the foetus and 'playing it safe.' But what makes *refraining* from running a trial with a prospect of direct benefit look like the safe choice? As my co-authors and I put it elsewhere, during pregnancy we have a tendency "to attend to the risks of intervening out of proportion to the clear risks, to both woman and foetus, of *failing* to intervene" (Lyerly et al. 2007, 982). We are invested in an image of the foetus as pure, and we interpret almost all interventions as morally charged attacks on this purity (Kukla

[11] This has been extensively documented and discussed. See, for instance, Randal (2002).

[12] See Lyerly and Mahowald (2001) and Rodrigues and Van Den Berg (2014). Chervenak and McCullough (2012) also talk about equipoise in the context of pregnancy research but I find their use of the term unrecognizable.

2005, 2006, 2010). In our earlier work, we interpreted this simply as a risk distortion – a bias produced by an irrational ideology of foetal purity and fragility. But I now think it is better understood as a specific kind of inductive risk balancing judgement in light of a particular cluster of cultural values around foetuses and pregnant women.

As Lyerly, Little, and Faden put it, "Cultural reasoning about risk in pregnancy, in short, tends to invoke the precautionary principle in a particularly unfettered way" (2008, 17). The precautionary principle is a kind of systematic inductive risk balancing judgement, which councils erring on the side of avoiding a risk from intervening in the face of uncertainty, thus inevitably settling for a higher risk from non-intervention. Almost any potential risk to a foetus from an intervention, including a purely theoretical one, will disturb equipoise and make a trial seem unethical. In contrast, in research and health care practices, we tolerate quite substantial evidence of risks to pregnant women and foetuses from non-intervention. Indeed, this approach to balancing inductive risks is built into our research policies and practices, which classify pregnant women as a vulnerable group for purposes of research, and interpret this as making non-intervention and non-participation in research the default 'safe' choice (see Ballantyne and Rogers 2016; Johnson 2016).

Our tendency towards a foetal protectionism that values foetal purity is culturally pervasive and to a large extent institutionalised. However, there are many other values and interests at play when it comes to assessing risks during pregnancy, and many of these will be dramatically more variable. As we saw in detail in the last section, one cannot establish whether one is in equipoise except relative to one's assessment of how bad a mistake in either direction would be. And when it comes to the kinds of risks we face in research on pregnant women, there are likely to be big differences in these judgements between different kinds of stakeholders – for instance, between drug company executives, paediatricians, obstetricians, and pregnant women themselves. Paediatricians and obstetricians, for instance, are likely to have different stakes and values when it comes to care and interventions during pregnancy. Rodrigues and Van Den Berg write, "'Mamma doctors' and 'baby doctors' do not always have the same perspectives on risk, treatment, and outcome … In general, pediatric surgeons, the ones actually performing the operations, tend to be more 'tolerant' of maternal risk and less 'tolerant' of maternal refusal of interventions that are recommended for foetal benefit" (2014, 410). We can understand this as an unsurprising bias towards their own patients in each case, as long as we remember that there is no unbiased 'objective' perspective from which to value various risks.

Meanwhile, pregnant women may vary widely in their judgements of how bad or good various outcomes would be. For instance, incontinence or weight gain might be a minor risk for many women, but a life-destroying outcome for a ballerina; future infertility might be a major tragedy for a 25-year-old woman planning a large family, but of no concern to a 45-year-old with no interest in having future children. As well, women will not equally disvalue the symptoms of untreated illnesses, such as depression or low energy. When it comes to foetal well-being, potential mothers will not be univocal in their judgements about the badness of outcomes such as deafness, mild limb deformity, and so forth. For that matter, judgements will vary widely as to the badness of miscarriage. For some pregnant women, this is an outcome so maximally tragic that almost any increase in the risk of foetal death is intolerable. For others, this is a manageable risk that can be traded off more easily for other potential benefits.[13]

My point here is not just the relatively familiar one that values influence risk assessments, but rather that no judgement of equipoise of the sort we need to make before assessing the acceptability of a trial can be established independently of such values, some of which are highly variable and others of which are culturally pervasive. When we judge that it is an open question whether a treatment path is worth pursuing, we do so relative to an evidence bar for when the question would be settled. This bar can only be set by way of an inductive risk balancing process, which compares the risk of an overly precipitous false positive with that of an overly cautious false negative; this comparison depends in turn on the specific valuation of these outcomes.

We can take away at least two important lessons so far. First, we won't be able to assess the acceptability of clinical trials absent careful thought about whose inductive risk judgements we care about, and how we can coherently integrate divergent ones. Second, since many of the values that shape epistemic judgements around pregnancy are deeply culturally fraught and shaped by ideology, we need to engage in a critical hermeneutics of the values at play, and not just leave them unquestioned.

10.4 Case Study: Dexamethasone for Foetuses with Congenital Adrenal Hyperplasia

The various epistemological and ethical issues I have discussed so far are thrown into sharp relief by a recent controversial case of research involving pregnant women. Foetuses with Congenital Adrenal Hyperplasia (CAH) are 'at risk' for being born with intersexual anatomy. Girls with CAH also seem to be less likely to end up in, or to yearn for, traditional heterosexual family roles. Researchers – most notably Maria New and her colleagues at Mount Sinai School of Medicine[14] – have been experimenting with foetal administration of dexamethasone, not for the purposes of curing or preventing CAH, but specifically for the purpose of producing more gender-conforming babies and adults or, as she puts it, more 'female-typical development.'

Feminist bioethicists have thoroughly and vociferously criticised this research. In particular, Alice Dreger and colleagues (2010) have condemned New's ethical

[13] See Lyerly et al. (2007) and (2009) for more in-depth discussions of these sorts of differences in risk assessment.

[14] See Elton (2010) and Dreger et al. (2010), (2012) for discussions of the centrality of New's team to dexamethasone treatment.

motivations, her methodology, and her recruitment practices. They document the history of shoddy science and questionable outcomes concerning dexamethasone research. For instance, in a meta-analysis, only four of over a thousand clinical trials were deemed of high enough quality for inclusion. Further, none of these four trials had a follow-up trial and there was no anonymising so the evidential quality and scope was low, at best (Dreger et al. 2010, 2012). Indeed, current knowledge of the long-term risks and potential benefits of dexamethasone for pregnant women or children is nearly non-existent. Yet New has not only defended continued research on dexamethasone administration to pregnant women whose foetuses have CAH, but she has urged that it be adopted as the standard of care (Dreger et al. 2010, 2012).

Why are New and her supporters and patients willing to take such an interventionist approach in this research, imposing unknown risks on foetuses without even a claimed prospect of medical benefit, whereas normally we err on the side of anti-interventionist protectionism when it comes to foetuses? I propose that we can answer this question effectively by critically examining the inductive risk balancing judgements at work and the values and interests that undergird those judgements. Let us assume that New and colleagues believe that their research is genuinely justifiable. Presumably, then, they think that the truth of the *hypothesis* that dexamethasone is an effective treatment for foetuses who will develop CAH is either genuinely unresolved, or ought by now to be accepted (but deserves more research so as to convince an overly conservative medical community). Normally, with the weak evidence base available, the possibility of short and long term harms to women and babies, the lack of any expected medical benefit, and our general unwillingness to conduct interventionist research involving foetuses, this would be an odd epistemic stance. But consider New's own story about the potential benefits of dexamethasone and the risks of CAH:

> Without prenatal therapy, masculinization of external genitalia in females is *potentially devastating*. It carries the risk of wrong sex assignment at birth, difficult reconstructive surgery, and subsequent long-term effects on quality of life. Gender related behaviors, namely childhood play, peer association, career and leisure time preferences in adolescence and adulthood, maternalism, aggression, and sexual orientation become masculinized in 46,XX girls and women with 21HOD deficiency. ... Genital sensitivity impairment and difficulties in sexual function in women who underwent genitoplasty early in life have likewise been reported. We anticipate that prenatal dexamethasone therapy will reduce the well-documented behavioral masculinization and difficulties related to reconstructive surgeries.[...]
>
> The challenge here is ... to see what could be done to restore this baby to the normal female appearance which would be compatible with her parents presenting her as a girl, with her eventually becoming somebody's wife, and having normal sexual development, and becoming a mother. (quoted in Dreger et al. 2012, 282, my emphasis)

New here makes clear that she takes gender-nonconforming behaviour and family life to be a devastating outcome. Becoming a wife and mother and presenting and developing in a 'normal' feminine fashion are directly associated, for her, with quality of life. This is a value judgement on her part rather than any sort of scientific fact.

Among the negative outcomes she points to in children with CAH are compromises in genital sensitivity and in sexual function. This is ironic, since these are a potential consequence of reconstructive surgery, which she also treats as an *effect* of untreated atypical development. But of course surgery, and in turn compromises of sensitivity and sexual function, are effects of untreated CAH only if we judge that atypical gender development is so terrible that surgery is a necessary solution. We can't, without unforgivable circularity, count the surgery itself and its own negative effects as one of the terrible effects of the atypical development!

My point is not that New's views on gender identity and acceptable family arrangements are hopelessly outdated and offensive, although that point is worth making. Rather, it is that New has a specific value system that treats specific outcomes as distinctively bad. If one really does see gender non conformism as an unmitigated, life-destroying disaster that will lead to risky surgery, among other harms, then one has a high stake in accepting a hypothesis about how to prevent such an outcome when the hypothesis might be true. In turn, then, one's inductive risk balancing process will favour a low evidence bar, which minimises false negatives and tolerates false positives. From this point of view, given her value system, her reaction to the evidence is actually quite reasonable. She would need exceptionally strong evidence of harms to women or children or of the ineffectiveness of the intervention before her equipoise would be rationally disturbed in a way that would preclude continuing the research.

New has substituted typical overcautiousness concerning false positives with an overzealous attempt to avoid false negatives. It is not entirely surprising that New's values concerning gender identity and norms – an area fraught with tenacious ideology and emotional complication – managed to invert our typical inductive risk balancing. In this case as in others, we need to probe and critically assess the values that drive methodological decisions about when research involving pregnant women is appropriate.

10.5 Conclusions and Steps Towards Solutions

When it comes to research involving pregnant women, whose equipoise matters? How shall we settle on an appropriate balance of inductive risks? Even if we only look to the epistemic state of stakeholders who have a reasonably complete, undistorted, comprehending grasp of the available evidence, my argument suggests that there will still be considerable disagreement as to when we have the right kind of open question on our hands – especially in a vexed and intensely value-ridden domain such as pregnancy. This disagreement cannot be resolved or understood by appeal to objective scientific principles, but only by critical engagement with values and interests.

Pregnant women are both direct recipients of any interventions that might be done in the course of research during pregnancy, and also the stewards of their foetuses, who do not have value systems of their own. For these reasons, it seems that women's own values and epistemic perspectives should have an especially large role in settling whether a trial addresses a genuinely open question of the right sort. This is already quite different from the standard form of community equipoise that looks only to physicians' opinions. Other stakeholders whose values would seem to be legitimately relevant include obstetricians and gynaecologists, paediatricians, public health professionals, and health economists. Particular projects might need input from disability rights activists, mental health professionals, oncologists, or any number of other types of stakeholders.

When we think about whose equipoise matters, we need to separate out the question of whose *values* matter from that of who is *properly informed* and able to understand the evidence. Hence including the perspectives of relevant stakeholders is not enough; we also have to *build* a group of stakeholders who, at a minimum, understand the current state of the evidence and its practical significance well enough for their assessments of it to be epistemologically relevant. Distorted views of the evidence, blinded by unrealistic hopes, poor statistical reasoning, or whatever else, do not count as epistemically legitimate contributions to scientific methodology. We should not just add women's (or anyone else's) opinions to the mix in the name of patient self-determination, without regard for how well-informed they are about the evidence or how competent they are at assessing its strength. This means that careful communication of information has to happen before a trial is fully developed and approved, and not just at the stage when participants are being recruited.

We have seen detailed epistemological reasons why the standard version of the principle of equipoise – which looks to clinical equipoise amidst the community of clinical researchers – is insufficient. Rodrigues and Van Den Berg write: "If clinical equipoise, by eliminating individual researcher's bias, aims at safeguarding the rights and well-being of prospective research subjects, it should also avoid the introduction of a 'professional bias'" (2014, 411). Different professions tend to have their own special investments and values that don't necessarily generalise. But calling this professional bias can be misleading, since as we have seen, there is no one objective, interest-free way to balance inductive risks and judge equipoise. Since there is no reason to think that clinical researchers' values and interests are the ones that ought to uniquely control the inductive risk judgements involved, uncertainty in this community is not clearly necessary or sufficient for appropriate equipoise. This is especially (but not only) so in a domain such as research on pregnant women, where values are likely to diverge in important ways, ideology is rife, and pregnant women are the distinctive ethical stewards of the half of the research participants who are foetuses.

Ubel and Silbergleit have recently proposed substituting the notion of clinical equipoise with that of 'behavioural equipoise'. They write, "According to this standard, the moral justifiability of a proposed clinical trial would be based not only on traditional standards of scientific evidence, such as the results of previous trials, but also on the beliefs and behaviors of clinicians who are likely to care for the patient population in question" (Ubel and Silbergleit 2011, 3). They are interested primarily in cases in which researchers may be convinced of the superiority of a treatment, but practitioners are not. Their point is that it may be important to run a trial designed to address behavioural resistance to implementing a treatment, even if researchers who are not the ones delivering it are already convinced. Ubel and Silbergleit are not focused on research on pregnant women, and their point is narrower than mine, but I take them to be making a kindred claim. The epistemological and ethical point of checking for equipoise, on their account and mine, is to design trials that will practically intervene on people's inductive inferences in helpful ways. Like me, they want to separate cases where people are just misinterpreting evidence from cases where genuine differences in stakes and values lead to different responses to the same evidence:

> Behavioral equipoise should not be a vehicle for performing excessively risky research when we already know the answer. That is, acknowledging behavioral equipoise is not intended to be capitulation to irrational and stubborn obstructionists somehow unwilling to be persuaded by what more enlightened experts know to be true and beneficial. It is rather a way to embrace a larger sense of uncertainty about what constitutes potential harm to patients. (Ubel and Silbergleit 2011, 6)

Ubel and Silbergleit, like me, acknowledge that subjective values will *rationally* play a role in determining whether we are willing to accept or reject a hypothesis, or stay in a state of equipoise, on the basis of the existing evidence.[15] Thus, we cannot just perform trials when a hypothesis is 'objectively' uncertain; we need to shape our trials around resolving the actual, practical uncertainty that matters.

I suggest that behavioural equipoise is a necessary condition on the acceptability of a trial, where I interpret this as equipoise among relevant stakeholders who understand the evidence, but whose balance of inductive risk does not yet allow them to resolve in favour of, or against, a treatment path. I think that Ubel and Silbergleit still restrict their scope too much by broadening from researchers to clinicians. We also need to worry about, for instance, patients who will refuse an intervention because their own equipoise has not been tipped, or policy makers who are not ready to act on a body of evidence. Behavioural equipoise is especially relevant in the case of pregnant women, who often will have internalised extreme versions of the precautionary principle when it comes to caring for their own health, and often will be loath to take even a medication deemed safe and overall beneficial in the eyes of researchers.[16]

I have defended equipoise as an important epistemological and ethical consideration when it comes to deciding whether to run (or continue) a trial. I have analysed why judgements of equipoise will vary, and when that variance is due to ineliminable differences in values and interests, rather than to different understandings of the evidence or to distorted reasoning. Because of these differences in values and interests, it will always be a substantial question whose equipoise matters, and our

[15] They add on the important point that people's prior background beliefs will properly affect how quickly they are willing to accept or reject a hypothesis, because their Bayesian priors will shape their probability updating.

[16] In certain contexts with specific ideological charges and cultural meanings, however, internalised values may push women towards, rather than away from, interventions; see the dexamethasone case discussed above, as well as the probiotics case identified in fn. 3 (see Ballantyne et al. 2016).

answer to this will need to be richer and broader than the standard roster of answers in the literature. In the case of research involving pregnant women, deeply held cultural tropes and ideologies will shape people's balancing of inductive risks. Hence we need careful critical attention to how people's judgements of equipoise or its lack are being formed. Furthermore, pregnant women will likely vary substantially in how they value different outcomes, so we need to be especially careful about generalisations and assumptions concerning what count as acceptable and unacceptable risks for research participants. Finally, because pregnant women are both the ultimate keepers of their own bodies and also the proper stewards of their foetuses, they have a special double investment in the care they receive and the interventions they undergo. For this reason, their well-informed judgements of inductive risk and equipoise (or its lack) should hold special weight. This requires communicating with them about the current state of the evidence and eliciting their values and judgements at the stage of trial design. Pregnant women (and others) need to be included in the process of developing trials that reflect appropriate equipoise, and not just given an informed choice whether to participate in trials that are already up and running.

References

Ballantyne, A., C. Barthow, and K. Wickens. 2016. Research into lifestyle changes in pregnancy. In *Clinical research involving pregnant women*, eds. F. Baylis and A. Ballantyne, 211–225. Cham: Springer.

Baylis, F., and R. MacQuarrie. 2016. Why physicians and women should want pregnant women included in clinical trials. In *Clinical research involving pregnant women*, eds. F. Baylis and A. Ballantyne, 17–31. Cham: Springer.

Borgerson, K. 2014. Redundant, secretive, and isolated: When are clinical trials scientifically valid? *Kennedy Institute of Ethics Journal* 24(4): 385–411.

Chervenak, F.A., and L.B. McCullough. 2012. Ethics of fetal surgery. In *International Academy of Perinatal Medicine annual meeting: Maternal-fetal conflict of interest*, ed. CNGOF (Collège National des Gynécologues et Obstétriciens Français), 25–36. Paris: CNGOF. http://www.cngof.asso.fr/D_TELE/2012_AIPM.pdf. Accessed 19 May 2015.

Churchman, C.W. 1948. Statistics, pragmatics, and induction. *Philosophy of Science* 15: 249–268.

Douglas, H. 2000. Inductive risk and values in science. *Philosophy of Science* 67(4): 559–579.

Dreger, A., E. Feder, and A. Tamar-Mattis. 2010. Preventing homosexuality (and uppity women) in the womb? *Hastings center report bioethics forum*, 6/29/10. http://www.thehastingscenter.org/Bioethicsforum/Post.aspx?id=4754andamp;blogid=140andterms=preventing+homosexuality+in+the+womb+and+%23filename+*.html. Accessed 26 Mar 2015.

Dreger, A., E. Feder, and A. Tamar-Mattis. 2012. Prenatal dexamethasone for congenital adrenal hyperplasia: An ethics canary in the modern medical mine. *Bioethical Inquiry* 9: 277–294.

Elton, C. 2010. A prenatal treatment raises questions of medical ethics. *Time Magazine*. June 18.

Freedman, B. 1987. Equipoise and the ethics of clinical research. *New England Journal of Medicine* 317(3): 141–145.

Hempel, C. 1965. Science and human values. In *Aspects of scientific explanation and other essays in the philosophy of science*, 81–96. New York: The Free Press.

Johnson, L.S.M. 2016. When hypothetical vulnerability becomes actual: Research participation and the autonomy of pregnant women. In *Clinical research involving pregnant women*, eds. F. Baylis and A. Ballantyne, 161–178. Cham: Springer.

Karlawish, J.H., and J. Lantos. 1997. Community equipoise and the architecture of clinical research. *Cambridge Quarterly of Healthcare Ethics* 6: 385–396.

Kitcher, P. 2011. *Science in a democratic society*. Amherst: Prometheus Books.

Kukla, R. 2005. *Mass hysteria: Medicine, culture and mothers' bodies*. Lanham: Rowman and Littlefield.

Kukla, R. 2006. Pregnant bodies as public spaces. In *Motherhood and space: Configurations of the maternal through politics, home, and the body*, eds. C. Wiedmer and S. Hardy, 283–305. New York: Palgrave MacMillan.

Kukla, R. 2007. Resituating the principle of equipoise: Justice and access to care in non-ideal conditions. *Kennedy Institute of Ethics Journal* 17(3): 171–202.

Kukla, R. 2010. The ethics and cultural politics of reproductive health warnings: A case study of proposition 65. *Health, Risk and Society* 12(4): 323–334.

Langston, L. 2016. Better safe than sorry: Risk, stigma, and research during pregnancy. In *Clinical research involving pregnant women*, eds. F. Baylis and A. Ballantyne, 33–50. Cham: Springer.

Lyerly, A.D., and M.B. Mahowald. 2001. Maternal-fetal surgery: The fallacy of abstraction and the problem of equipoise. *Health Care Analysis* 9: 151–165.

Lyerly, A.D., L.M. Mitchell, E.M. Armstrong, L.H. Harris, R. Kukla, M. Kuppermann, and M. Little. 2007. Risk, values, and decision making surrounding pregnancy. *Obstetrics and Gynecology* 109(4): 979–984.

Lyerly, A.D., M. Little, and R. Faden. 2008. The second wave: Toward responsible inclusion of pregnant women in research. *International Journal for Feminist Approaches to Bioethics* 1(2): 5–22.

Lyerly, A.D., L. Mitchell, E.M. Armstrong, L. Harris, R. Kukla, M. Kuppermann, and M. Little. 2009. Risk and the pregnant body. *Hastings Center Report* 39(6): 34–42.

Lyerly, A.D., M. Little, and R. Faden. 2011. Reframing the framework: Toward fair inclusion of pregnant women as participants in research. *The American Journal of Bioethics* 11(5): 50–52.

Miller, F.G., and S. Joffe. 2011. Equipoise and the dilemma of randomized clinical trials. *New England Journal of Medicine* 364(5): 476–480.

Mitchell, A.A., S.M. Gilboa, M.M. Werler, K.E. Kelley, C. Louik, and S. Hernández Díaz. 2011. National birth defects prevention study. Medication use during pregnancy, with particular focus on prescription drugs: 1976–2008. *American Journal of Obstetrics and Gynecology* 205(1): 51.e1–51.e8.

Passamani, B.E. 1991. Clinical trials: Are they ethical? *New England Journal of Medicine* 324(22): 1590–1591.

Prout, M., and S. Fish. 2001. Participation of women in clinical trials for drug therapies: A context for the current controversy. *Medscape General Medicine* 3: 4. http://www.medscape.com/viewarticle/408956. Accessed 8 Apr 2016.

Randal, J. 2002. The end of an era: Study reveals harms of hormone replacement therapy. *Journal of the National Cancer Institute* 19(15): 1116–1118.

Rodrigues, H.C.M.L., and P.P. Van Den Berg. 2014. Randomized clinical trials of maternal-fetal surgery: A challenge to clinical equipoise. *Bioethics* 28(8): 405–413.

Rudner, R. 1953. The scientist qua scientist makes value judgments. *Philosophy of Science* 20(1): 1–6.

Ubel, P., and R. Silbergleit. 2011. Behavioral equipoise: A way to resolve ethical stalemates in clinical research. *The American Journal of Bioethics* 11(2): 1–8.

Veatch, R. 2007. The irrelevance of equipoise. *Journal of Medicine and Philosophy* 32(2): 167–183.

Weber, M. 1949. *The methodology of the social sciences*. Trans. and eds. E.A. Shils, and H.V. Finch. Glencoe: Free Press.

Wilholt, T. 2009. Bias and values in scientific research. *Studies in History and Philosophy of Science* 40: 92–101.

Chapter 11
Does My Bias Look Big in This?

David Healy and Derelie Mangin

Abstract *Randomised controlled trials (RCTs) are thought to be the gold standard in evidence. This review of their origins and adoption, highlights commonly ignored shortcomings with RCTs. If RCTs are used indiscriminately, their adverse effects may outweigh their benefits. This chapter focuses on antidepressants and how RCTs give the wrong message about safety, efficacy, and effectiveness. The arguments hold true in principle for all treatments, including all treatments for pregnant women. The received wisdom since thalidomide, that we should rarely if ever use drugs in pregnancy, increasingly is being eroded by arguments in support of the use of RCTs. In the case of antidepressants, this has made them among the most commonly prescribed drugs in pregnancy.*

There is a presumption that objectivity comes from the procedures of an RCT. We argue that objectivity comes from collective scrutiny of publicly available data and, in the case of pregnancy, this mandates the creation of pregnancy registries to generate sound evidence on the basis of which to make treatment decisions for pregnant women and women of child-bearing years.

There is a general belief that (RCTs) provide the best possible evidence regarding treatments and that RCTs are the only way to avoid biased judgements on the safety, efficacy and effectiveness of treatments. Allied to a position that pregnant women deserve access to the best quality evidence, this belief mandates an increased use of RCTs in pregnancy and in women of child-bearing years.

In contrast, articles on the value of observational studies invariably include disclaimers as to validity of such studies. The explicit message is that with a great deal of care, it might be possible to reduce the amount of bias to which such studies inevitably will be subject, but observational studies will never approach the quality of disinterested evidence that stems from RCTs.

D. Healy, MD (✉)
North Wales Department of Psychological Medicine, Cardiff University, Cardiff, Wales, UK
e-mail: david.healy54@googlemail.com

D. Mangin, MBChB, DPH, FRNZCGP
Department of Family Medicine, McMaster University, Hamilton, ON, Canada

University of Otago, Christchurch, New Zealand

F. Baylis, A. Ballantyne (eds.), *Clinical Research Involving Pregnant Women*,
Research Ethics Forum 3, DOI 10.1007/978-3-319-26512-4_11

We argue that RCTs are mechanical exercises, that in general they ***do not*** provide good quality evidence, that they are commonly ineradicably confounded, and that an unthinking application of RCTs in pregnancy would be a mistake. If women who are (or who might become) pregnant want the best quality evidence, researchers need to carefully think about the design of appropriate studies. Moreover, if researchers want to reduce bias, the focus should be on collective scrutiny of publicly available data, and a key source of such data will be pregnancy registries. In our view, objectivity comes from collective scrutiny and not mechanical exercises.

This chapter focusses on antidepressants and how RCTs give the wrong message on safety, efficacy, and effectiveness, but the arguments hold true in principle for all treatments.

11.1 The Origin of Randomised Trials

Ronald Fisher created the modern RCT in the 1920s, when investigating the effect of fertilisers. Many factors can confound fertiliser studies such as differences in soil drainage, exposure to wind or sunlight, and a myriad of soil elements. These known factors can be controlled for, but Fisher's insight was to control for unknown confounding factors by randomising fertilisers under study to alternate soil patches. Fisher tied significance testing to randomisation. If experimenters got the same result every time they repeated the intervention, they had designed a good experiment. There was a *Quod Erat Demonstrandum* quality to this strategy – shave a bit off one side of a coin and you can expect heads to come up nineteen times out of twenty. Randomisation in this sense is about leaving nothing to chance and it allows experimenters to show that they know what they are doing. This insight on what Fisher meant by an RCT has slipped out of view (Fisher 1935; Savage 1976; Marks 1997).

Randomisation was first used in medicine in Bradford Hill's trial of streptomycin for tuberculosis (Medical Research Council 1948). Earlier, non-randomised trials had established all that is known about streptomycin for tuberculosis – that streptomycin works in the short term, that the germ becomes resistant over time, and that treatment comes with significant risks such as ototoxicity (Toth 1998). As no new knowledge was gained from the RCT, Hill's streptomycin trial put randomisation rather than streptomycin on the map. If 'efficacy' means that trials accomplish something, while 'effectiveness' means that they work for the intended purpose, then Hill's trial demonstrated the efficacy of trials, but did not establish their effectiveness.

Early Doubts About RCTs

The primacy of RCTs today as a method of evaluation stems not from greater rational or logical coherence, but from events centring on women and pregnancy – the thalidomide tragedy. The birth defects linked to the use of a sleeping pill, thalidomide, created a political imperative to be seen to be doing something to make patients safer (see Langston 2016). As a result, in 1962 a change was made to the provisions of the US Food and Drugs Act requiring companies to demonstrate the effectiveness of new compounds, with an understanding that this would be done through placebo-controlled RCTs.

While RCTs initially appeared to be a way to contain pharmaceutical company claims, by the mid-1960s Hill (1966) noted that company salesmen were deploying RCT evidence to encourage doctors to use their company's products. At that time, Hill suggested that if RCTs ever became the only way to evaluate drugs that "the pendulum would not just have swung too far [away from physician judgement] it would have come off its hook" (Hill 1966, 113).

In 1962, at the time the US Food and Drug Act was amended, RCTs were a novel evaluation method whose suitability for the task at hand (i.e., improving patient safety) was uncertain (Healy 2012). Indeed, as early as the 1950s, some recognised that the philosophical basis of RCTs was uncertain; there was no agreement on the meaning of statistical significance, and no logical basis for randomisation had been elaborated (shortcomings that remain the case today) (Gigerenzer et al. 1990; Toth 1998). A powerful symbol of this uncertainty is the fact that in 1962, only one drug had demonstrated safety and effectiveness through a placebo-controlled RCT prior to marketing – thalidomide (Lasagna 1960).

Finally, there is a crisis today within drug development that sits poorly with claims that RCTs are an effective evaluation method. As Table 11.1 shows, most major drug groups were introduced in the 1950s without the benefit of RCTs, and the drugs that were introduced during this time remain more effective than drugs that have since come to market using RCTs. Empirically, therefore, it appears that RCTs are not necessary in developing an effective drug arsenal.

11.2 The Placebo Effect

RCTs of fertilisers are not controlled with placebos. The first RCTs in medicine were not placebo-controlled. The first placebo-controlled trials in medicine were not RCTs.

The marriage of RCTs and placebos gives the impression that a further set of confounding factors (biases) is being controlled for. But placebos introduce a systematic bias. An active drug simply needs to beat a placebo on some dimension for others to claim that it is effective. This can be achieved for ever weaker drugs by powering trials accordingly. Manipulations of this sort mean that recent antihypertensives, hypoglycemics and antidepressants are often less effective than drugs introduced without RCTs before 1962 (Table 11.1). Previously, the effects of a drug treatment on a patient had to be visible; today the effects can be invisible at the individual patient level, underpinning distinctions between clinical and statistical significance.

Table 11.1 Drug effectiveness with and without RCTs

Drugs introduced pre 1962	Exemplars of pre-1962 medicines	Post-1962 medicines: more effective or not
Analgesics	Morphine, Paracetamol	No
Antibiotics	Penicillins, Tetracyclines	No
Anticonvulsants	Barbiturates, Valproate	Possible
	Phenytoin	
Antidepressants	Tricyclics, MAOIs	No
Antihistamines	Chlorphenamine, Diphenhydramine	No
Antihypertensives	Thiazides	No
Antipsychotics	Clozapine, Haloperidol	No
Chemotherapies	Nitrogen mustards	Yes
	Cisplatin	
Contraceptives	Second generation COC	No
Diuretics	Furosemide	No
Hypoglycaemics	Metformin	No
Steroids	Prednisone	No
Stimulants	Dexamphetamine	No
	Methylphenidate	
Tranquilisers	Diazepam	No
Vaccines	Polio, Smallpox	No

RCTs & Efficacy

Currently, the design of RCTs to establish efficacy and effectiveness for an antidepressant, involves testing an active drug against a placebo in trials lasting six weeks, using rating scales as outcome measures. There is academic debate about whether a statistically significant finding of doubtful clinical significance constitutes efficacy. But in practice, regulators like the US Food and Drug Administration (FDA) license drugs for use on the basis of such data, and so for almost everyone debates about either efficacy or effectiveness are academic. The drugs are assumed to be effective and are put into ever increasing use as few doctors and no politicians want to deny patients, especially pregnant women, effective treatments.

Consider the following thought experiment. A company does a series of RCTs on alcohol as an antidepressant. It uses its current abilities to hide data from RCTs, to only publish data from selected RCTs, and to ghost-write all publications. Using these strategies, the company could achieve an identical outcome as is achieved with RCTs of SSRIs (selective serotonin reuptake inhibitors) as antidepressants. It is clear, however, that designating alcohol as an antidepressant would not be a good outcome for pregnant women (or any other patient group). With this thought experiment, we are able to pit other knowledge about the use of alcohol during pregnancy against knowledge that might stem from RCTs of alcohol as a possible effective antidepressant. That is, we can put the RCT knowledge in perspective. But in most other cases we can't do this, and we assume that because the claims have come through an RCT, as in the case of comparable claims about SSRIs, that this is good quality data. In real life, for instance, the example of burgeoning stimulant use driven by RCTs suggests that we are not able to deploy wisdom or even common sense against the treatment imperative that flows from RCTs.

The received wisdom since thalidomide, that we should rarely if ever use drugs in pregnancy, is being eroded by RCTs that have made drugs such as antidepressants among the most commonly prescribed drugs in pregnancy. If safety in pregnancy improved in the 1960s, following the thalidomide tragedy, this had nothing to do with the introduction of RCTs, and everything to do with the fact that pregnant women and their doctors became reluctant to use medications during pregnancy.

RCT Crisis

It is important to distinguish the failings of RCTs that we have linked to a crisis in drug development, from a much more commonly referred-to crisis concerning the conduct of RCTs. This latter crisis is linked to the use of surrogate outcomes, in trials of inadequate duration, against a regulatory background that will license products on the basis of biased trial data.[1] Our goal in this chapter is not to offer another list of the many failings of the conduct of RCTs. We recognise that RCTs have an important place in therapeutics but, we maintain that even if they are carried out impeccably, their adverse effects may outweigh their benefits. Adapting Muir Gray's dictum that all screening is harmful, we might say that all RCTs are harmful but, in some instances, there are also benefits that warrant taking the unavoidable risks involved (Raffle and Gray 2007).

Mediculture or Medicine?

Our argument in brief is: People and their diseases, and the treatment of those diseases, are not uni-dimensional in the same way as Fisher's soil patches and growing crops. As a result, transforming a chemical into a medicine is a different matter to demonstrating a chemical is an effective fertiliser.

For example, a fertiliser has only one action we need pay heed to, but a medicine may have a hundred effects all of which need attention. It is not problematic to designate a primary effect in an RCT of fertilisers and ignore other effects. The fact that a small proportion of ears of corn might die prematurely because of the fertiliser is of no consequence. But medicine is critically concerned with potential benefit to an

[1] For a discussion of licensing in the face of inadequate, contested data see Healy's discussion of a decision to license Zoloft on the basis of ghost-written publications stemming from these two positive RCTs, when there were ten or more negative RCTs (Healy 2012).

individual patient, and average effects are only useful insofar as they might be of help to an individual patient. Average effects that obscure potential harm to an individual patient entail risks that may not be worth taking. Clinical practice wants to manage heterogeneity, not act as though it doesn't exist. This is especially true in pregnancy (see Baylis and MacQuarrie 2016).

Randomisation aims at eliminating sources of objective bias. But in the case of SSRI trials, for instance, the possible effects of these drugs on mood are designated as primary effects, when such effects are less likely than effects on sex and bowel function. The trial process then means that these more common effects are ignored. Is bias being eliminated here or systematised?

11.3 Antidepressants and Suicide

A Thought Experiment

The confounders that randomisation can introduce in testing a medication (that are not confounders in testing a fertiliser), can be drawn out through two examples involving antidepressants and suicide in depression. The lessons about confounders, however, apply to all drug groups and all drug effects.

Imipramine, the first antidepressant, was launched in 1958 without RCTs. In 1959, at a meeting convened to discuss its effects, several clinicians reported having witnessed patient agitation after exposure to Imipramine, how the agitation cleared after stopping the Imipramine, and then reappeared on re-exposure – a medical testing protocol known as challenge-dechallenge-rechallenge. These clinicians decided that, wonderful though Imipramine was for many patients, it could trigger suicidal and homicidal ideation in some people (Davies 1964). The challenge-dechallenge-rechallenge protocol offers as convincing a demonstration of cause and effect, as the statistical significance testing of the type advocated by Fisher. Both tests show greater replicability than is found with the statistically significant findings reported from most RCTs today.

Imipramine and related tricyclic antidepressants are serotonin reuptake inhibitors. They are more clinically potent than most SSRIs, 'beating' SSRIs in patients with melancholia (Healy 1998). Melancholic patients are 80 times more likely to commit suicide than mildly depressed patients (Hagnell et al. 1981). Accordingly, comparing Imipramine and placebo in an RCT of melancholic patients would likely show less suicides and suicidal acts on Imipramine than on placebo. The relative risk might be as low as 0.5. Thus, a drug that causes suicide will also appear to protect against suicide, in some clinical trials.

In contrast, various meta-analyses of suicides and suicidal acts in SSRI and post-SSRI RCTs indicate a relative risk that SSRIs will cause suicide and suicidal acts of roughly 2.0 (Fergusson et al. 2005). This different outcome results, in part, from the fact that SSRIs weaker than Imipramine were tested in people who were at less risk

Table 11.2 Suicidal acts in Major Depressive Disorder trials (MDD) trials

Major Depressive Disorder trials (MDD)	Paroxetine	Placebo	Relative risk
Suicidal acts/Patients	11/2943	0/1671	Inf (1.3, inf)

of suicide. As a result, the rate of suicidal acts on placebo is reduced – making the risk from SSRIs more noticeable. When a drug such as Imipramine is put into this assay system, it would show the same excess of suicidal acts.

It is common to hear claims that RCTs demonstrate cause and effect. But this thought experiment shows that if a trial is not designed to look at an issue, it cannot show cause and effect with respect to that issue. These RCTs of SSRIs say nothing about causality, except insofar as there could not be an excess of suicides and suicidal acts if the SSRIs don't cause suicide. Better evidence that SSRIs can cause suicide in some people comes from Teicher's 1990 paper on Prozac and suicide that demonstrated the challenge-dechallenge-rechallenge relationships (Teicher et al. 1990). Another implication of the thought experiment is that RCTs do not generate reliable data on frequency. Even in studies designed to look at antidepressants and suicide, we cannot in fact have any idea from RCTs how often antidepressants might trigger suicidality in clinical practice. The next example will buttress this point.

Actual Experiments

In the early 1990s, SmithKline undertook a study of paroxetine (Study 106) in patients with Intermittent Brief Depressive Disorders (IBDD). The study terminated early, and was never published. The rate of suicidal acts on paroxetine was threefold higher than on placebo.[2] SmithKline then undertook study 057 in a similar group of patients (Verkes et al. 1998). Neither trial supported using paroxetine for IBDD.

In April 2006, GlaxoSmithKline issued a press release that presented the following data for patients in paroxetine Major Depressive Disorder (MDD) trials (Table 11.2).

The MDD patients on paroxetine showed a significant increase in the risk of suicidal acts as compared with placebo (GlaxoSmithKline 2006). The press release also contained data on suicidal acts from the previous IBDD trials (Studies 106 and 057), despite the fact that these studies did not support using paroxetine for IBDD. When the data from all three studies were aggregated, surprisingly the risk of suicidal acts on paroxetine in depression trials vanished (see Table 11.3).

It is possible to add 16 more suicidal acts to the paroxetine IBBD column in Table 11.3 (viz., 48/147), increasing the relative risk of an adverse event on paroxetine to 1.4 (viz., the combined paroxetine suicidal act number increases to 59 (viz.,

[2] Data available upon request from David Healy.

Table 11.3 Suicidal acts in Major Depressive Disorder trials (MDD) and Intermittent Brief Depressive Disorders (IBDD) trials

	Paroxetine	Placebo	Relative risk
MDD Trials Suicidal Acts/Patients	11/2943	0/1671	Inf (1.3, inf)
IBDD Trials Suicidal Acts/Patients (Studies 106 and 057)	32/147	35/151	0.9
MDD and IBDD Trials Combined Suicidal Acts/Patients	43/3090	35/1822	0.7

43 + 16)), and still get the same apparently protective outcome overall with paroxetine (paroxetine 59/3090; placebo 35/1822). This paradoxical outcome is predictable. Knowing what a drug can do makes it possible to design placebo-controlled RCTs that use a problem the drug actually causes, to hide that same problem. In this respect, a medicine is unlike a fertiliser.

It is clearly bad meta-analytic technique to lump the datasets for the IBDD and the MDD trials, but the example points to a deeper problem for RCTs undertaken in heterogeneous clinical populations – from back pain to Parkinson's disease. Just as IBDD patients can meet the criteria for MDD, so too diverse patients with back pain or Parkinson's disease or Type II diabetes can meet criteria for different illnesses. Provided there is more than one IBDD patient entered into MDD trials, randomisation will ensure these patients will hide the effect of an SSRI on suicidal acts. Similarly, drugs for back pains, parkinsonian syndromes, or diabetic states of one type may mask what may be beneficial treatment effects on other types of back pain, Parkinson's disease, or diabetes.

The only way to overcome this bias and get a result that would have Fisher agreeing that we know what we are doing is, in fact, to understand the pathophysiology of the clinical condition we are treating and the pharmacogenetics of the drug we are using. It is only then that RCTs would take on the quality of a demonstration that was Fisher's original intention.

11.4 Pharmagnosia

Unlike fertilisers, medicines have a hundred or more effects. When we design an experiment employing randomisation to manage the unknown unknowns for one of these effects, we risk generating ignorance about ignorance regarding most of what the drug does. The process is akin to hypnosis, where holding a subject's attention to one focus can lead him/her to miss more important material out of focus, especially when for the sake of 'objectivity' patient reports are essentially ignored.

In the case of the SSRIs, the choice of endpoint was dictated by business considerations. This meant powering studies to produce a statistically significant outcome on rating scales that measure clinical changes in a very rough fashion. But because the Hamilton Ratings Scale for Depression (HAMD) was the primary endpoint there was a focus on scores from this, data on sexual functioning and the other 99 effects these drugs have, were either not collected or poorly collected, letting companies claim afterwards that less than 5 % of those taking SSRIs had a disturbance of sexual functioning, when the true figure is closer to 100 %.

Thus, trial design has generated an agnosia for most of the effects of these drugs. This agnosia has been compounded by a rhetoric that gives the impression that since these drugs have been through RCTs, most of what needs to be known about them is known. We become ignorant of effects that are termed adverse solely because they are not the primary effects being looked at and, in so doing, we compromise safety and reduce the rate of discovery of new drugs.

Atom-Agnosia

The clinical encounter is a relationship, and good care involves close attention to the individual (the 'atom'). RCTs have affected this relationship by treating individual variation as inconsequential. As a result, clinical encounters have become an industrial process, like agriculture, that aims at implementing impersonal algorithms and guidelines, leaving clinicians practising mediculture rather than medicine.

In agriculture, RCTs work and (perhaps until the advent of genetically modified crops), there was little attempt to hide the data. In contrast, in mediculture, RCTs don't work, and treatment guidelines are based on data that are often miscoded and always inaccessible, in ghost-written publications from trials that are not designed to detect many of the significant effects of treatment. The key point is that the individual has vanished, and it is becoming progressively more difficult for any of us to form a genuine relationship with a doctor.

Efficacy and Effectiveness of RCTs

The problems from pharmacognosia to randomisation-induced confounders are not an inconvenience that stem from some oddity to do with antidepressants or suicide. They are intrinsic to RCTs within medicine and can be expected every time a treatment and an illness produce similar or superficially similar outcomes – whether a benefit or a harm.

Similar scenarios unfold with cardiac rhythm problems in trials of anti-arrhythmics, with breathing difficulties in trials of anti-asthmatics, and with vaccines and viral infections that cause brain or other damage. Controlled trials show that ACE-inhibitors improve renal function in patients on diabetes but, in a proportion of cases, they can make renal function worse by aggravating renal artery stenosis. The interaction between the heart attack producing effects of both diabetes and rosiglitazone obscure the adverse effects of rosiglitazone (Cohen 2010). Exenatide and sitagliptin produce pancreatitis, but diabetes can too (Cohen 2013). These problems happen as much with the effects of treatments termed benefits as with those termed harms.

RCTs were introduced to the regulatory apparatus in an attempt to enhance safety by demonstrating effectiveness. The only time RCTs are unambiguously effective at enhancing safety is when they demonstrate that a drug is inefficacious or ineffective. An example of this is the Women's Health Initiative study of hormone replacement therapy, where a proposed reduction in cardiovascular risk specifically, and mortality in general, was not found and indeed the opposite was found (Women's Health Initiative 2002). With this kind of outcome, the risks inherent in RCTs of a medicine are warranted because the demonstrated effects led to a reduced rather than an increased exposure to unknown effects (see Kukla 2016). It is worth noting that the hormone replacement therapy studies were helpful, but not because they generated the right answer. The negative answer they generated put the onus to back up claims being made where it belonged – namely on those who might make money from vulnerable people.

11.5 The Best Evidence

Evidence-based medicine has become synonymous with RCTs even though such trials fail to tell the physician what she wants to know which is which drug is best for Mr Jones or Ms Smith – not what happens to a non-existent average person (Lasagna 1998).

The verdict of RCTs is often pitted against clinical judgement, despite the fact that an RCT may not be able to show an antidepressant causes suicidality, whereas the exercise of clinical judgement within an RCT can. Clinicians and patients often can distinguish between depression-induced and drug-induced suicidality.

Patients can distinguish between the beneficial effect of a drug and the effect of that benefit on outcomes as when, for instance, patients make it clear that an SSRI is producing a useful emotional numbing but this is not leading to recovery. This information is important, if doctors want to introduce another drug with a different mode of action into the mix, or want to stop the original drug and start another.

As things stand, because RCTs de facto discourage the engagement of doctors with patients, they obscure any specific effects that quite different therapeutic principles might have. In the case of antidepressants or hypoglycemics, RCTs make diverse drugs acting selectively on different systems look exactly the same. This leads to patients being put on drug cocktails because all have been shown to 'work' without any effort to match a therapeutic principle to a patient's needs. This clinical approach prevents doctors finding the right drug for the patient in front of them, and blocks the possibility of insights into the nature of the syndromes they are treating.

If we are to reverse this, there needs to be a focus on safety rather than efficacy. A focus on safety will increase the chance that a pregnant woman will end up on a drug that in fact works for her and, of even greater importance, she will only end up on a treatment she values.

Registries and Objectivity

Given that the challenge-dechallenge-rechallenge protocol is not a reasonable option for pregnant women, there is an urgent need for comprehensive pregnancy registries. The only explanation for the lack of such registries would seem to be the fact that until recently, drug use in pregnancy was minimal. As this changes, women (including pregnant women), and administrators, midwives, doctors and nurses working in antenatal and postnatal services need to work together to put such registries in place (see Ballantyne and Rogers 2016).

Part of the appeal behind RCTs is that when contrasted to individual judgements by patients or doctors, they appear to offer objectivity. But as this exposition of some of the limitations of RCTs reveals, RCTs are largely a mechanical process. This means objectivity must come from elsewhere. In our opinion, objectivity generally results from the ability to bring many points of view to bear on an issue – it results from a collective exercise.

The challenge of objectively establishing whether a treatment causes birth defects is one of the greatest challenges in science, given that the challenge-dechallenge-rechallenge strategy is unavailable and RCTs are not likely to be helpful. When it comes to minimising any bias in registry data, objectivity is most likely to be achieved if we have the greatest possible input from the widest range of sources – including from those whose bias might be thought to be biggest because of the outfits they are currently in (institutions they work for).

References

Ballantyne, A., and W. Rogers. 2016. Pregnancy, vulnerability, and the risk of exploitation in clinical research. In *Clinical research involving pregnant women*, eds. F. Baylis and A. Ballantyne, 139–159. Cham: Springer.

Baylis, F., and R. MacQuarrie. 2016. Why physicians and women should want pregnant women included in clinical trials. In *Clinical research involving pregnant women*, eds. F. Baylis and A. Ballantyne, 17–31. Cham: Springer.

Cohen, D. 2010. Rosiglitazone. What went wrong? *BMJ* 341: c4848.

Cohen, D. 2013. European drugs agency clashes with scientists over safety of GLP-1 drugs. *BMJ* 347: f4838.

Davies, E.B. (ed.). 1964. *Depression: Proceedings of the symposium held at Cambridge, 22–26 September 1959*. Cambridge: Cambridge University Press.

Fergusson, D., S. Doucette, K. Cranley-Glass, S. Shapiro, D. Healy, P. Hebert, and B. Hutton. 2005. The association between suicide attempts and selective serotonin reuptake inhibitors: Systematic review of randomized controlled trials. *BMJ* 330(7488): 396–399.

Fisher, R.A. 1935. *The design of experiments*. London: Macmillan.

Gigerenzer, G., Z. Swijtink, T. Porter, L. Daston, J. Beatty, and L. Kruger. 1990. *The empire of chance: How probability changed science and everyday life*. Cambridge: Cambridge University Press.

GlaxoSmithKline, 2006. Paroxetine adult suicidality analysis. GlaxoSmithKline. Updated 5 April 2006; originally cited 8 August 2007. On file with and available from the authors.

Hagnell, O., J. Lanke, and B. Rorsman. 1981. Suicide rates in the Lundby study: Mental illness as a risk factor for suicide. *Neuropsychobiology* 7(5): 248–253.

Healy, D. 1998. *The antidepressant era*. Cambridge, MA: Harvard University Press.

Healy, D. 2012. *Pharmageddon*. Berkeley: University of California Press.

Hill, A.B. 1966. Reflections on controlled trial. *Annals of Rheumatic Diseases* 25(2): 107–133.

Kukla, R. 2016. Equipoise, uncertainty, and inductive risk in research involving pregnant women. In *Clinical research involving pregnant women*, eds. F. Baylis and A. Ballantyne, 179–196. Cham: Springer.

Langston, L. 2016. Better safe than sorry: Risk, stigma, and research during pregnancy. In *Clinical research involving pregnant women*, eds. F. Baylis and A. Ballantyne, 33–50. Cham: Springer.

Lasagna, L. 1960. Thalidomide – A new non-barbiturate sleep-inducing drug. *Journal of Chronic Diseases* 11(6): 627–631.

Lasagna, L. 1998. Back to the future. Evaluation and drug development. In *The psychopharmacologists*, vol. 2, ed. D. Healy, 135–166. London: Arnold.

Marks, H.M. 1997. *The progress of experiment: Science and therapeutic reform in the United States, 1900–1990*. Cambridge: Cambridge University Press.

Medical Research Council. 1948. Streptomycin treatment of pulmonary tuberculosis. *BMJ* 2(4582): 769–782.

Raffle, A., and M. Gray. 2007. *Screening: Evidence and practice*. Oxford: Oxford University Press.

Savage, L.J. 1976. On rereading R.A. Fisher. *The Annals of Statistics* 4(3): 441–500.

Teicher, M.H., C. Glod, and J.O. Cole. 1990. Emergence of intense suicidal preoccupation during fluoxetine treatment. *The American Journal of Psychiatry* 147(2): 107–210.

Toth, B. 1998. Clinical trials in British medicine 1858–1948, with special reference to the development of the randomised controlled trial. Bristol University PhD dissertation.

Verkes, R.J., R.C. Van der Mast, M.W. Hengeveld, J.P. Tuyl, A.H. Zwindermann, and G.M. Van Kempen. 1998. Reduction by paroxetine of suicidal behavior in patients with repeated suicide attempts but not major depression. *The American Journal of Psychiatry* 155(4): 543–547.

Women's Health Initiative (Writing group for the Womens' Health Initiative Investigators). 2002. Risks and benefits of estrogen plus progestin in healthy postmenopausal women: Principal results from the Women's Health Initiative randomized controlled trial. *JAMA* 288(3): 321–333.

Part IV

Chapter 12
Research into Lifestyle Changes in Pregnancy

Angela Ballantyne, Christine Barthow, and Kristin Wickens

Abstract *Clinical research undertaken in pregnant women is limited due to the belief that pregnant women and their foetuses are a vulnerable group. As a result, much of the research that does occur in this population focuses primarily on aspects of obstetric practice, preterm labour and foetal safety. There are increasing calls to broaden the research agenda to include a much wider range of conditions and study both short and long-term maternal and foetal outcomes. For example, this includes research topics such as the impact of maternal mental health, asthma, oral health, and hypertension during pregnancy on both maternal and infant outcomes. The use of probiotic supplements is an interesting example of an intervention in pregnancy-related research because they are widely considered dietary supplements rather than medications, and therefore fall into the category of 'lifestyle' interventions. Other lifestyle factors include use of complementary and over-the-counter medications, behaviours (including exercise or smoking), stress, and other dietary factors.*

This chapter provides an overview of some examples of lifestyle interventions aimed at pregnant women and current research into probiotic supplement use during pregnancy. We then focus on a specific study "Probiotics in Pregnancy: Improving health during pregnancy and preventing infant eczema and allergy" (PiP Study) in Wellington and Auckland, New Zealand. This study usefully illustrates some of the methodological issues associated with recruiting and running studies using a lifestyle intervention for pregnant women. The conclusion looks at future priority areas for this research agenda, and considers barriers to conducting lifestyle research during pregnancy.

It has been argued that pregnant women remain one of the most underserved populations in clinical research (Baylis 2010). This is likely to be the result of a range of inter-related factors including: liability concerns on the part of manufacturer; restrictive regulatory environments; researchers' concerns about the vulnerability of

---

A. Ballantyne, BSc, PhD (✉)
Department of Primary Health Care and General Practice, University of Otago, Wellington, New Zealand
e-mail: angela.ballantyne@otago.ac.nz

C. Barthow, RCompNs, MA(appl) • K. Wickens, PhD
Department of Medicine, University of Otago, Wellington, New Zealand

F. Baylis, A. Ballantyne (eds.), *Clinical Research Involving Pregnant Women*, Research Ethics Forum 3, DOI 10.1007/978-3-319-26512-4_12

pregnant women and their foetuses; ethical guidelines stating that research that can be undertaken in other populations should not be done in vulnerable populations; reluctance of health care providers to recruit pregnant women; and the risk aversion of pregnant women (and their families, and communities) (Kukla 2005) steering them away from research participation. As a result, clinical research with pregnant women is limited, and research that does occur focuses primarily on aspects of obstetric practice, preterm labour, and foetal safety. There are increasing calls to broaden the research agenda to include conditions that affect both maternal and foetal interests and outcomes, particularly around mental health, asthma, oral health, and hypertension during pregnancy.

The use of probiotic supplements is an interesting example of an intervention in pregnancy-related research because they are widely considered dietary supplements rather than medications, they may be used to improve both maternal and foetal health, and they fall into the category of 'lifestyle' interventions. Probiotics are "live micro-organisms that, when administered in adequate amounts, confer a health benefit on the host" (Joint FAO/WHO Working Group 2002). Probiotics are commonly found in foods (such as fermented milks or yogurts), or sold as a dietary supplement over-the-counter (without a prescription) at pharmacies, health food stores or supermarkets. Probiotic supplements vary considerably in their microbial composition and number (dosage) of viable bacteria (Barrett et al. 2014). In the United States (US), a product containing probiotics can be placed along a continuum of FDA regulatory classifications including – drug, new drug, food, food additive, dietary supplement – each having implications on the nature and degree of regulatory requirements for clinical research (Degnan 2012). This uncertainty has led to a dearth of probiotic studies from the US.

Egger and colleagues (2008) define lifestyle interventions as the application of environmental, behavioural, medical, and motivational principles to the management of lifestyle related health problems in a clinical setting. This includes interventions that focus on nutrition, physical activity, stress management, sleep management, smoking cessation, personal hygiene, and a variety of other non-drug modalities (Egger et al. 2008). Lifestyle interventions also include use of other complementary and over-the-counter products.

Most studies of lifestyle interventions during pregnancy measure both maternal and neonatal outcomes, but the primary focus of the studies remain neonatal health outcomes. In publications of lifestyle intervention studies, neonatal outcomes and child health are usually introduced first and discussed more extensively than maternal outcomes, and tend to be used as the primary justification for the research.

Research into the effect and treatment of obesity during pregnancy provide an example of lifestyle-related research. The increased prevalence of obesity in Western high-income countries has led to significant health care issues in obstetric practice. Being overweight or obese is a risk factor for developing gestational diabetes mellitus (Han et al. 2012). Women who suffer from gestational diabetes mellitus are at increased risk of developing type 2 diabetes at some point following the index pregnancy (Kim et al. 2002). Maternal obesity is correlated with larger birth weight (over 4.5kgs), and observational studies show an association between infant birth weight and the subsequent risk of childhood and adulthood obesity (Dodd 2014). Infants are at increased risk of adverse outcomes including being large-for-gestational age, macrosomia, and birth trauma. Nutrition management is the primary treatment tool and is supported by a small amount of good-quality evidence (Buchanan et al. 2012). Treatment can be intensified with the use of insulin, glyburide, and metformin for high-risk groups. While some studies have investigated the impact of exercise modification during pregnancy, a recent Cochrane review concluded that existing research is insufficient to guide practice, and larger, well-designed randomised trials are needed (Han et al. 2012).

Nicotine replacement therapy during pregnancy is another example of lifestyle-related research. Worldwide 10–35 % of pregnant women smoke (De Long et al. 2014). Smoking cessation is recommended for the health of the pregnant women and their foetuses. Yet most manufacturers of nicotine replacement products do not officially endorse use during pregnancy. The safety and efficacy of nicotine replacement products for smoking cessation in pregnant women has not been clearly demonstrated (De Long 2014), with almost all evidence-based on observational studies (Dhalwani et al. 2014). Despite the lack of evidence, experts have tended to recommend the use of nicotine replacement products during pregnancy on the grounds that they are less harmful than smoking (WHO 2013). The recently published Smoking, Nicotine and Pregnancy (SNAP) trial was the first randomised, placebo-controlled trial of nicotine replacement therapy in pregnancy, with a 2-year post-natal follow up. This trial enrolled 1,050 women, and found that nicotine replacement patches had no enduring, significant effect on smoking during pregnancy. However, the trial also found that 72.6 % (323/445) of 2-year-olds born to women who used these patches showed no developmental impairment, compared to 65.5 % (290/443) in the placebo group (Cooper et al. 2014).

One obstacle is the variability in the specific methodology of the lifestyle interventions – different diets, exercise regimes, etc. Conducting meta-analyses of these studies is therefore difficult, and translating promising results into clinical practice is especially challenging.

Even when there is good evidence about the relationship between specific behaviours and maternal and foetal outcomes, there is an additional hurdle of communicating and motivating pregnant women to change behaviours. Adherence to folate supplementation during pregnancy remains low despite the simple nature of the intervention, and the strong evidence of the protective action of folate against spina bifida (Callaway et al. 2009). For conditions where the evidence is weaker or conflicting, and the intervention more complicated, the task of working with pregnant women to encourage healthy behaviours is challenging.

Consider, for example, the confusion around fish consumption during pregnancy in the United States. Pregnant women in the US are advised to limit the overall amount and to avoid specific types of fish consumed during pregnancy and lactation in order to reduce the risk of mercury-related adverse effects to the foetus/baby. Dietary intake of omega-3 fatty acids by pregnant and postpartum women in the US falls short of recommended safe levels (Benisek et al. 2000). Furthermore, some research indicates that women who exceed the recommended weekly intake of fish had babies who showed enhanced infant brain development, suggesting that public health advice to limit fish intake is detrimental (Hibbeln et al. 2007). Recent research into the use of omega-3 fatty acid as a treatment for depression in the general population (in most cases an adjunct to anti-depressant treatment) has shown promising results (Freeman 2006). But the interpretation and meta-analysis of these studies is hindered by the use of different particular omega-3 fatty acids or their combinations, and by dosing differences between studies (Freeman 2006).

12.1 Existing Research on Probiotic Supplements

Over recent years, an increasing body of research has provided evidence of the complexity of human microbiota and its impact on human health (Tojo et al. 2014). While much is unknown, evidence supports the widespread use of probiotic supplements in the prevention of antibiotic-associated diarrhoea and clostridium difficile (Pattani et al. 2013; Tojo et al. 2014). A review of studies also supports the use of probiotic supplements in premature infants weighing between 1,000 and 1,500 g at risk of necrotising enterocolitis (Alfahel and Anabrees 2014). Research into the efficacy of probiotic supplements for the prevention of allergic disease, gestational diabetes mellitus, and bacterial vaginal infections is less conclusive.

Worldwide rates of allergic disease have increased dramatically since the beginning of the twentieth century. In Europe, allergies are now the leading cause of chronic disease. Most forms of allergic disease begin in childhood, with many children first developing eczema, then later developing asthma and allergic rhinitis. New Zealand has high rates of infant eczema with 40 % of children developing the skin condition by 15 months (Silvers et al. 2008). Eczema is associated with significant morbidity and lower quality of life for both infant and family, and can be associated with poorer school achievement, depression, and negative impact on social development (Papadopoulos et al. 2012). As well as eczema, other illnesses including IgE-mediated food allergies, asthma, and allergic rhinitis are associated with atopic sensitisation and cause considerable and potentially long-term burdens on children, families, and health care services.

There is increasing evidence about the role of probiotic supplements as an intervention to prevent the development of eczema and allergic disease in infants. Almost all studies have intervened late in pregnancy, usually around 35 weeks gestation, perhaps because intervention later in pregnancy was considered safer than intervention at an earlier gestational age. Probiotic supplement use from about 35 weeks gestation onward has been shown to reduce the development of eczema in children, but there is less evidence for an effect on underlying atopic sensitisation, which predisposes children to allergic disease. Our own research has shown that *Lactobacillus rhamnosus* HN001 supplementation from 35 weeks gestation, and in infants to age 2 years, halved the rates of eczema by 2, 4 and 6 years (Wickens et al. 2008, 2012, 2013), and may also have had an impact on rates of atopic sensitisation by 6 years (Wickens et al. 2013). These findings for eczema are consistent with

Finnish probiotic supplement studies (Kalliomaki et al. 2001, 2003, 2007) using a similar probiotic – *Lactobacillus rhamnosus* GG. However, not all studies using *Lactobacillus rhamnosus* GG (Kopp et al. 2008) or other probiotic species have confirmed these findings (Osborne and Sinn 2007). Further research is required before clear advice can be given to pregnant women regarding these interventions during pregnancy to prevent allergic disease. For example, questions remain regarding the choice of probiotic species, strain, dose, timing, and duration of intervention, and whether the intervention should be used in pregnant women, infants or both.

Probiotic supplementation may also benefit maternal health outcomes during pregnancy in women with gestational diabetes mellitus, which affects health directly and predisposes pregnant women to further complications (such as gestational hypertension and preeclampsia) (Laitinen et al. 2009). Gestational diabetes mellitus also has an impact on foetal/infant health by increasing risks of congenital defects (Poston et al. 2011), macrosomia, complicated birth (shoulder dystocia), and postnatal hypoglycaemia (Tieu et al. 2014). In addition, it is now recognised that nutrition and glucose levels during pregnancy may have a long-term impact on infant metabolic and immune-inflammatory conditions that may become evident in later life (Laitinen et al. 2009). In a Finnish study (Luoto et al. 2010), probiotic supplement use from the 1st trimester of pregnancy was associated with beneficial outcomes for gestational diabetes mellitus. In women given dietary advice plus probiotic supplementation, the prevalence of gestational diabetes mellitus was 13 %, compared to 36 % in a group given diet advice only, and 34 % in a control group with no intervention ($p=0.003$). The authors suggest that this effect may be due to probiotics contributing to glucose regulation during pregnancy (Laitinen et al. 2009). In this same study population, probiotic supplementation taken from the 1st trimester was associated with half the risk of adiposity, defined as having a waist circumference ≥80 cm, at 6 months post-partum ($p=0.03$) (Ilmonen et al. 2011). If probiotics can influence maternal weight gain during pregnancy, we can expect improvements in all obesity-related conditions that occur in pregnancy including gestational diabetes mellitus, with potential benefits also for the future child.

Vaginal colonisation with Group B Streptococcus is frequently present but may be asymptomatic in pregnant women; 50 % of babies born to colonised women acquire the infection, and 1–2 % of colonised infants become seriously ill with potentially fatal consequences (Chambers et al. 2004). Lactobacilli have been shown to have inhibitory effects on Group B streptococcus growth in vitro (Acikgoz et al. 2005; Zarate and Nader-Macias 2006) and are less frequently found in vaginal swabs in pregnant women with Group B streptococcus (Moghaddam 2010). Although the popular literature supports the use of probiotics in the prevention of Group B streptococcus, we are aware of only one trial examining an oral probiotic effect on Group B streptococcus prevention in pregnancy (NCT01479478 is currently recruiting, see clinicaltrials.gov). As with Group B streptococcus, bacterial vaginosis is associated with a depletion of the normal vaginal lactobacillus population (Yudin and Money 2008; Moghaddam 2010). Women who have symptomatic bacterial vaginosis experience discomfort from an unpleasant odorous vaginal discharge, and may require antibiotic treatments. In addition, pregnancy complications such as maternal infection, late miscarriage, premature rupture of membranes, spontaneous abortion, chorioamnionitis, and preterm birth with neonatal morbidities related to prematurity are associated with this condition (Yudin and Money 2008). Although lactobacilli have been found to be effective in treating bacterial vaginosis (Abad and Safdar 2009), there is no current evidence that probiotics could prevent this condition in pregnant women.

The lack of well-powered studies investigating the effect of probiotic supplementation on these outcomes in pregnant women indicates the need for double-blind randomised placebo-controlled trials as the next step in determining the role of probiotics (see Healy and Mangin 2016).

12.2 Probiotics in Pregnancy Study: Improving Health During Pregnancy and Preventing Infant Eczema and Allergy (PiP Study)

The Probiotics in Pregnancy Study (PiP Study) aims to improve maternal health during pregnancy by reducing gestational diabetes mellitus, bacterial vaginosis and group B streptococcal vaginal colonisation before birth, and prevent the development of eczema and atopic sensitisation in infants by age 12 months. This study is a two centre, double-blind randomised placebo-controlled trial in Wellington and Auckland, New Zealand that commenced in 2013 with an expected completion date in 2016. In the study, pregnant women take either a daily probiotic supplement (*Lactobacillus rhamnosus* HN001) or placebo from 14 to 16 weeks gestation until 6 months after birth while breastfeeding.

This study differs from previous work by the same research team (Wickens et al. 2008, 2012, 2013) in that infant exposure to the probiotic is through the placenta and maternal breast milk only, and the maternal intervention begins early in the 2nd trimester rather than at 35 weeks gestation.

Probiotic supplements have been found to be very safe among the general healthy population (Boyle et al. 2006), although there are some identified risk factors for probiotic sepsis. As a result, probiotics are not recommended for use among those who are immunocompromised, debilitated, have malignancy or cardiac valvular disease (Boyle et al. 2006). When conducting research with probiotic supplements, there is a need for appropriate screening tools prior to study inclusion and documentation of the safety profiles related to specific probiotic organisms. The probiotic organism trialled in the PiP study *Lactobacillus rhamnosus* HN001 has GRAS (Generally Regarded As Safe) status in the US. Apart from general cautions about probiotic supplement use as indicated above, there is no biological mechanism or evidence suggesting that probiotics can cause harm in pregnant women, yet health benefits for pregnant women and their infants may be extensive.

Recruitment and Retention of Study Participants

There is conflicting evidence regarding the most successful recruitment strategies for research studies in primary care. A systematic review of recruitment strategies concluded that recruitment via clinical practitioners is the most successful strategy for primary health research (Ngune et al. 2012); however, others suggest that a range of barriers exist in this method of recruitment (Rendell et al. 2007). Recruitment in PiP proved to be a major hurdle to the research, much more so than originally anticipated. The story of the PiP team's recruitment efforts highlights several barriers to conducting research with pregnant women. The study aimed to recruit 440 pregnant women between 14 and 16 weeks gestation and 390 participants were expected to complete the study.

In New Zealand, the Midwifery and Maternity Providers Organisation was set up in 1997 to support midwifery practices and maternity services. Most midwives are registered with this organisation. The original study recruitment plan, approved by the Multi-region Ethics Committee (Wellington, New Zealand), was for staff at the Midwifery and Maternity Providers Organisation to mail study information to women in early pregnancy who were listed in their database. Contact details for the pregnant women would not be revealed to the researchers unless the woman chose to contact the research team. However, this plan was revoked when it was discovered that at the time of providing the registration data to the midwives, the pregnant women were not informed that their data could be used to notify them of potential research studies.

A further recruitment initiative planned to identify pregnant women on general practitioners' databases using electronic queries and to invite pregnant women to contact research staff about study participation. However, newly confirmed pregnancies are not coded by all general practitioners in the same manner. For example, gestational age at the time a pregnancy is confirmed are not recorded in a uniform manner, and queries would not be able to reliably and automatically detect and remove names of women where the consultation was for a termination of pregnancy. In addition, in New Zealand not all women see a general practitioner for confirmation of pregnancy. This approach to recruitment was not used.

The next recruitment strategy was to invite lead maternity carers[1] and general practitioners to provide information about the study directly to pregnant women in their care early in their pregnancy. This strategy generated a low proportion of all pregnant women interested in the study (General practitioners 5%, Midwives

[1] In New Zealand, maternity care and delivery is managed through a system of lead maternity carers. These carers provide maternity care and support through the pregnancy, co-ordinate with other healthcare providers, organise scans and tests, manage the labour, and provide care for the first six weeks of the infant's life. Most lead maternity carers are midwives, though some doctors and obstetricians may also have this role. Lead maternity carers identify conditions that may require specialist care and refer pregnant women to appropriate specialists when necessary.

10.5 %)[2] and involved considerable time investment on the part of researchers who had to meet with individual clinicians to discuss the study. In some cases, all of the clinicians in the practice had to endorse trial participation before the clinic would help. Increasing the efficiency of this strategy would have involved more time than researchers had available. Moreover, even among those clinicians who supported the study, referrals were low, possibly reflecting the fact that providing information about trial participation is an extra demand on top of clinical care.

Due to the slow recruitment rates further strategies were developed. One such strategy involved enclosing study information brochures in information packs routinely given to women by clinicians and midwives in early pregnancy. Ten percent of women who sought further information about the study came through this source. Additional recruitment strategies included emailing employees of large organisations (11 %), using web based approaches such as Facebook (8 %), and arranging for sonographers (6 %) and phlebotomists (7 %) to hand out study brochures at the time of early scans and the first antenatal blood collection. Surprisingly 13 % of expressions of interest came through word of mouth (friends or other study participants). Smaller numbers of trial participants learned of the study through radio advertising, newspaper articles, as well as pamphlets and posters placed at hospitals, clinics, crèches, libraries, and pharmacies.

Once women had contacted the research team, a further barrier to enrolment was the women's desire to enhance their infant's health by taking non-study probiotic supplements rather than risk being randomly allocated to the placebo arm. Perceived health benefits of probiotic supplements resulted from previous research (Wickens et al. 2008, 2012, 2013), as well as a wider awareness and availability of these products over-the-counter, and (ironically) from PiP study promotion. This was a notable and repeated reason for women declining study involvement. Opting to use over-the-counter probiotic supplements for themselves or their baby was the reason for the ineligibility in 32 % of women at the Wellington site (46/142). Some women even asked the researchers to provide the name and details of the probiotic supplement used in the study so that they could buy the study product. In addition, anecdotal evidence suggests that pregnant women are choosing to take probiotic supplements during the pregnancy, and for this reason may never have contacted the study team for screening. The plethora of probiotic supplements readily available over-the-counter means that many women will have opted to take ineffective or unproven products (for example, evidence for a probiotic preventive effect against allergy still needs confirmation for specific brands of probiotics), or inadequate doses at considerable personal expense in an effort to avoid placebo allocation (see Baylis and MacQuarrie 2016).

Women/infant dyads remain in the study for 18 months from recruitment until the infant reaches 1 year of age. Four hundred and twenty three eligible participants were initially enrolled in the study. Mid-study 389 women/infant dyads were active

[2] Recruitment data provided in this section reflect recruitment in Wellington alone. That is, the data do not reflect recruitment in Auckland.

participants (92 %),[3] 22 were inactive (i.e., no longer taking study capsules) (5.2 %), and 12 had withdrawn (2.8 %) mainly due to perceived side-effects of the study capsule and/or concerns about infant health.

During the study a number of questionnaires, measures, and samples were requested. This included maternal vaginal swabs at baseline and 35 weeks for bacterial vaginosis and Group B streptococcus, an oral glucose tolerance test, collection of cord blood and tissue samples, maternal and infant faecal samples, infant skin swabs and buccal samples. Both biological parents and the infant were skin-prick tested to determine if they are sensitised to a range of allergens.

As the study originated outside of maternity services, it was essential to communicate, co-ordinate and co-operate with lead maternity carers throughout the process of study set-up and during the field work. Some tests such as the oral glucose tolerance test and Group B streptococcus swabs were relevant to the clinical care of the women. These tests were ordered by the study, with results copied to the lead maternity carers who maintain responsibility for clinical care. Sharing these test results with the lead maternity carers decreased the burden on women by preventing duplicate testing. Lead maternity carers plan glucose tolerance testing based on a woman's past history and current clinical features, meaning that in practice there is variability in when glucose testing is ordered. Study protocols needed to be flexible and take account of this variability. Despite the complexity of this co-ordinated care approach, there is reason to believe that it has resulted in smoother clinical care for the pregnant women and improved retention in the study.

Maintaining clear boundaries between clinical care and research was important throughout the study and, in particular, in relation to the collection of cord blood and cord tissue samples. These samples were collected in order to enable further work to understand the mechanisms of probiotic action if the intervention arm proved to be effective. Typically, lead maternity carers establish an understanding of women's preferences and discuss birth plans late in pregnancy. There is considerable variation in women's and lead maternity carers' preferences regarding management of the 3rd stage of labour (placental delivery), with some preferring no interventions until the cord has stopped pulsing after birth, while others prefer active management. Differing approaches to this stage of labour have an impact on quantity and quality of cord blood samples achieved. During the informed consent process, researchers explained that the collection of samples for study purposes would only occur after any samples needed for clinical care had been obtained, and that women's personal preferences for birthing plans would be prioritised before the collection of study samples. Women were encouraged to discuss their birthing plans with their lead maternity carers, including the risks and potential benefits of different approaches to the 3rd stage of labour. In conjunction with this approach, lead maternity carers were supported in understanding study processes through teaching sessions and personal communications from research staff. This approach respected the relationship between pregnant women and lead maternity carers, and supported

[3] This data is combined data from Wellington and Auckland.

the role of lead maternity carers in determining a woman's birthing preferences and guiding appropriate clinical care.

12.3 Challenges of Doing Lifestyle Research

In this section, we discuss some of the challenges associated with lifestyle research during pregnancy, particularly with dietary supplements, and offer some advice for future research.

General Problems

It is difficult to conduct meta-analysis of lifestyle interventions because of heterogeneity in the product or intervention. A medical product, even a generic version such as paracetamol (i.e., acetaminophen), comes in a standardised dose. Reviews of this research can compare and contrast interventions. By comparison, probiotics products vary enormously in their composition (type and number of bacteria), storage (bench, fridge, shelf life), and delivery mechanism (supplement, added to food processed products, naturally occurring in the food items). As a result, the amount of live active bacteria in a probiotic product varies considerably. Because of the relatively unregulated nature of supplements compared to medicines, probiotics may be present in processed food that research participants are consuming without their awareness. For example, probiotics are increasingly being added to infant formula, infant cereal formulas, nutrition bars and fruit drinks, despite inconclusive evidence regarding the type of probiotics, dosage, or positive health effect. It follows, that in addition to the research probiotic supplement, participants may be consuming various other probiotic products. To correct for non-study probiotic consumption, a range of data such as yogurt consumption, infant formula containing probiotics and use of non-study probiotic supplements are collected in the PiP study. The same challenge applies to research into Omega-3 fatty acids, and vitamins. Lifestyle behavioural interventions such as diet and exercise suffer from even greater heterogeneity.

Specific to Pregnancy

Furthermore, there are challenges associated with doing lifestyle research with pregnant populations. Risk aversion in pregnancy may make pregnant women reluctant to join clinical trials, especially randomised placebo controlled trials (see Langston 2016; Kukla 2016). If a pregnant woman is diagnosed with a clinical condition and there is a trial for a medical intervention not otherwise available, there is greater incentive for the woman to participate in the trial. However, if the product is available over-the-counter and the research aims to prevent rather than treat a medical condition, there is less incentive to accept the possibility of randomisation.[4] This was certainly the case in the PiP study, where many women stated a preference for purchasing their own over-the-counter probiotic supplements, as opposed to risking blind allocation to the placebo arm. Pregnant women may also erroneously believe that it is safer to buy a marketed product over-the-counter, rather than receive a carefully designed product in the monitored environment of a research study.

Pregnancy is a physically hard and draining process for many women. Participating in research where there is no obvious medical need (no therapeutic intent) may seem overly demanding for many women, especially if they can simply buy a perceived equivalent product over-the-counter.

Priorities for Future Research Into Lifestyle Changes During Pregnancy

We should systematically investigate women's experience of, and views about, research interventions (see Wild and Biller-Andorno 2016). Even if studies show positive or promising results for maternal health, uptake in the population will depend on pregnant women's experience of interventions and their ability to integrate these new behaviours into their existing, often busy, lives. Pregnancy is hard work. Even a non-complicated healthy pregnancy adds another dimension to a woman's life, which may already include combining paid work, unpaid domestic work, and childcare. Ordinary pregnancy is associated with a decline in physical and mental health for most pregnant women. An Australian study showed the following rates of pregnancy complications in the general population: exhaustion (87 %), nausea (64 %), back pain (46 %), constipation (44 %), and severe headaches/migraines (30 %) (Gartland et al. 2010). Given the challenges of pregnancy, it is a difficult time to be trying to implement new, healthier habits. Informing a pregnant woman that she should exercise more or in a different way, change her diet, or add supplements, will only work if there is robust evidence to support benefits of the interventions that are proposed, the relevant information is presented in an understandable manner, and women are motivated and supported to change. In the case of over-the-counter supplements, even where there is robust evidence of effectiveness, cost may also be a significant limiting factor for uptake for some pregnant women. Pregnancy is associated with increased motivation to adopt healthy behaviours, especially reduction in smoking, alcohol, and coffee consumption (Crozier et al. 2009). Yet consumption of folic acid supplements remains lower than advised

[4] This could be a challenge outside of pregnancy as well. People might be reluctant to participate in research with interventions that are already available to them outside of research when there is a chance that they might be randomised to the placebo group. Thank you to the reviewer for pointing this out.

(Callaway et al. 2009). Development of new habits requires mental effort and can takes weeks to establish. The mental effort required to adopt new habits around smoking and alcohol may deplete the will power available to change other behaviours.

Pregnant women who volunteer to participate in a lifestyle study (especially a randomised placebo controlled trial that offers no guarantee to access a desired intervention), are likely to be especially motivated, and are unlikely to be those suffering most acutely from their pregnancies. Participants in the PiP study who were diagnosed with gestational diabetes mellitus were often those that chose to withdraw from the study due to the extra stress associated with this diagnosis. The fact that many pregnant women complied with the intervention during the study, does not mean that pregnant women in the general population will have sufficient support to instigate new habits. Lifestyle intervention studies should be systemically collecting qualitative data about pregnant women's experiences of the intervention. Furthermore, there is a need for studies that explore the most effective means of communicating the vast, confusing, and often conflicting advice about healthy habits during pregnancy. Research should investigate pregnant women and their partners' decision-making with regard to lifestyle interventions and strategies that support women to adopt new behaviours.

More generally, researchers should foster an attitude of partnership with pregnant women. Pregnant women receive a huge amount of often unsolicited advice. Kukla (2005) has described how the notion of the 'unruly mother' has influenced the perception of pregnant women as the primary source of harm and damage to the 'innocent foetus'. Kukla reports that when searching a modern academic library catalogue with the term 'pregnancy', 80 % of the subheadings were associated with toxins pregnant women must avoid – alcohol, tobacco, drugs, unhealthy food, and too much food (Kukla 2005). This near exclusive focus on pregnant women's potentially harmful behaviours obscures other significant threats to foetal health such as poverty, racism, and male violence (Aizer 2011). A paternalistic and critical approach to pregnant women's behaviours persists amongst some researchers: "Many women engage in health risk behaviors during pregnancy.... [this intervention] appears to be a feasible and convenient approach to addressing the multiple health behavior change needs of pregnant women" (Davis et al. 2014, 1165). Future research into lifestyle and behaviour factors during pregnancy should frame pregnant women as active partners, rather than passive recipients of medical advice.

In conclusion, there is the need for robust well-designed research regarding lifestyle interventions during pregnancy. Such research needs to be designed with pregnant women as active partners, and address both maternal and infant outcomes. In conjunction with this research, there is an urgent need to expand the research agenda to explore pregnant women's views of involvement in research during pregnancy, what they need to support their involvement, and how they make decisions and act upon the outcomes of such research.

References

Abad, C.L., and N. Safdar. 2009. The role of Lactobacillus probiotics in the treatment or prevention of urogenital infections – a systematic review. *Journal of Chemotherapy* 21(3): 243–252.

Acikgoz, Z., S. Gamberzade, S. Gocer, and P. Ceylan. 2005. Inhibitor effect of vaginal Lactobacilli on group B streptococcus. *Mikrobiyoloji Bülteni* 39(1): 17–23.

Aizer, A. 2011. Poverty, violence and health: The impact of domestic violence during pregnancy on newborn health. *The Journal of Human Resources* 46(3): 518–538.

Alfahel, K., and J. Anabrees. 2014. Probiotics for prevention of necrotizing enterocolitis in preterm infants. *Cochrane Database of Systematic Reviews* 4: CD005496.

Barrett, H.L., M. Dekker Nitert, L.S. Conwell, and L.K. Callaway. 2014. Probiotics for preventing gestational diabetes. *Cochrane Database of Systematic Reviews* 2: CD009951.

Baylis, F. 2010. Pregnant women deserve better. *Nature* 465(7299): 689–690.

Baylis, F., and R. MacQuarrie. 2016. Why physicians and women should want pregnant women included in clinical trials. In *Clinical research involving pregnant women*, eds. F. Baylis and A. Ballantyne, 17–31. Cham: Springer.

Benisek, D., J. Shabert, and R. Skornik. 2000. Dietary intake of polyunsaturated fatty acids by pregnant or lactating women in the United States. *Obstetrics and Gynecology* 95(4): S77–S78.

Boyle, R.J., R.M. Robins-Browne, and M.L.K. Tang. 2006. Probiotic use in clinical practice: What are the risks? *The American Journal of Clinical Nutrition* 83(6): 1256–1264.

Buchanan, T.A., A.H. Xiang, and K.A. Page. 2012. Gestational diabetes mellitus: Risks and management during and after pregnancy. *Nature Reviews Endocrinology* 8(11): 639–649.

Callaway, L., P.B. Colditz, and N.M. Fisk. 2009. Folic acid supplementation and spontaneous preterm birth: Adding grist to the mill? *PLoS Medicine* 6(5): e1000077.

Chambers, S., M. Baker, F. Frizelle, K. Sexton, D. Lennon, P. Oster, et al. 2004. The prevention of early-onset neonatal group B streptococcus infection: technical report from the New Zealand GBS consensus working party. *New Zealand Medical Journal* 117: 1–19.

Cooper, S., S. Lewis, J.G. Thornton, N. Marlow, K. Watts, J. Britton, M.J. Grainge, J. Taggar, H. Essex, S. Parrott, A. Dickinson, R. Whitemore, and T. Coleman; Smoking, Nicotine and Pregnancy Trial Team. 2014. The SNAP trial: A randomised placebo-controlled trial of nicotine replacement therapy in pregnancy – Clinical effectiveness and safety until 2 years after delivery, with economic evaluation. *Health Technology Assessment* 18(54): 1–128.

Crozier, S.R., S.M. Robinson, S.E. Borland, K.M. Godfrey, C. Cooper, H.M. Inskip, and SWS Study Group. 2009. Do women change their health behaviors in pregnancy? Findings from the Southampton Women's Survey. *Paediatric and Perinatal Epidemiology* 23(5): 446–453.

Davis, A.M., K.A. Wambach, E.L. Nelson, C. Odar, T. Lillis, A. McKinley, and M. Gallagher. 2014. Health behavior change in pregnant women: A two-phase study. *Telemedicine Journal and E-Health* 20(12): 1165–1169.

De Long, N.E., N.G. Barra, D.B. Hardy, and A.C. Holloway. 2014. Is it safe to use smoking cessation therapeutics during pregnancy? *Expert Opinion on Drug Safety* 13(12): 1721–1731.

Degnan, F.H. 2012. Clinical studies involving probiotics: When FDA's investigational new drug rubric applies-and when it may not. *Gut Microbes* 3(6): 485–489.

Dhalwani, N.N., L. Szatkowski, T. Coleman, L. Fiaschi, and L.J. Tata. 2014. Prescribing of nicotine replacement therapy in and around pregnancy: A population-based study using primary care data. *The British Journal of General Practice* 64(626): e554–e560.

Dodd, J.M. 2014. Dietary and lifestyle advice for pregnant women who are overweight or obese: The LIMIT randomized trial. *Annals of Nutrition and Metabolism* 64(3–4): 197–202.

Egger, G., A. Binns, and S. Rossner. 2008. *Lifestyle medicine*. Sydney: McGraw-Hill.

Freeman, M.P. 2006. Omega-3 fatty acids and perinatal depression: A review of the literature and recommendations for future research. *Prostaglandins, Leukotrienes, and Essential Fatty Acids* 75(4–5): 291–297.

Gartland, D., S. Brown, S. Donath, and S. Perlen. 2010. Women's health in early pregnancy: Findings from an Australian nulliparous cohort study. *The Australian and New Zealand Journal of Obstetrics and Gynaecology* 50(5): 413–418.

Han, S., P. Middleton, and C.A. Crowther. 2012. Exercise for pregnant women for preventing gestational diabetes mellitus. *Cochrane Database of Systematic Reviews* 7: CD009021.

Healy, D., and D. Mangin. 2016. Does my bias look big in this? In *Clinical research involving pregnant women*, eds. F. Baylis and A. Ballantyne, 197–208. Cham: Springer.

Hibbeln, J.R., J.M. Davis, C. Steer, P. Emmett, I. Rogers, C. Williams, and J. Golding. 2007. Maternal seafood consumption in pregnancy and neurodevelopmental outcomes in childhood (ALSPAC study): An observational cohort study. *Lancet* 369(9561): 578–585.

Ilmonen, J., E. Isolauri, T. Poussa, and K. Laitinen. 2011. Impact of dietary counselling and probiotic intervention on maternal anthropometric measurements during and after pregnancy: A randomized placebo-controlled trial. *Clinical Nutrition* 30(2): 156–164.

Joint FAO/WHO Working Group. 2002. *Report on drafting guidelines for the evaluation of probiotics in food: London, Ontario, Canada, April 30 and May 1, 2002*. http://www.who.int/foodsafety/fs_management/en/probiotic_guidelines.pdf. Accessed 8 Apr 2016.

Kalliomaki, M., S. Salminen, H. Arvilommi, P. Kero, P. Koskinen, and E. Isolauri. 2001. Probiotics in primary prevention of atopic disease: A randomised placebo-controlled trial. *Lancet* 357(9262): 1076–1079.

Kalliomaki, M., S. Salminen, T. Poussa, H. Arvilommi, and E. Isolauri. 2003. Probiotics and prevention of atopic disease: 4-year follow-up of a randomised placebo-controlled trial. *Lancet* 361(9372): 1869–1871.

Kalliomaki, M., S. Salminen, T. Poussa, and E. Isolauri. 2007. Probiotics during the first 7 years of life: A cumulative risk reduction of eczema in a randomized, placebo-controlled trial. *Journal of Allergy and Clinical Immunology* 119(4): 1019–1021.

Kim, C., K. Newton, and R.H. Knopp. 2002. Gestational diabetes and the incidence of type 2 diabetes: A systematic review. *Diabetes Care* 25(10): 1862–1868.

Kopp, M.V., L. Hennemuth, A. Heinzmann, and R. Urbanek. 2008. Randomized, placebo-controlled trial of probiotics for primary prevention: No clinical effects of Lactobacillus GG supplementation. *Pediatrics* 121(4): e850–e856.

Kukla, R. 2005. *Mass hysteria: Medicine, culture, and mothers' bodies*. Oxford: Rowman and Littlefield Publishers.

Kukla, R. 2016. Equipoise, uncertainty, and inductive risk in research involving pregnant women. In *Clinical research involving pregnant women*, eds. F. Baylis and A. Ballantyne, 179–196. Cham: Springer.

Laitinen, K., T. Poussa, E. Isolauri, and the Nutrition, Mucosal Immunology and Intestinal Microbiotia Group. 2009. Probiotics and dietary counselling contribute to glucose regulation during and after pregnancy: A randomized controlled trial. *The British Journal of Nutrition* 101(11): 1679–1687.

Langston, L. 2016. Better safe than sorry: Risk, stigma, and research during pregnancy. In *Clinical research involving pregnant women*, eds. F. Baylis and A. Ballantyne, 33–50. Cham: Springer.

Luoto, R., K. Laitinen, M. Nermes, and E. Isolauri. 2010. Impact of maternal probiotic-supplemented dietary counselling on pregnancy outcome and prenatal and postnatal growth: A double-blind, placebo-controlled study. *The British Journal of Nutrition* 103(12): 1792–1799.

Moghaddam, N. 2010. Recto-vaginal colonization with group B streptococcus in pregnant women referred to a hospital in Iran and its effect on Lactobacillus normal flora. *Journal of Biological Sciences* 10(2): 166–169.

Ngune, I., M. Jiwa, A. Dadich, J. Lotriet, and D. Sriram. 2012. Effective recruitment strategies in primary care research: A systematic review. *Quality in Primary Care* 20(2): 115–123.

Osborn, D., and J. Sinn. 2007. Probiotics in infants for prevention of allergic disease and food hypersensitivity. *Cochrane Database of Systematic Reviews* 4: CD006475.

Papadopoulos, N.G., I. Agache, S. Bavbek, B.M. Bilo, F. Braido, V. Cardona, et al. 2012. Research needs in allergy: An EAACI position paper, in collaboration with EFA. *Clinical and Translational Allergy* 2(1): 21.

Pattani, R., V.A. Palda, S.W. Hwang, and P.S. Shah. 2013. Probiotics for the prevention of antibiotic-associated diarrhoea and clostridium difficile. *Open Medicine* 7(2): e56–e67.

Poston, L., L.F. Harthoorn, and E.M. van der Beek, on behalf of contributors to the ILSI Europe Workshop. 2011. Obesity in pregnancy: Implications for the mother and lifelong health of the child. A consensus statement. *Pediatric Research* 69(2): 175–180.

Rendell, J.M., R.D. Merritt, and J.R. Geddes. 2007. Incentives and disincentives to participation by clinicians in randomised controlled trials. *Cochrane Database of Systematic Reviews* 2: MR000021.

Silvers, K., C. Frampton, K. Wickens, M. Epton, P. Pattemore, T. Ingham, D. Fishwick, J. Crane, I. Town, and the New Zealand Asthma and Allergy Cohort Study Group. 2008. Breastfeeding protects against adverse respiratory outcomes at 15 months of age. *Maternal and Child Nutrition* 5(3): 243–250.

Tieu, J., A.J. McPhee, C.A. Crowther, and P. Middleton. 2014. Screening and subsequent management for gestational diabetes for improving maternal and infant health. *Cochrane Database of Systematic Reviews* 2: CD007222.

Tojo, R., A. Suárez, M.G. Clemente, C.G. de Los Reyes-Gavilán, A. Margolles, M. Gueimonde, and P. Ruas-Madiedo. 2014. Intestinal microbiota in health and disease: Role of bifidobacteria in gut homeostasis. *World Journal of Gastroenterology* 20(41): 15163–15176.

Wickens, K., P.N. Black, T.V. Stanley, E. Mitchell, P. Fitzharris, G.W. Tannock, G. Purdie, J. Crane, Probiotic Study Group. 2008. A differential effect of 2 probiotics in the prevention of eczema and atopy: A double-blind, randomized, placebo-controlled trial. *The Journal of Allergy and Clinical Immunology* 122(4): 788–794.

Wickens, K., P. Black, T.V. Stanley, E. Mitchell, C. Barthow, P. Fitzharris, et al. 2012. A protective effect of Lactobacillus rhamnosus HN001 against eczema in the first 2 years of life persists to age 4 years. *Clinical and Experimental Allergy* 42(7): 1071–1079.

Wickens, K., T.V. Stanley, E. Mitchell, C. Barthow, P. Fitzharris, G. Purdie, R. Siebers, P. Black, and J. Crane. 2013. Early supplementation with Lactobacillus rhamnosus HN001 reduces eczema prevalence to 6 years: Does it also reduce atopic sensitization? *Clinical and Experimental Allergy* 43(9): 1048–1057.

Wild, V., and N. Biller-Andorno. 2016. Pregnant women's views about participation in clinical research. In *Clinical research involving pregnant women*, eds. F. Baylis and A. Ballantyne, 119–136. Cham: Springer.

World Health Organization. 2013. *WHO recommendations for the prevention and management of tobacco use and second-hand exposure in pregnancy*. Geneva: World Health Organization.

Yudin, M.H., and D.M. Money. 2008. Screening and management of bacterial vaginosis in pregnancy. *Journal of Obstetrics and Gynaecology Canada* 30: 702–708.

Zarate, G., and M.E. Nader-Macias. 2006. Influence of probiotic vaginal lactobacilli on in vitro adhesion of urogenital pathogens to vaginal epithelial cells. *Letters in Applied Microbiology* 43(2): 174–180.

Chapter 13
Ethics and Research with Pregnant Women: Lessons from HIV/AIDS

Margaret Olivia Little, Anne Drapkin Lyerly, Anna C. Mastroianni, and Ruth R. Faden

Abstract *HIV/AIDS is among the most serious diseases confronting women who are pregnant. It is also one of the few areas of research involving humans where there is a long track record of research involving pregnant women. Yet the HIV/AIDS research community has struggled to expand the research agenda from research to prevent mother-to-child transmission of HIV to research encompassing issues pertaining to the pregnant woman's own health. Research questions of interest include: which antiretrovirals are safest and most effective for pregnant women; how best to pursue preventive regimes for pregnant women who are not infected; or, how to treat HIV's deadly co-infections, such as tuberculosis (TB), during pregnancy. In this chapter, we describe two key lessons about research in pregnancy from the context of HIV/AIDS: first, why addressing the health needs of pregnant women, not just the needs of their offspring, is so critical; and second, why doing so is immediately possible, even as we work to resolve certain ethical and regulatory debates, particularly about when it is appropriate to impose foetal risk without the prospect of foetal benefit. In particular, the HIV/AIDS context shows how treatment or prevention of maternal disease often entails not just risk – but the prospect of benefit – to the foetus; and creative trial designs can advance no-benefit studies without imposing foetal risk in the first place. For all the challenges that research with pregnant women entails, the HIV/AIDS context reveals that it is possible to conduct a wide range of important research during pregnancy that is both ethically responsible and consonant with US regulations.*

M.O. Little, BPhil, PhD (✉)
Kennedy Institute of Ethics, Georgetown University, Washington, DC, USA
e-mail: littlem@georgetown.edu

A.D. Lyerly, MD, MA
Center for Bioethics, Department of Social Medicine, University of North Carolina at Chapel Hill, Chapel Hill, NC, USA

A.C. Mastroianni, JD, MPH
University of Washington School of Law, Seattle, WA, USA

R.R. Faden, PhD, MPH
Johns Hopkins Berman Institute of Bioethics, Baltimore, MD, USA

F. Baylis, A. Ballantyne (eds.), *Clinical Research Involving Pregnant Women*, Research Ethics Forum 3, DOI 10.1007/978-3-319-26512-4_13

Of the many serious diseases that can confront women while pregnant, HIV/AIDS is surely among the most significant. With 16 million women living with HIV worldwide (UNAIDS 2014), and millions more at high risk of infection, a striking number will face decisions about HIV treatment or prevention while pregnant. To give an idea of the scale, 1.5 million women living with HIV give birth annually (UNAIDS 2014). HIV incidence among pregnant women in sub-Saharan Africa is comparable to non-pregnant high-risk groups, such as sex workers, HIV-discordant couples, and men who have sex with men (Drake et al. 2014). Add to that the burdens of treating HIV's deadly co-infections such as drug-resistant TB during pregnancy, and the magnitude of the issue is clear.

HIV/AIDS is also a particularly interesting research context for those who want to advance the evidence base for the prevention and treatment of serious illness during pregnancy. For one thing, HIV/AIDS is one of the few areas of research where there is a long track record of research involving pregnant women. Indeed, one of the major foci of HIV research in the last quarter century has centred on just this population, as researchers worked to study interventions to prevent transmission of HIV from pregnant women to their offspring (mother-to-child transmission). Large studies have tested various regimens to find the most effective mode for administering antiretroviral medication to women during pregnancy, labour and delivery, and breastfeeding, with inspiring results. Without any intervention, the rate of mother-to-child transmission is estimated to be 15–45 % (WHO 2015). Antiretroviral medication taken during pregnancy in accordance with evidence-based guidelines can reduce this risk to less than 1 % (CDC 2014). As a result of clinical research and public health efforts, the birth of a child with HIV is now "very rare" in developed countries such as the United States (Kelland 2010), and the elimination of mother-to-child transmission in developing countries is increasingly being identified as a "realistic public health goal" (WHO 2014).

When it comes to HIV, then, the pregnant research participant is far from taboo. Indeed, even pharmaceutical companies – widely seen as perhaps the most difficult actors to engage in research with pregnant women – have joined the research effort here as part of public-private partnerships. HIV/AIDS is thus an impressive demonstration of what political and civic will can do to encourage research with pregnant women.[1]

[1] The scope and urgency of the pandemic have resulted in unusually high levels of public funding and robust public-private partnerships, including private pharmaceutical companies, with a notable emphasis on trials that have included research with pregnant women for the purposes of preventing mother-to-child transmission. In 2012, PEPFAR (the President's Emergency Plan for AIDS Research) accounted for 73 % of all bilateral aid for HIV, 49 % of all international assistance, and 23 % of total HIV funding. Multilateral organisations, including the Global Fund to Fight AIDS, Tuberculosis and Malaria and the World Bank provided 28 % of international assistance for HIV, while the private sector, including the Bill and Melinda Gates Foundation, have contributed more than 5 % (AVERT 2014). Mirroring the public response, the pharmaceutical industry has invested significantly, reporting in 2014 that 44 medicines and vaccines are currently being developed with 94 active trials in the US alone and many more worldwide, including several trials that have included pregnant women (PhRMA 2014).

At the same time, the HIV/AIDS research community has struggled to expand research during pregnancy to encompass issues pertaining to the pregnant woman's *own* health. Compared to the robust evidence base for the goal of preventing mother-to-child transmission, little is known about which antiretrovirals are safest and most effective for women to use during pregnancy. Just as critically, little is known about how best to pursue preventive regimes for pregnant women who are not infected. And little is known about how to treat HIV's deadly co-infections, such as TB, during pregnancy.

In an effort to provide concrete guidance for ethically advancing the evidence base for pregnant women at risk for or infected with HIV, we initiated the PHASES Project – Pregnancy and HIV/AIDS: Seeking Equitable Study. Funded in 2013 by the US National Institutes of Allergy and Infectious Disease (NIAID), this project aims to explore the ethical need for research into pregnant women's *own* health in the pressing contexts of HIV prevention and treatment with particular attention to the ethical reticence around, and the ethical paths forward in, advancing such research. As part of our initial work, we conducted a series of meetings and discussions with HIV investigators, ethics and law scholars, toxicologists, and institutional review board (IRB) officials, and explored patterns in the research that has been conducted.

In this chapter, we describe two key lessons about research in pregnancy that can be taken from the context of HIV/AIDS: first, why addressing the health needs of pregnant women themselves, and not just the needs of their offspring, is so critical; and second, how clarity around foetal risk and potential benefit, as well as creative trial designs, show that it is possible to move forward now (even as we await resolution of certain ethical and regulatory debates). For all the challenges that the conduct of research with pregnant women entails, the HIV/AIDS context reveals that it is immediately possible to conduct a wide range of important research during pregnancy that is both ethically responsible and consonant with US regulations.

13.1 The Critical Need for Research with Pregnant Women: The Case of HIV/AIDS

If anyone understands that it is wrong to interrupt medical treatment just because of pregnancy, it is those who work in, or live with, HIV/AIDS. As opposed to other areas of medicine in which treatment interruption is considered and debated, patients and practitioners alike agree that pregnancy is no time for suspension of antiretroviral medication.

One reason for the continuation of medical treatment is the key role that medication plays in the prevention of transmission from infected pregnant women to the children they bear. But the health of women quickly emerges as its own critical reason for continued medical treatment. Without appropriate medication, viral loads can rapidly increase, bringing AIDS-related complications in their wake. Without appropriate medication, co-infections can take a profound toll: TB, for instance, is a leading cause of maternal death, and a significant proportion of those who die from the disease are co-infected with HIV. TB and HIV are 'inextricably linked': poorly controlled HIV is associated with higher rates of death from TB, as are delays in treatment of TB (Mofenson and Laughon 2007).

Further, it turns out that decisions made about the treatment of HIV in pregnancy can cast a long shadow with implications that endure long after the pregnancy ends. For example, when pregnant women have taken potent antiretroviral medication in short courses to prevent mother-to-child transmission during pregnancy, the same medications are less effective when those women eventually need to use them for their own health. When women are choosing from a limited arsenal of antiretroviral drugs, as is often the case in developing countries, the result can be the worsening of the disease or death (Gray 2008).

Simply shifting to older, more established antiretroviral drugs when considering treatment of HIV/AIDS during pregnancy is not an effective alternative. Newer antiretroviral medications have critical advantages. In particular, game-changing agents have emerged to treat what is known as 'refractory' HIV/AIDS – disease that is unresponsive to first-line drugs. In the complex world of HIV/AIDS, finding a treatment that works for a given patient, and has side-effects that the patient can tolerate, can be a matter of trial and error; it is critical that health care providers have a portfolio of options to try. Alongside this, new appreciation has emerged of the critical importance of patients, including pregnant women, *continuing* the drug regimen that has been found to work for them (AIDSInfo 2014). Finally, women who were themselves infected since birth are now emerging as a new cohort of obstetrical patients. Having used antiretroviral medication since childhood, they often require novel antiretrovirals to stay healthy.

Yet there is little evidence about the efficacy and risk profiles in pregnancy of HIV/AIDS medications for women's health, especially newer ones. Information on treating pregnant women with novel antiretroviral drugs is primarily empirical. We lack systematic data to inform dosing over the course of pregnancy (Mirochnick and Capparelli 2004), and have only limited data on the safety, efficacy, or pregnancy-specific dosing of the vast majority of these newer drugs.

The situation is similar for the treatment of HIV's life-threatening co-infections such as TB. While a new class of antitubercular drugs is revolutionising the treatment of drug-resistant TB, we lack data to guide dosing and treatment decisions during pregnancy – and this, despite the fact that drug-resistant TB associated with HIV is both a major cause of morbidity and mortality in the babies born to these women (Churchyard et al. 2007; Gupta et al. 2007). Likewise the absence of adequate data on first-line antitubercular drugs in pregnancy has led to conflicting national and international treatment guidelines.[2] It is perhaps no surprise that a sig-

[2] For instance, the World Health Organization (WHO) recommends the drug pyrazinamide (PZA) during pregnancy as first line therapy, but the US Centers for Disease Control (CDC) do not due to inadequate data on foetal effects (Gupta et al 2016).

nificant proportion of the women who die from TB in pregnancy are co-infected with HIV.

Why are pregnancy-specific data so critical here? First, it is widely known that pregnancy can substantially alter the way that drugs are processed by the body (pharmacokinetics) and the way that the body reacts to drugs (pharmacodynamics) (Andrew et al. 2007). In the context of HIV treatment, these physiological changes may result in decreased drug efficacy, with resulting suboptimal maternal viral suppression and mother-to-child transmission (Patterson et al. 2013). At the same time, unintentional overdosing may increase maternal and foetal toxicity (Simon et al. 2011). We need to understand the pharmacokinetic and pharmacodynamic values of treatments and preventives in the pregnant body across the trimesters of pregnancy (Patterson et al. 2013), as well as its various 'compartments' – blood, genital tract, cord blood, and amniotic fluid (Yeh et al. 2009) – to make sure that pregnant women are actually being given effective and appropriate doses.

Second, we need to understand the specific risk profiles that medicines prescribed in pregnancy have for the foetus. Such information is critical to guiding recommendations for initiating and continuing antiretroviral therapy in pregnant women in the high-stakes context of mother-to-child transmission. It is also important, in the other direction, because uncertainty about foetal safety can lead to reticence to prescribe medication that is indicated and safe.

In 2012, for instance, the World Health Organization called attention to the problems that uncertainty around medication generates for the responsible treatment of HIV in pregnancy. Limited data had indicated an increased risk of congenital defects in the foetus (specifically, in closure of the spinal cord or 'neural tube') associated with efavirenz, an antiretroviral medication widely used in resource-limited settings. Further research, however, indicated that these fears of teratogenicity were unfounded (Ekouevi et al. 2011; Ford et al. 2011), and that the treatment turned to as a substitute, nevirapine, may carry greater risk of toxicity and treatment failure (Shearer et al. 2014). Guessing about foetal safety is bad for pregnant women – and for foetuses. In short, we need effective treatment for pregnant women living with HIV – both for the health of women themselves and for the children they will bear.

We also need good interventions aimed at preventing women who are pregnant from getting HIV/AIDS in the first place. It turns out that pregnancy brings an increased risk of becoming infected with HIV. For one, pregnancy and HIV 'share' a risk factor, namely unprotected intercourse: sex without a condom places a woman at risk for getting pregnant *and* places her at risk of becoming infected with a sexually transmitted infection such as HIV. Further, and hauntingly, we are now finding that pregnancy itself appears to increase the risk of infection. Pregnancy can make negotiation of condom use yet more difficult, as the contraceptive rationale is no longer present (Mugo et al. 2011). Further, pregnancy induces physiological changes including changes in the way the cells of the vagina and cervix are structured, and changes in the way the immune system functions. These changes make women more susceptible to infection (Gray et al. 2005; Moodley et al. 2009).

It also turns out that preventing infection during pregnancy is important to foetal outcomes in ways that were not earlier appreciated. Dramatic new evidence has emerged to indicate that when women become infected with HIV *during* pregnancy, versus coming to pregnancy already infected, the risk of the virus being transmitted to the foetus increases at least fivefold. This is due to both the high viral load that accompanies new infection, as well as missed opportunities for diagnosis and treatment before delivery. In a recent study of nearly 10,000 women in South Africa, among women who seroconverted (were infected with HIV) during pregnancy, 11 % passed the disease on to their children, compared with 2 % of women who were seropositive prior to becoming pregnant (Dinh et al. 2015).[3] Achieving this dramatic decrease in the rate of seroconversion matters. For as path-breaking as antiretroviral interventions for the prevention of mother-to-child transmission have been, we do not want to rely solely on this to protect foetal health: not all pregnant women who become infected will have access to the regimes, and none are 100 % effective (Chi et al. 2012). The preferred way of keeping foetuses free from the virus remains keeping women from becoming infected in the first place. For many reasons, then, pregnancy represents a critical period in which women need access, not only to effective treatment, but to the best available preventive measures.

A crucial weapon in that preventive arsenal is microbicides, which are emerging as a potentially groundbreaking approach to prevention for women. Microbicides are gels or other topicals that can be applied inside the vagina to protect against sexually transmitted infections. In the last few years, microbicides have shown considerable promise in preventing sexual transmission of HIV, with one large study demonstrating a 39 % reduction in HIV infection risk among women using an investigational gel (Abdool Karim et al. 2010). Part of the exciting value of microbicides is that they bypass the need to negotiate with partners, since their use does not require the cooperation, consent, or even knowledge of a partner for use. All told, mathematical models suggest that widespread use of effective microbicides *by pregnant women* could reduce population-level incidence of HIV by 40 % in women, 15 % in men, and reduce perinatally acquired HIV by 8–15 % over a 10-year period (Dimitrov et al. 2013).

Despite the promise of microbicides in preventing sexual transmission of HIV, the vast majority of microbicide studies have excluded pregnant women (HIV Prevention Trials Network 2006; Microbicide Trials Network 2008; Abdool Karim et al. 2010; Baeten et al. 2012). Research participants who become pregnant have been asked to discontinue microbicide use due to safety concerns. This strategy has been adopted despite the fact that the active ingredient – Tenofovir – is the same drug that, in oral form (associated with much higher serum concentrations, see Krakower and Mayer 2011), has been studied widely in pregnant women to prevent mother-to-child transmission, and is recommended in the US Perinatal Guidelines as the treatment of choice for pregnant women co-infected with HIV and hepatitis (Wang et al. 2013). When it comes to microbicide studies, though, there is a tendency to portray pregnancy as a 'complication' or 'adverse event' or 'major meth-

[3] Other studies have estimated even higher transmission rates. For example, in a cohort analysis of HIV-exposed births in New York City between 2002 and 2006, Birkhead et al. found that maternal acquisition of HIV during pregnancy was associated with a 15-fold increase (2010).

odological challenge' to the conduct of studies (Raymond et al. 2007; Sibeko et al. 2011), rather than a priority area for research.

But a priority area it should be. We need to know whether and how well microbicides prevent HIV in the context of pregnancy. Pregnancy brings with it significant effects on the vagina, which potentially alter the gel's absorption. Research is thus required to determine the proper dosing of microbicides across trimesters to ensure dosing can be achieved at therapeutic levels. Further, we need to determine the risks that a particular formulation might entail for the foetus. Without information about whether microbicides cross the placenta, in what concentrations, and to what effect, it is not possible to make granular recommendations on their use during pregnancy, such as whether they should be used in areas of lower infection rates or just in high risk areas or in contexts of serodiscordant couples.

Time and again in our discussions with members of the HIV research community,[4] we have heard researchers describe the devastating consequences of inadequate data for addressing the health needs of women living with or at risk for HIV. For all of the critical advances made in research with pregnant women for preventing mother-to-child transmission, there are deep gaps in the evidence base for pursuing prevention and treatment of HIV for pregnant women themselves.

It is worth pausing to consider why this pattern continues. The 'reticence gap' as we call it – namely, the tendency for research with pregnant women to proceed, where it does, with a focus on foetal outcomes and not the women's own health – is not unique to HIV/AIDS. Consider, for instance, the National Institutes of Health Maternal-Fetal Medicine Units Network, funded since 1986 to address "clinical questions in maternal-fetal medicine and obstetrics." This admirable network has conducted more than 45 randomised controlled trials, cohort studies, and registries with pregnant women.

But the vast majority of previous studies, and all current studies focus primarily or exclusively on foetal, neonatal, or child outcomes of maternal disease or obstetric intervention. Below are some examples of the research questions these studies are designed to address: whether administration of steroids to women late in pregnancy reduces the need for *neonatal respiratory support*; whether different types of treatment of pregnant women for mild gestational diabetes affects the rate of *obesity in their children*; whether administration of CMV (cytomegalovirus) immune globulin to pregnant women reduces the rate of *congenital CMV infection*; and whether treating women for subclinical hypothyroidism improves the *cognitive abilities in children*. (MFMU 2015) Though maternal outcomes are sometimes measured, this is often not the case; protecting the health of the offspring appears to predominantly drive and shape the research protocols.[5] If we are going to advance the health of

[4] In the first year of the PHASES Project we held several group and individual discussions with HIV researchers and clinicians to get a sense of the priorities, barriers, and opportunities they saw around HIV research in pregnant women. Formal summary of these findings is in progress.

[5] We observed this tendency previously in the National Children's Study, the largest US study to investigate the effects of the environment on children's health, from before birth to age 21. Of the 100,000 children to be studied, a cumulative 90% were to be enrolled as foetuses, while their

women during and after pregnancy, more research studies will need to be designed to directly address women's own health needs and not just those of their babies.

13.2 Researching Pregnant Women's Health Needs: Ethically Impossible?

However urgent the public health need, the importance of a research question is not enough to justify its pursuit: if research is not ethical, it cannot proceed. Without assurances about what is ethical and appropriate when conducting research on pregnant women, even those researchers willing in principle to conduct such research can experience profound uneasiness about the permissibility of conducting studies needed to treat the populations they serve.[6] Central to these concerns are questions about what risk to the foetus is ethically acceptable.

There is no doubt that research with pregnant women is ethically complex. The reason is not the traditional depiction – ensconced in advisory language in the US regulations – of pregnant women as a 'vulnerable population'. That term is increasingly acknowledged to be both misleading and offensive as applied to pregnant women (see Ballantyne and Rogers 2016; Johnson 2016). Others also classified as vulnerable, including prisoners, children, and individuals with mental disabilities, are populations that either by capacity or by context are compromised in some respect that affects autonomous decision-making. Pregnancy does not itself limit the ability to reason or to act autonomously.

The reason that research with pregnant women is ethically complex, instead, is because of the risks that it may impose on the foetus. This is an issue on which regulators and researchers alike clearly continue to struggle: when is the imposition of foetal risk in research acceptable?

Everyone should agree that this is a critically important question. Whatever one believes about the current moral status of the foetus – an issue on which people will disagree – for pregnancies that are continued, what happens *in utero* to the foetus can profoundly affect the welfare of the child who may be born – a child who has not, and cannot, consent to participation in the research (see Ashcroft 2016; Kukla 2016).

Nor does concern for this issue denigrate the autonomy of pregnant women. As legal scholar Vanessa Merton aptly noted, "It is hard to find better decision makers [about foetal welfare] than these women" (1996, 233; see also Johnson 2016; Wild and Biller-Andorno 2016). While that point is critical, the research context brings with it specific issues that warrant added layers of regulation. In paediatric research,

---

intended mothers were in the first trimester of pregnancy. In other words, tens of thousands of pregnant women would be studied, but primarily in terms of the effect their bodies and environments had on the health of their children (Lyerly et al. 2009).

[6] Our discussions with HIV investigators – many of whom had considered, proposed, or conducted research with pregnant women – widely evoked these concerns.

for instance, we do not simply defer to parents to decide when it is acceptable for their children to become research participants. Given that children cannot themselves consent or volunteer for research, regulations tightly monitor and restrict what constitutes acceptable research risk.

What risk, then, is ethically acceptable to impose on the foetus in the context of research? Here is what the key paragraph of the US regulations states:

> Pregnant women or fetuses may be involved in research if ... the risk to the fetus is caused solely by interventions or procedures that hold out the prospect of direct benefit for the woman or fetus; or, if there is no such prospect of direct benefit, the risk to the fetus is not greater than minimal and the purpose of the research is the development of important biomedical knowledge which cannot be obtained by any other means. (DHHS 2009, 45 CFR 46.204(b))[7]

Here we see the regulations organised around a familiar, important distinction regarding the acceptable imposition of risk: namely, whether participation in the trial carries the prospect of direct medical benefit. Prospect of direct medical benefit studies – sometimes called 'therapeutic research' – are studies in which the intervention being tested may directly benefit the research participant. There is only the *prospect* of direct medical benefit, however, both because there is not yet confirmation of efficacy (that being one of the points of clinical research); and because, for trials with control arms, a given participant may not receive the agent being studied or an alternative intervention of proven benefit.

In contrast, studies with no prospect of direct medical benefit are those in which the possibility of benefit cannot reasonably be attributed. These studies include many early phase trials in which researchers have intentionally minimised dose as a strategy to answer specific questions about safety; studies marked by too little evidence to reach a threshold where benefit can be reasonably expected, even if benefits happily turn out to accrue; and studies whose focus is to better understand a point of biology rather than to test a potential preventive or therapeutic intervention. With studies that have no prospect of direct medical benefit, enrolment is purely for the (laudable) value of advancing biomedical knowledge to the potential benefit of future populations and patients.

Given historical examples of research abuses (e.g. Presidential Commission for the Study of Bioethical Issues 2011), special scrutiny attends research with no prospect of direct medical benefit. In particular, and critical for the context of pregnancy, for those who cannot consent, the risk that can be imposed in such studies is strongly capped. The idea here is that researchers, or family members, should not be able to authorise the involvement of those who cannot consent to even laudable research if there is no prospect of direct medical benefit and the research carries more than very low or 'minimal' risk to them. Minimal risk is defined elsewhere in the regulations as risks "not greater in and of themselves than those ordinarily encountered in daily

[7] Subpart B – Additional Protections for Pregnant Women, Human Fetuses and Neonates Involved in Research is part of DHHS regulations, and directly regulates DHHS/FWA funded research. That said, it has been highly influential as guidance governing other contexts, including any medications that seek FDA approval.

life or during the performance of routine physical or psychological examinations or tests" (DHHS 2009, 45 CFR 46.102(i)).

Guidance around the imposition of foetal risk in research with pregnant women, then, is organised around a regulatory disjunct. Each disjunct brings key ethical questions. Consider first the regulatory language regarding studies with the prospect of direct medical benefit. According to the regulations, foetal risk can be justified by procedures or interventions where there is the prospect of direct medical benefit for the woman *or* the foetus. In other words, foetal risk can be justified not only by the prospect of foetal benefit, but by the prospect of maternal benefit. This is different from other research contexts, and has raised questions and unease for many. How, ethically, can risk be distributed across – or balanced between – these two parties? Is it really acceptable in research to justify risk to one party for the benefit of another? Isn't this treating the foetus – or the child it would become – as an object or a 'mere means'?

Turn next to the regulatory language regarding studies that carry no prospect of direct medical benefit for either the foetus or the pregnant woman. According to US regulations, such studies must meet the standard of no more than minimal risk to the foetus. But what really qualifies as 'minimal risk'? Although the concept at first blush seems intuitively clear, a well-developed literature reveals that there is nothing intuitive about it. Depending on which risks we think of, the risks of everyday life are either under- or over- inclusive and thus not a helpful guide to determining an acceptable level of research risk (Wendler 2005; Strong 2011a, b). IRBs give the concept of 'minimal risk' radically varying interpretations (Shah et al. 2004; Westra et al. 2011; see also Ells and Lyster 2016). As with paediatric and other populations that cannot offer meaningful consent to research, unresolved issues about what counts as minimal risk continue to be particularly vexing.

Together, these two concerns can have a profoundly chilling effect on research involving pregnant women. Without an understanding of where the lines should be drawn, and what tradeoffs are permissible, researchers and IRBs can end up avoiding studies involving pregnant women altogether. More deeply, once we turn to research aimed at pregnant women's own health outcomes, it becomes impossible to overlook the elephant in the room. Such research seems to ask us precisely to justify risk to the foetus for the prospect of benefit to another. Put together, these two concerns leave some with a nagging sense that research into pregnant women's own health must somehow, however regrettably, be ethically impossible.

We concur that regulations contain areas of vagueness, and that more work needs to be done to develop consensus frameworks to address these questions. However, we do not believe that these gaps mean that research involving pregnant women, either in general or for their own health, need be put on hold. And, the HIV/AIDS context can help us see why. On reflection, it has lessons to offer about studies on both sides of the regulatory disjunct alluded to above. Even before we have full resolution of those important ethical questions, there is much that can be done when we look more closely at how to think through the prospect of foetal benefit from maternal participation, and what might be accomplished by pursuing creative trial designs when no prospect of foetal benefit can be reasonably anticipated.

13.3 Prospect of Direct Medical Benefit: Whose Benefit? Whose Risk?

We begin with studies that carry a prospect of direct medical benefit, and the question of whether maternal benefit from participating in a trial can justify any risk to the foetus.

Some believe that the prospect of maternal benefit from participation in research cannot justify any increment of risk to the foetus. For instance, Carson Strong, citing an analogy to paediatrics (since the foetus, like a child, is not a consenting partner to research), argues that the risk to the foetus of participation in the research cannot be justified by the benefit of participation to the pregnant woman (Strong 2011a). If the research carries no prospect of direct medical benefit for the foetus, he argues, then no matter what the prospect of benefit to the pregnant woman might be, the study must return to the more restrictive standard of imposing no more than minimal risk to the foetus. According to this view, any foetal risk above 'minimal' can be justified only by a sufficient prospect of direct medical benefit to the foetus.

We find this reasoning deeply problematic. There is a critical difference between research in pregnancy and research in paediatrics. In the case of pregnancy, there is no way for the pregnant woman to accrue the potential benefit of research participation without leaving the foetus behind; and pregnancy, we believe, should not entail forfeiting important potential health benefits, including access to trials that may carry health- or life-saving potential.

But the point we want to make here is a different one. Debates over whether foetal risk can be justified by the prospect of benefit to participating pregnant women – and, if so, how much – will not be resolved any time soon. Fortunately, a great deal of research involving pregnant women carries with it the prospect of direct medical benefit to the foetus and can proceed ethically without waiting for that resolution. The reason is as simple as it is critical: when we appreciate the real world effects of serious maternal disease on foetal health, it turns out that studies involving the prospect of direct medical benefit to participating pregnant women often *entail* the prospect of direct medical benefit for the foetus (see Wild and Biller-Andorno 2016).

Consider, for example, the large microbicide trials aimed at investigating how to empower pregnant women to decrease their risk of acquiring HIV from intercourse. As noted above, there is a critical need for research to determine effective and responsible dosing of microbicides for use in pregnancy. The core reason for such research ultimately is to improve health outcomes for pregnant women: to prevent pregnant women from getting a devastating disease during a time of heightened susceptibility to infection in the event of exposure to HIV.

Participation by pregnant women in such trials carries the prospect of direct medical benefit to their foetuses. We remember that women are at increased risk of getting HIV when they are pregnant. Furthermore, when infected during pregnancy, women are also at least five times more likely to transmit the virus to their foetus. Regimes for the prevention of mother-to-child transmission are neither universally available nor foolproof. It is good for pregnant women not to get infected with HIV, and this is also good for their foetuses. This logic carries a critical lesson for the research context. If participating in a trial carries the prospect of reducing the woman's chance of getting infected with HIV, it thereby carries the prospect of reducing the foetus' chance of the same.

Turn now to the need for research into effective modes of treating pregnant women living with HIV. The core point of such research is to ensure that women living with the virus are effectively treated during the variabilities of pregnancy. Optimising such treatment will directly benefit the foetus. Poorly controlled HIV in pregnancy is associated with numerous outcomes of relevance to foetal/paediatric health. The sequelae of gestating in the body of a woman with poorly controlled HIV include higher risks of stillbirth, preterm delivery, small for gestational age infants, and neonatal death (Brocklehurst and French 1998; Chen et al. 2012). That is, the threat to the foetus of living in a woman with HIV is not just the risk of becoming infected with HIV, but of developing in a physiological context that is highly compromised by maternal disease.

This clearly demonstrates that a research protocol that carries the prospect of better control for pregnant women's HIV *is* a protocol that carries the prospect of direct medical benefit to their foetuses. Given the risks that maternal disease itself can entail for the foetus, the prospect of helping the pregnant woman's HIV brings with it the prospect of directly benefitting the foetus and the child the foetus may become.

What these examples point to is a second critical disanalogy between research with pregnant women and paediatric research. Paediatrics involves a physically separated parent and child; pregnancy involves deep physiological intertwinement. Living in and from the woman's body, the foetus can be, and often is, directly medically impacted by maternal disease.

The issue is especially vivid with HIV/AIDS, but is by no means limited to this medical context. Take, for example, depression. Familiar headlines report small increased risks associated with anti-depressant medications in pregnancy (Huybrechts et al. 2015; Spencer 2015). Yet, it is widely known that leaving serious depression untreated is associated with preterm birth, small for gestational age infants, and low birthweight (ACOG 2008). Recent studies indicate that untreated depression in pregnancy may have longer-term health implications for children, including even depression when these children reach adolescence (Pearson et al. 2013; see also Healy and Mangin 2016).

More dramatically, maternal disease can be a teratogen (Wilson 2007). While the term usually evokes drugs like thalidomide, chemicals, or radiation to which a pregnant woman (and thus her foetus) might be exposed, maternal disease itself – whether infectious or metabolic – can have deeply harmful effects on foetal development. For instance, pre-gestational diabetes (diabetes whose onset occurs prior to pregnancy, as in Type I or Type II diabetes mellitus) has been identified as one of the 'most challenging' medical conditions to manage during pregnancy (ACOG 2005). Not only does pregnancy accelerate a range of diabetes-related complications related to women's health (such as retinopathy which can lead to blindness), but poorly controlled diabetes is a serious threat to foetal health. Most dramatically, when blood sugar far exceeds a healthy range in the first trimester, the rate of major congenital abnormalities reaches 20–25 % – a risk level comparable to thalidomide. Elevated blood sugar from uncontrolled maternal diabetes, it has been said, is among the *worst* teratogens (Allen and Armson 2007).

All of this has critical implications for assessing acceptable foetal risk for trials designed to test interventions for maternal health. If participation in a trial offers the prospect of direct medical benefit to pregnant women, very often that intervention entails specific and quantifiable medical benefits to the foetus – namely, by reducing the effects of maternal disease on its health. IRBs still need to determine whether the potential risk to the foetus is acceptable relative to the potential benefit – whether, in essence, participation in the trial would involve net benefit to the foetus. This, though, is the traditional question posed by any research with the potential for direct medical benefit, and needs to be analysed in the usual way.

Even before getting into knotty questions about whether maternal benefit from research participation can justify imposition of risk onto the foetus (and if so, how much), there are real possibilities that pregnant women's participation in research can offer the prospect of net foetal benefit. Thus researchers proposing a trial, and IRBs vetting it, should consult what is known about foetal risks entailed *by the specific maternal disease state* under discussion – even if those risks don't seem to have as obvious a mechanism of action as a virus being transmitted in the vivid case of HIV. The goal is to determine potential foetal benefits, not just risks, associated with women's research participation, and to determine how much research risk is justified by the potential benefits.

To clarify, once again, we are not endorsing the view that potential foetal benefit is the only way in which risk to the foetus can be justified in research. If the potential benefit to the pregnant woman is strong enough, and the stakes are high enough, we believe that potential maternal benefit, even in the absence of any prospect of potential direct foetal benefit, can justify some measure of risk to the foetus. Our point here is that quite apart from this claim, research designed around the prospect of maternal benefit can often ethically proceed based on the entailed potential benefit to the foetus. On even the most conservative interpretation of Subpart B's guidance on trials with the prospect of direct medical benefit, there is wide berth for research designed around maternal health in which foetal risk is justified by the prospect of potential foetal benefit.

13.4 No Prospect of Direct Medical Benefit and Minimal Risk

What, then, of those trials that carry no prospect of direct medical benefit to either the pregnant woman or the foetus? Such trials can be critically important, using small exposures to confirm whether an agent crosses the placental barrier, or to provide critical pharmacokinetic data for those dosing questions. According to US regulations, as noted above, these trials can impose no more than minimal risk onto the foetus. But the interpretation of minimal risk is widely regarded to be a vexing one, both in general, and in particular when considering those that cannot meaningfully volunteer for research.

Serious issues here include how we should think about the 'risks of everyday life' for a foetus, taking into consideration both the risks noted above of living in a body with serious disease, and the ever present environmental health risks. As regards this second category of risks, consider, for example, recent studies documenting over 200 industrial chemicals and pollutants in umbilical cord blood including pesticides, consumer product ingredients, and wastes from burning coal, gasoline, and garbage (EWG 2005). When it comes to the perception of risk in pregnancy, we appear to have framing bias in our tendency to think that pharmacologic agents must be above minimal risk – however small and limited the dose, however robust the preclinical data, however well understood the mode of action, or whatever the stage of pregnancy.

Instead of addressing those points here, though, we want to pursue a different lesson. Even without solving those outstanding issues, there is much research that can be done. Vanguard research around women's health needs in HIV/AIDS point the way to creative strategies for gaining critically needed data working within conservative interpretations of minimal risk. We draw attention here to two such strategies. The first strategy is around *when* in pregnancy research can be done in order to meet conservative understandings of minimal risk. The second strategy involves 'opportunistic' research where there is no research risk to the foetus.

For the first strategy, we take as our example a progressive vanguard project that has begun to study microbicides in a small cohort of pregnant women. These researchers began with a trial (MTN-002) involving 16 pregnant women not infected by HIV who were given a single dose of vaginal tenofovir gel a few hours before a scheduled Caesarean; drug concentrations were then measured in the woman's blood, the foetal cord blood, and the amniotic, placental, and endometrial tissues. The researchers found that the microbicide was well-tolerated, had very low exposure to the foetus (about 40 × lower than after an oral dose of the same medication), showed up in extremely low concentrations in maternal serum (about 100 × lower than after an oral dose), and had no adverse effects on the woman or foetus (Beigi et al. 2011).

Clearly this study had no prospect of direct medical benefit. The single dose of vaginal tenofovir in an uninfected woman hours before a planned Caesarean was not about preventing infection in the woman or the foetus (to put it mildly, these hours are not reasonably a window of heterosexual transmission risk). This trial was simply about gathering critical basic information about absorption rates and the like.

Just as clearly, this was a study that meets even conservative standards of minimal risk. The exposure was very small, in both dose and length of time; the active ingredient had been well studied in pregnancy in large oral doses (with far higher foetal exposures), for the prevention of mother-to-child transmission. This was not a first-in-human or even a first-in-foetus experience with a drug.

The subsequent stages of the research project are also worth mentioning. With reassuring results indicating very low foetal exposure, the study then moved to a cohort of women who were very late in pregnancy to minimise any hypothesised risks, such as prematurity from gel use. MTN-008 involved use of a microbicide gel daily for 1 week among 45 women at 37–39 weeks gestation (and 16 women who were breastfeeding). One of the theoretical questions about the gel was whether it might have an effect of increasing preterm labour. In moving next to women who were 37–39 weeks gestation, the research could explore that question with women whose infants would be least harmed by any such effect, if it emerged. After yet more reassuring data came in, a cohort at 34–36 weeks gestation was then enrolled (Microbicide Trials Network). (This third study has not yet been published, but early reports indicate that again the drug was well tolerated with low levels of foetal exposure.)

For the second strategy, we turn to a study of an important novel treatment for HIV/AIDS, the antiretroviral agent raltegravir. Raltegravir is one of the newer drugs in a class known as 'integrase inhibitors'. These drugs target a distinct step in the HIV life cycle, and can be taken in combination with other types of antiretroviral medications to reduce the likelihood of the virus developing resistance. Raltegravir, already recommended as initial therapy in non-pregnant individuals, has been identified as potentially important in the context of pregnancy because of its ability to rapidly reduce levels of virus in the blood – which might be particularly important to preventing mother-to-child transmission in women with untreated disease who present for care late in pregnancy. In addition, raltegravir has been identified as a potentially important option for 'treatment experienced' pregnant women who may have developed HIV resistance to other medications.

Shortly after its approval, a group of researchers proposed a study of raltegravir that involved giving a single dose of the drug to pregnant women who were being effectively treated on a different antiretroviral regimen at three different points in time (second trimester, third trimester, and post-partum). Following administration of this single dose, the researchers would intensively measure the pharmacokinetics of the drug through multiple blood draws. The objective was to obtain pharmacokinetic data without extensive or prolonged foetal exposure to the drug, and without the pregnant woman needing to change a drug regimen that had been effective for her. An anticipated benefit of this research was that as the drug became more widely used in pregnant populations, dosing recommendations could be put in place. Like MTN-002, this trial held no prospect of direct medical benefit. Yet differing interpretations of minimal risk – specifically, whether a single dose of an antiretroviral medication that had not been used widely in pregnancy constituted more than minimal risk – led to a series of regulatory delays. Ultimately, the researchers shelved the proposal, despite what many considered its elegance and ethics.

Three years later the researchers took a different approach – studying raltegravir 'opportunistically'. By then the drug had made its way into clinical settings and pregnancy exposures – either empiric or unintentional – were widespread. Capitalising on this 'natural history' of antiretroviral uptake in pregnant populations, the IMPAACT[8] Network Protocol 1026a enrolled 42 pregnant women who were already receiving standard doses of raltegravir as a part of their clinical care. The study conducted intensive pharmacokinetic sampling at three time periods: between 20 and 26 weeks and between 30 and 36 weeks gestation, as well as between 6 and 12 weeks post-partum. Concerns about minimal risk were easily met, since any risk to the foetus was already assumed in the clinical setting, as part of a conscientious provider's best therapeutic decision for the care of the woman and foetus/child. Study risks were limited to the blood draws themselves (something that does not impact the foetus at all). The *research* risk to the foetus was nil. The researchers found decreased exposure at standard doses but high rates of viral suppression, suggesting that standard doses were appropriate (Watts et al. 2014). Similar studies have been conducted since (Blonk et al. 2015).

Once again, to clarify, we do not believe that opportunistic studies are the only ethical pathway for Phase I-style questions in pregnancy. Indeed, in our view, the interpretations of minimal risk that led to the regulatory delays in this case were overly restrictive. The initial proposed study was highly ethical, and proceeding with this research when first proposed would have allayed deep concerns about dosing. Our point here, as throughout the chapter, is that research with pregnant women need not be held hostage to difficult debates about minimal risk. In this case, a creative trial design allowed researchers to move forward even as the research community continues to debate interpretations of minimal risk.

13.5 Conclusion

At the end of the day, addressing the research community's residual ethical unease with research in pregnant women will require the development of a more complete and unified ethical framework, including when and how the prospect of maternal benefit can justify some increment of risk to the foetus, and better consensus around the interpretation of minimal risk. In this chapter, we have not attempted to craft such a framework. Instead, our point has been to emphasise that there is much we can do now, consistent with consensus interpretations of guidelines, to get deeply needed knowledge. The thought that research on pregnant women's own health outcomes must await adjudication of all of the ethical questions the regulations currently leave open is deeply misguided. There is much we can clearly do, consistent with both best ethical and scientific practices.

[8] International Maternal Pediatric Adolescent AIDS Clinical Trials.

References

Abdool Karim, Q., S.S. Abdool Karim, J.A. Frohlich, A.C. Grobler, C. Baxter, L.E. Mansoor, et al. 2010. Effectiveness and safety of tenofovir gel, an antiretroviral microbicide, for the prevention of HIV infection in women. *Science* 329(5996): 1168–1174.

ACOG (American College of Obstetricians and Gynecologists). 2005. ACOG practice bulletin no. 60: Pregestational diabetes mellitus. *Obstetrics and Gynecology* 105(3): 675–685.

ACOG (American College of Obstetricians and Gynecologists). 2008. ACOG practice bulletin no. 92: Use of psychiatric medications during pregnancy and lactation. *Obstetrics and Gynecology* 111(4): 1001–1020.

AIDSInfo. 2014. Recommendations for use of antiretroviral drugs in pregnant HIV-1 infected women for maternal health and interventions to reduce perinatal HIV transmission in the United States. https://aidsinfo.nih.gov/contentfiles/lvguidelines/perinatalgl.pdf. Accessed 8 Apr 2016.

Allen, V.M., and B.A. Armson for the SOGC Genetics Committee. 2007. SOGC clinical practice guideline – Teratology associated with pre-existing and gestational diabetes. *Journal of Obstetrics and Gynaecology Canada* 29(11): 927–944.

Andrew, M.A., M.R. Easterling, D.B. Carr, D. Shen, D.L. Buchanan, T. Rutherford, et al. 2007. Amoxicillin pharmacokinetics in pregnant women: Modeling and simulations of dosage strategies. *Clinical Pharmacology and Therapeutics* 81(4): 547–556.

Ashcroft, R. 2016. Ethical issues in a trial of maternal gene transfer to improve foetal growth. In *Clinical research involving pregnant women*, eds. F. Baylis and A. Ballantyne, 247–263. Cham: Springer.

AVERT. 2014. Funding for HIV and AIDS. http://www.avert.org/funding-hiv-and-aids.htm. Accessed 8 Apr 2016.

Baeten, J.M., D. Donnell, P. Ndase, N.R. Mugo, J.D. Campbell, J. Wangisi et al. 2012. Antiretroviral prophylaxis for HIV prevention in heterosexual men and women. *The New England Journal of Medicine* 367(5): 399–410.

Ballantyne, A., and W. Rogers. 2016. Pregnancy, vulnerability, and the risk of exploitation in clinical research. In *Clinical research involving pregnant women*, eds. F. Baylis and A. Ballantyne, 139–159. Cham: Springer.

Beigi, R., L. Noguchi, T. Parsons, I. Macio, R.P. Kunjara, N. Ayudhya, et al. 2011. Pharmacokinetics and placental transfer of single-dose tenofovir 1 % vaginal gel in term pregnancy. *The Journal of Infectious Diseases* 204(10): 1527–1531.

Birkhead, G.S., W.P. Pulver, B.L. Warren, S. Hackel, D. Rodríguez, and L. Smith 2010. Acquiring human immunodeficiency virus during pregnancy and mother-to-child transmission in New York. *Obstetrics and Gynecology* 115(6): 1247–1255.

Blonk, M., A. Colbers, C. Hidalgo-Tenorio, K. Kabeya, K. Weizsäcker, A. Haberl, et al. 2015. Raltegravir in HIV-1 infected pregnant women: Pharmacokinetics, safety and efficacy. *Clinical Infectious Diseases* 61(5): 809–816.

Brocklehurst, P., and R. French. 1998. The association between maternal HIV infection and perinatal outcome: A systematic review of the literature and meta-analysis. *British Journal of Obstetrics and Gynaecology* 105(8): 836–848.

CDC (Centers for Disease Control). 2014. HIV in pregnant women, infants and children. http://www.cdc.gov/hiv/risk/gender/pregnantwomen/facts/index.html. Accessed 8 Apr 2016.

Chen, J.Y., H.J. Ribaudo, S. Souda, N. Parekh, A. Ogwu, S. Lockman, et al. 2012. Highly active antiretroviral therapy and adverse birth outcomes among HIV-infected women in *Botswana. The Journal of Infectious Diseases* 206(11): 1695–1705.

Chi, B.H., M.R. Adler, O. Bolu, D. Mbori-Ngacha, D. K. Ekouevi, A. Gieselman, et al. 2012. Progress, challenges, and new opportunities for the prevention of mother-to-child transmission of HIV under the US President's Emergency Plan for AIDS Relief. *JAIDS* 15(60, Suppl 3): S78–S87.

Churchyard, G.J., F. Scano, A.D. Grant, and R.E. Chaisson. 2007. Tuberculosis preventive therapy in the era of HIV infection: Overview and research priorities. *The Journal of Infectious Diseases* 196(Suppl. 1): S52–S62.

DHHS (US Department of Health and Human Services). 2009. *Code of Federal Regulations: Title 45, Part 46, Protection of Human Subjects*. http://www.hhs.gov/ohrp/regulations-and-policy/regulations/45-cfr-46/index.html. Accessed 3 Oct 2016.

Dimitrov, D., M. Boily, J. Marrazzo, R. Beigi, and E.R. Brown. 2013. Population-level benefits from providing effective HIV prevention means to pregnant women in high prevalence settings. *PLoS ONE* 8(9): e73770.

Dinh, T., K.P. Delaney, A. Goga, D. Jackson, C. Lombard, S. Woldesenbet, et al. 2015. Impact of maternal HIV seroconversion during pregnancy on early mother to child transmission of HIV (MTCT) measured at 4–8 weeks postpartum in South Africa 2011–2012: A national population-based evaluation. *PLoS ONE* 10(5): e0125525.

Drake, A.L., A. Wagner, B. Richarson, and G. John-Stewart. 2014. Incident HIV during pregnancy and postpartum and risk of mother-to-child HIV transmission: A systematic review and meta-analysis. *PLoS Medicine* 11(2): e1001608.

Ekouevi, D.K., P.A. Coffie, E. Ouattara, R. Moh, C. Amani-Bosse, E. Mossou, et al. 2011. Pregnancy outcomes in women exposed to efavirenz and nevirapine: An appraisal of the IeDEA West Africa and ANRS databases, Abidjan, Côte d'Ivoire. *JAIDS* 56(2): 183–187.

Ells, C., and C. Lyster. 2016. Research ethics review of drug trials targeting medical conditions of pregnant women. In *Clinical research involving pregnant women*, eds. F. Baylis and A. Ballantyne, 95–118. Cham: Springer.

EWG (Environmental Working Group). 2005. Body burden: The pollution in newborns. A benchmark investigation of industrial chemicals, pollutants, and pesticides in umbilical cord blood. http://www.ewg.org/research/body-burden-pollution-newborns. Accessed 8 Apr 2016.

Ford, N., A. Calmy, and L. Mofenson. 2011. Safety of efavirenz in the first trimester of pregnancy: An updated systematic review and meta-analysis. *AIDS* 25(18): 2301–2304.

Gray, G.E. 2008. Walking the tightrope in prevention of mother-to-child transmission of HIV infection. *Clinical Infectious Diseases* 46(4): 622–624.

Gray, R.H., X. Li, G. Kigozi, D. Serwadda, H. Brahmbhatt, F. Wabwire-Mangen, et al. 2005. Increased risk of incident HIV during pregnancy in rakai, Uganda: A prospective study. *The Lancet* 366(9492): 1182–1188.

Gupta, A., U. Nayak, M. Ram, R. Bhosale, S. Patil, A. Basavraj, et al. 2007. Postpartum tuberculosis incidence and mortality among HIV-infected women and their infants in Pune, India, 2002–2005. *Clinical Infectious Diseases* 45(2): 241–249.

Gupta, A., J.S. Matthad, S.M. Abdel-Rahman, J.D. Albano, R. Botgros, V. Brown, et al. 2016. Toward earlier inclusion of pregnant and postpartum women in tuberculosis drug trials: Consensus statements from an international expert panel. TB drug trials: Consensus statements from an expert panel. *Clinical Infectious Diseases* 62(6): 761–769.

Healy, D., and D. Mangin. 2016. Does my bias look big in this? In *Clinical research involving pregnant women*, eds. F. Baylis and A. Ballantyne, 197–208. Cham: Springer.

HIV Prevention Trials Network. 2006. HPTN 035: Phase II/IIb safety and effectiveness study of the vaginal microbicides buffergel and 0.5% pro 2000/5 gel (p) for the prevention of HIV infection in women. http://www.hptn.org/research_studies/hptn035.asp. Accessed 8 Apr 2016.

Huybrechts, K.F., B.T. Bateman, K. Palmsten, R.J. Desai, E. Patorno, C. Gopalakrishnan, et al. 2015. Antidepressant use late in pregnancy and risk of persistent pulmonary hypertension of the newborn. *JAMA* 313(21): 2142–2151.

Johnson, L.S.M. 2016. When hypothetical vulnerability becomes actual: Research participation and the autonomy of pregnant women. In *Clinical research involving pregnant women*, eds. F. Baylis and A. Ballantyne, 161–178. Cham: Springer.

Kelland, K. 2010. UNICEF says HIV-free generation is achievable. *Reuters Health* (London), 30 November. http://www.reuters.com/article/2010/11/30/us-aids-unicef-idUSTRE6AT3IC20101130. Accessed 8 Apr 2016.

Krakower, D., and K.H. Mayer. 2011. Promising prevention approaches: Tenofovir gel and prophylactic use of antiretroviral edications. *Current HIV/AIDS Reports* 8(4): 241–248.

Kukla, R. 2016. Equipoise, uncertainty, and inductive risk in research involving pregnant women. In *Clinical research involving pregnant women*, eds. F. Baylis and A. Ballantyne, 179–196. Cham: Springer.

Lyerly, A., M. Little, and R. Faden. 2009. The National Children's Study: A golden opportunity to advance the health of pregnant women? *The American Journal of Public Health* 99(10): 1742–1745.

Maternal Fetal Medicine Units Network. 2015. *Research projects*. https://mfmu.bsc.gwu.edu/research-projects. Accessed 8 Apr 2016.

Merton, V. 1996. Ethical obstacles to women in biomedical research. In *Feminism and bioethics: Beyond reproduction*, ed. S. Wolf, 216–251. Oxford/New York: Oxford University Press.

Microbicide Trials Network. 2008. MTN-003: Phase 2B safety and effectiveness study of tenofovir 1 % gel, tenofovir disoproxil fumarate tablet and emtricitabine/tenofovir disoproxil fumarate tablet for the prevention of HIV infection in women. http://www.mtnstopshiv.org/sites/default/files/attachments/MTN-003_FINAL_Version_2.0_31DEC2010.pdf. Accessed 8 Apr 2016.

Mirochnick, M., and E. Capparelli. 2004. Pharmacokinetics of antiretrovirals in pregnant women. *Clinical Pharmacokinetics* 43(15): 1071–1087.

Mofenson, L., and B.E. Laughon. 2007. Human immunodeficiency virus, mycobacterium tuberculosis, and pregnancy: A deadly combination. *Clinical Infectious Diseases* 45(2): 250–253.

Moodley, D., T. Esterhuizen, T. Pather, V. Chetty, and L. Ngaleka. 2009. High HIV incidence during pregnancy: Compelling reason for repeat HIV testing. *AIDS* 23(10): 1255–1259.

Mugo, N.R., R. Heffron, D. Donnel, A. Wald, E.O. Were, H. Rees, et al. 2011. Increased risk of HIV-1 transmission in pregnancy: A prospective study among African HIV-1 serodiscordant couples. *AIDS* 25(15): 1887–1895.

Patterson, K.B., J.B. Dumond, H.A. Prince, A.J. Jenkins, K.K. Scarsi, R. Want, et al. 2013. Protein binding of lopinavir and ritonavir during 4 phases of pregnancy: Implications for treatment guidelines. *JAIDS* 63(1): 51–58.

Pearson, R.M., J. Evans, D. Kounali, G. Lewis, J. Heron, P.G. Ramchandani, et al. 2013. Maternal depression during pregnancy and the postnatal period: Risks and possible mechanisms for offspring depression at age 18 years. *JAMA Psychiatry* 70(12): 1312–1319.

PhRMA – America's Biopharmaceutical Research Companies. 2014. 2014 report – Medicines in development: HIV/AIDS. http://www.phrma.org/sites/default/files/pdf/2014-meds-in-dev-hiv--aids.pdf. Accessed 8 Apr 2016.

Presidential Commission for the Study of Bioethical Issues. 2011. "Ethically impossible": STD research in Guatamala from 1946 to 1948. http://bioethics.gov/sites/default/files/Ethically-Impossible_PCSBI.pdf. Accessed 8 Apr 2016.

Raymond, E.G., D. Taylor, W. Cates Jr., E.E. Tolley, D. Borasky, A. Cancel, et al. 2007. Pregnancy in effectiveness trials of HIV prevention agents. *Sexually Transmitted Diseases* 34(12): 1035–1039.

Shah, S., A. Whittle, B. Wilfond, G. Gensler, and D. Wendler. 2004. How do institutional review boards apply the federal risk and benefit standards for pediatric research? *JAMA* 291(4): 476–482.

Shearer, K., A.T. Brennan, M. Maskew, L. Long, R. Berhanu, I. Sanne, and M.P. Fox. 2014. The relation between efavirenz versus nevirapine and virologic failure in Johannesburg, South Africa. *Journal of the International AIDS Society* 17(1): 19065.

Sibeko, S., C. Baxter, N. Yende, Q. Abdool Karim, and S.S. Abdool Karim. 2011. Contraceptive choices, pregnancy rates, and outcomes in a microbicide trial. *Obstetrics and Gynecology* 118(4): 895–904.

Simon, A., J. Warszawski, D. Kariyawasam, J. Le Chenadec, V. Benhammou, P. Czernichow, et al. 2011. Association of prenatal and postnatal exposure to lopinavir-ritonavir and adrenal dysfunction among uninfected infants of HIV-infected mothers. *JAMA* 306(1): 70–78.

Spencer, B. 2015. Taking antidepressants during pregnancy increases risk of baby developing a rare circulation problem which causes permanent disabilities. *Daily Mail*, 2 June. http://www.dailymail.co.uk/health/article-3107709/Taking-antidepressants-pregnancy-increases-risk--baby-developing-rare-circulation-problem-causes-permanent-disabilities.html. Accessed 8 Apr 2016.

Strong, C. 2011a. How should risks and benefits be balanced in research involving pregnant women and fetuses? *IRB* 33(6): 1–5.

Strong, C. 2011b. Minimal risk in research involving pregnant women and fetuses. *The Journal of Law, Medicine & Ethics* 39(3): 529–538.

UNAIDS. 2014. 2014 progress report on the global plan. http://www.unaids.org/sites/default/files/documents/JC2681_2014-Global-Plan-progress_en.pdf. Accessed 8 Apr 2016.

Wang, L., A.P. Kourtis, S. Ellington, J. Legardy-Williams, and M. Bulterys. 2013. Safety of tenofovir during pregnancy for the mother and fetus: A systematic review. *Clinical Infectious Diseases* 57(12): 1773–1781.

Watts, H.D., A. Stek, B.M. Best, J. Wang, E.V. Capparelli, T.R. Cressey, et al. 2014. Raltegravir pharmacokinetics during pregnancy. *JAIDS* 67(4): 375–381.

Wendler, D. 2005. Protecting subjects who cannot give consent: Toward a better standard for "minimal" risks. *Hastings Center Report* 35(5): 37–43.

Westra, A.E., J.M. Wit, R.M. Sukhai, and I.D. de Beaufort. 2011. How best to define the concept of minimal risk. *The Journal of Pediatrics* 159(3): 496–500.

WHO (World Health Organization). 2014. Global guidance on criteria and processes for validation: Elimination of maternal to child transmission (EMTCT) of HIV and syphilis. http://apps.who.int/iris/bitstream/10665/112858/1/9789241505888_eng.pdf?ua=1&ua=1. Accessed 8 Apr 2016.

WHO (World Health Organization). 2015. Mother to child transmission of HIV. http://www.who.int/hiv/topics/mtct/en/. Accessed 8 Apr 2016.

Wild, V., and N. Biller-Andorno. 2016. Pregnant women's views about participation in clinical research. In *Clinical research involving pregnant women*, eds. F. Baylis and A. Ballantyne, 119–136. Cham: Springer.

Wilson, R.D., and The SOGC Genetics Committee. 2007. Principles of human teratology: Drug, chemical and infectious exposure. SOGC clinical practice guideline. *Journal of Obstetrics and Gynaecology Canada* 29(11): 911–917.

Yeh, R.F., N.L. Rezk, A.D. Kashuba, J.B. Dumond, H.L. Tappouni, H.C. Tien, et al. 2009. Genital tract, cord blood, and amniotic fluid exposures of seven antiretroviral drugs during and after pregnancy in human immunodeficiency virus type 1-infected women. *Antimicrobial Agents and Chemotherapy* 53(6): 2367–2374.

Chapter 14
Ethical Issues in a Trial of Maternal Gene Transfer to Improve Foetal Growth

Richard Ashcroft

Abstract *This chapter investigates the ethical issues associated with a proposed European gene transfer trial which would recruit pregnant women with severe early onset foetal growth restriction (Details of the research programme may be found on the programme website,* http://everrest-fp7.eu. *The clinical trial is building on ongoing preclinical studies, and is likely to begin recruitment towards the end of 2016. The trial has been registered with the US ClinicalTrials.gov (NCT02097667) and the UK Clinical Research Network (ID 15717)). The gene transfer is targeted on the pregnant woman, rather than her foetus, and seeks to improve blood flow and promote vascularisation of the uterus. Transfer of the vector across the placenta to the foetal blood supply should not occur in clinically significant quantities, and no modification of the foetal genome is planned or foreseen. In this chapter, I consider first whether such an intervention is in theory ethically acceptable. I then consider whether a trial of such an intervention meets regulatory guidelines. The core ethical issue here is how to weigh maternal and foetal interests. There is scant regulatory or legal guidance on this point. Here I assess the one- and two-patient models of foetal interests that dominate the ethics literature and find that neither model adequately captures the uniqueness of the pregnancy. The interests of the pregnant woman and her foetus are, in the vast majority of cases, co-dependent and intertwined. In this proposed trial (as in many other trials in conditions affecting foetal growth and development), the primary risk is potential exploitation of the pregnant woman's vulnerability, derived from her desire to undertake significant risks in order to gain potential benefits for the foetus.*

Foetal Growth Restriction (FGR) is a condition in which the foetus' growth slows or stops. It is one of the most common and complex problems in modern obstetrics (ACOG 2013). While some babies are naturally small, FGR occurs when a decrease in the foetal growth rate prevents an infant from obtaining its complete genetic growth potential (Alberry and Soothill 2007). FGR affects about 8 out of 100 pregnancies and is associated with an increased risk of perinatal mortality and morbidity, in addition to long-term health risks for foetuses that survive. There is currently

R. Ashcroft, MA, PhD, FHEA, FRSB (✉)
Queen Mary, University of London, School of Law, London, UK
e-mail: r.ashcroft@qmul.ac.uk

F. Baylis, A. Ballantyne (eds.), *Clinical Research Involving Pregnant Women*,
Research Ethics Forum 3, DOI 10.1007/978-3-319-26512-4_14

Table 14.1 EVERREST trial – research summary[a]

The EVERREST project is aiming to develop a treatment for foetal growth restriction. The most common cause is a lack of sufficient blood flowing to the womb via the mother's circulation. This results in a lack of nutrients and oxygen to the developing foetus. Our previous research has shown that increasing blood flow to the womb using localized maternal VEGF gene therapy can improve foetal growth. The aim of the project is to carry out the first trial of this therapy in pregnant women whose babies are most severely affected by foetal growth restriction to test out its safety and efficacy. Potential benefits from the research could include reduced stillbirths and neonatal deaths, and improved neonatal and long-term outcomes in pregnancies affected by severe early onset foetal growth restriction.

[a]Quoted without amendment from the EVERREST public website http://everrest-fp7.eu/ (accessed 8 June 2015)

no way to predict which foetuses will be affected, and no effective treatment. This chapter analyses the ethical issues raised by the proposed European gene transfer trial investigating a novel intervention for women with severe early onset FGR (see Table 14.1). The gene transfer seeks to improve blood flow and promote vascularisation of the uterus thereby increasing nutrition for the foetus. I focus here on two ethical questions: whether the gene transfer intervention is ethically acceptable and whether the proposed trial meets regulatory standards. Answering these questions requires consideration of the unique nature of pregnancy and to co-dependent and inter-related nature of foetal and maternal interests.

Although moderately effective preventive therapy is available for FRG in cases where a woman has had an affected pregnancy in the past, most affected pregnancies are only detected at the routine scan normally scheduled for week 18–20 of gestation. Where a foetus is detected as affected, there is currently no effective treatment at this stage or up to delivery. The EVERREST trial intervention is the administration of gene transfer to a pregnant woman, where at, or around, 20 weeks gestation her foetus is found to have seriously restricted growth, and where other causes (for example genetic defects in the foetus) of the growth restriction have been ruled out. The underlying pathology in these cases is placental insufficiency, where the blood supply to the developing pregnant uterus is insufficient, leading to poor nutrient and oxygen supply to the placenta. In these cases the intention is that by increasing uterine blood supply (by vasodilation and new vascularisation of the uterus) foetal growth may be boosted. By promoting foetal growth and improving the prospects that pregnancy will run to normal term, it is expected that the foetus will survive to be born alive (rather than lost to very late miscarriage or stillbirth), and will be born with a birth weight in or near the normal range, with greatly increased survival prospects and reduced likelihood (and severity) of any disability related to foetal growth restriction. The intervention is administered to the woman on the maternal side of the uterus/placenta junction and affects maternal vascularisation. The intervention should not cross into the foetal blood supply. As such, it is properly a maternal, not a foetal, intervention, and its direct effects are on the woman rather than the foetus.

The language used to describe the woman (woman, mother, prospective mother, and so on) is open to debate. In general, women presenting at this stage of pregnancy consider themselves to have established pregnancies that they expect to lead to birth. To this extent, they are relatively comfortable considering themselves mothers or mothers-to-be. On the other hand, one available option at this stage of pregnancy is a termination of pregnancy. Women choosing or considering this option may not consider themselves to be mothers or mothers-to-be – but equally, they may. Framing and forming a narrative that makes sense to them is a central part of both personal self-understanding and clinical care.

The key ethical, legal, and social issues arising in a trial of gene transfer to ameliorate foetal growth restriction in pregnant women are as follows:

(a) The ethical status of offering a potentially risky medical intervention to a pregnant woman that confers no benefit to maternal health but that may improve the health and survival prospects of her foetus or foetuses.
(b) The ethical status of offering an intervention that may result in the birth of babies who otherwise would have died in utero, who suffer from moderate to severe disability.
(c) The ethical status of offering participation in a trial in which the intervention is offered on an experimental basis to a pregnant woman, where there is no proven non-experimental treatment, and doing so in light of the above issues.

I consider (a) and (b) in the first part of this chapter, and then turn to point (c) in the second part.

14.1 Is the EVERREST Intervention Ethically Acceptable? The Ethical Status of Offering Medical Intervention to Pregnant Women for the Benefit of the Foetus

Most of the literature on maternal-foetal interests concerns treatment during pregnancy rather than research. This is in large part because pregnant women are routinely excluded from research so there has not been a pressing need for clinicians, researchers, or ethicists to consider these questions in the context of research. It should be noted, however, that 'treatment' during pregnancy is less evidence-based than medical treatment for most populations, owing to the lack of research on the basis of which treatment options are offered. So the degree of risk and uncertainty applying to decisions regarding treatment during pregnancy are in many ways similar to the degree of uncertainty regarding risks and potential benefits of research. I begin here with a review of the literature on maternal-foetal interests and potential trade-offs during clinical treatment.

A basic assumption in obstetrics and midwifery is that women who are pregnant desire healthy babies. Apart from the situation where the woman who is pregnant is seeking a termination of pregnancy, most women will take at least some steps to ensure that the pregnancy goes to term, that the foetus develops well in the womb, and that the foetus has good prospects for a healthy life during and after birth. Some of the steps pregnant women will take will be focussed on their own health during pregnancy; some will be focussed on their own health and the health of the developing foetus; and some will be focussed on the health of the developing foetus, independently of (or even, possibly, adverse to) their own health or interests. An example of a case in which a woman may willingly risk her health for the sake of her future child is where the pregnant woman may be diagnosed with cancer during pregnancy, and chooses to delay chemotherapy until after the baby is born. However, it is central to medical ethics, and in most jurisdictions to medical law as well, that the primary patient in pregnancy is the woman herself, not the foetus.

Despite widespread consensus that the woman is the primary patient, there is some debate in the literature as to whether there is one patient, two patients or a unique 'double-unit' that does not neatly fit into dominant conceptions of single-unit autonomous personhood. I do not address the double-unit model in this chapter in any detail as I have found it more helpful to use the one- and two-patient models 'diagnostically' to tease out the ethical issues. However, it will be clear that I favour a version of the one-patient model that uses an account of relational autonomy to reflect the importance to the woman of her pregnancy and status as a potential mother (MacKenzie and Stoljar 1999; Wild 2012). In this chapter, I do not use the term 'foetus as a patient', although this is sometimes used in the literature, because this terminology risks prejudging the issue of whether there are two patients or just one. The point is the choice of a conceptual model in the clinical relationship: should we act *as if* there are two patients or just one, and *do* we regulate *as if* there are two patients or just one (Dickenson 2002; Hervey and McHale 2004; Mahowald 2006; Mason 2007; Chervenak and McCullough 2011[1]).

One Autonomous Person, One Patient

The simplest analysis considers that there is only one patient: the woman herself. Until the foetus is born alive, it forms part of the woman's body, and only becomes a separate moral entity – a person – once it is born alive. This is, approximately, the position adopted by European law. There are some difficulties that arise in holding this view consistently: for example, what does this model imply for the medical care of a woman who is both pregnant but irreversibly comatose? In the more usual case that the woman has capacity, does the lack of moral status of the foetus mean that the woman can do anything she likes which may impact directly on the development of the foetus? These are difficult moral questions, quite apart from the issues that this model implies for the proper legal regulation of pregnancy.

One way to frame answers to these questions is to assert that the foetus does matter, morally speaking, but that only the woman has the full moral status that goes with moral agency – the ability to make decisions, the capacity to value actions and

[1] The peer commentaries on this article (pp.50-58 of the same issue) and the author's response (pp. W3-W7) are also relevant.

states of affairs, and the right to respect for one's autonomy. In light of respect for autonomy, we can say that while we cannot oblige or compel a woman to undergo medical treatment or research for the sake of her foetus alone, any more than we can oblige or compel a woman to undergo medical treatment or research for her own sake, we can also respect a woman's choice to act (or refrain from acting) in the interests of her developing foetus. Thus, should she choose to risk her own health, or undergo treatment or research that will benefit her foetus but not (necessarily) herself directly, this is a choice that should be respected and may be acted upon. More importantly perhaps, as a community we have positive duties to support safe and healthy pregnancies, for the sake of women and their foetuses.

Another puzzle in the one-patient model is whether the woman, though free to choose, and protected in her choices, is nevertheless under an obligation to her foetus to do her best to ensure that it is born healthy and with a good chance of life, so much as this lies within her power.[2] Some authors do construct an argument along these lines. In its weakest form, they may argue that a woman has a duty to refrain from specific behaviours that do – or may – impose serious harm on the foetus (or the future child), such as heavy drinking, smoking, or regular use of recreational drugs known to affect foetal development (Coutts 1990).[3] Some authors go a little further, and argue that there is a duty to engage in some specific behaviours that promote or protect foetal development (for instance, taking folate supplements in the diet). In each case, the authors advancing such arguments are pointing to specific harms that may be avoided by behavioural or dietary modifications on the part of the pregnant woman. Some authors go further than this, and argue that there is a general duty to minimise harm to the developing foetus or future child, or even that there is a general duty to maximise the ability of the future child to flourish over its future lifetime. All of these different arguments face conceptual and ethical difficulties. However, it is important to note that they do not override the priority given to autonomy in the one-patient model, but seek to engage the autonomous woman with proposals as to what, as a responsible person with a concern for the welfare of her foetus, she should consider doing.

The problem in the present context is that none of these welfare-of-the-future-child arguments assist us in determining whether to offer (or undergo) an intervention that might limit the effects of, or ameliorate, foetal growth restriction (Jackson 2002). These arguments simply do not fix for us whether the right thing to do is not to try to have a child at all, or, once pregnant, to have a termination if the foetus fails to thrive in utero, or to have an intervention that might promote foetal growth but potentially still leave the foetus small and prone to serious illness or disability after birth, or simply to let nature take its course and see what happens. The reason is that the comparisons of relative welfare are intractable. We cannot confidently say that it would be better for a particular child not to be born, or conversely that it would be better to have a chance of life than none. For the most part, we can say that it would be better to be less disabled, rather than more, at least in terms of the welfare of the

[2] The classical starting point being now Savulescu (2001).

[3] This helpful summary of the literature covers all the positions described in this paragraph.

future child. But if it cannot be known to any degree of certainty how much benefit the intervention to improve foetal growth would confer, then we cannot say much, if anything, about whether there is a duty to offer such an intervention, or a duty to undergo it as a patient, on the one-patient model. The most we can probably say is this: that it would be good to have the option to use this intervention because it increases the choices available to pregnant women, and thus it would be good to have research to this end – so long as this research is also organised so as to increase the options available to pregnant women.

The Two-Patient Model

Many obstetricians, and some others, consider that in care of the pregnant woman they are looking after two patients, the woman, and the foetus. A weak version of this thesis is that there are two patients, but that if the interests of the two conflict, the interests of the woman take priority. A strong version of this thesis is that no such tie-break applies: both sets of interests are of equal strength and importance. The central thesis of the two-patient model is that we have obligations to the foetus independent of any obligations to the woman, and independent of any obligations (if such there are) of the woman to her foetus.

On the two-patient model, it is clearer that there will be obligations to do whatever can feasibly be done to promote foetal development, with a view to the long-term interests of the potential future child, so long as these actions are consistent with the interests of the pregnant woman. Thus, provided the pregnant woman does not come to harm, the two-patient model authorises treatment that will benefit the foetus. It is this view that (for its proponents) licenses obstetricians to take a more paternalistic view of women's decision-making *in the interests of the foetal patient*. The more difficult case arises where the interests of woman and foetus clearly and explicitly conflict (for example, where a forced caesarean section is being proposed). The idea is that if the foetus is a patient in its own right, then circumstances may arise in which the woman and obstetrician may disagree about how best to protect the foetus, and the obstetrician may simply impose his or her view as the more expert or better informed one. It is exactly this sort of reasoning, quite apart from the physical and social vulnerability of the woman in the doctor-patient relationship, that the one-patient model resists.

A preliminary point must be made. Even on the two-patient model, we have no *legal* power to impose treatment for the benefit of the foetus upon a woman who does not consent to it. The reason is that any such treatment involves an invasion of the woman's body, which, absent her consent, would be an assault. There have been cases where a woman has been offered a Caesarean section in order to save the life of her foetus, and she has refused; and where doctors (sometimes with the agreement of a lower court) have sought to overrule her refusal by having her declared incapacitated or mentally disordered. In the United Kingdom, at least, all such cases have failed at appeal, if not before, and in any case what this line of cases shows is that where the woman is autonomous we have no power to force her to undergo treatment for the sake of her foetus (hence the attempts to find a way round this by having women found to be non-autonomous) (Munby 2010). One thing the two-patient model does clarify, on its own terms, is that where the woman *does* lack autonomy, then treating the foetus could be permissible so long as doing so did not go against the best interests (and previously stated wishes) of the woman.

A second preliminary point should also be made. In the single-patient model, the question of whether to offer treatment for the benefit of the foetus is not grounded objectively in the interests of the foetus, but in the interests of the woman and her own values, beliefs and preferences as to what choices she wishes to make for the sake of her foetus – including whether to continue the pregnancy. In the two-patient model, in theory we must consider the 'best interests of the foetus' as well as the autonomy of the pregnant woman. Notionally, as when applying the best interests of the child test in the law relating to medical care of children, these interests can be assessed independently of what the woman says they are.

In practice, it would be very unlikely that an assessment of these interests would lead to seeking to impose treatment against the wishes of the pregnant woman (though in some US jurisdictions pregnant women have had limitations on their behaviour or even treatment mandated in the interests of the foetus). But it might be that assessment of the interests of the foetus could lead doctors *not* to offer treatment that the pregnant woman might request (Ross 1998). The two-patient model would authorise research ethics committees (Institutional Review Board in the United States) to reject research that pregnant women might want to participate in, on the grounds of foetal interests.

As noted above, many pregnant women are deeply invested in the well-being of their foetus. And many will consider the foetus as their child-to-be, their baby, well before birth. For example, the EVERREST trial concerns pregnancies with a high risk of stillbirth or very low birthweight babies. If the woman has a history of miscarriage or premature birth, and has decided to try to have a baby nonetheless, she may have a relatively strong emotional investment in seeking to have a safe pregnancy and a healthy outcome for herself and her foetus. Even in the more frequent case where foetal growth restriction is detected in a first pregnancy or in a woman who has had previous, unproblematic pregnancies, the detection of a problem will occur at around 20 weeks, and at this stage the woman may have come strongly to feel that her foetus is her unborn baby or child, with a distinct identity, especially as this is the point in pregnancy at which the woman can start to feel the foetus move. To that extent, the woman may well appear to adopt the two-patient model herself. However, the two-patient model requires an objective external judgement about foetal interests, independent of the pregnant woman's views. If the foetus is a patient, on this model, its status as such is independent of what the woman thinks about it. All other views of foetal status accord some significance to the woman's own judgement of the matter, be this on the one-patient model, or some model involving relationality or the emergence of moral status over the course of pregnancy and birth.

Outside the very specific context of abortion law, there is no serious disagreement on the principle that maternal autonomy and physical integrity must be respected. The consent of the pregnant woman to foetus-directed interventions is thus accepted as a necessary condition of research participation or treatment. The difference between the one- and two-patient models comes down in practice to two different answers to the question of whether it is acceptable to offer a potentially risky trial to a pregnant woman where she will derive no direct health benefit herself, and where the intended beneficiary of the intervention is the foetus, even though the foetus will gain that benefit as an indirect consequence of its impact on the body of the woman on the 'maternal side' of the placenta. She may derive the psychological benefit that comes from believing that she has tried to do everything she can to ensure a good outcome for her foetus, but it is debatable whether this personal psychological benefit is the primary health benefit intended in the intervention. Moreover, whether risky physical interventions are justifiable where their primary health benefit is psychological is a vexed open question in bioethics more generally. The one-patient answer is that such intervention is permissible, though not obligatory, if the pregnant woman considers that she would rather proceed with the pregnancy and try to ensure that her baby, when born, is as healthy as possible, and that 'as healthy as possible' amounts to a decent and reasonably flourishing state of health and mental and physical ability. The one-patient answer leaves open the question of whether such an outcome is possible. It also leaves open the question of whether – from the pregnant woman's point of view – no intervention or even termination of pregnancy would be better than the likely outcome of intervention. The two-patient answer depends less on the pregnant woman's personal judgement of future well-being, and takes the benefit of intervention more or less at face value, even where the incremental benefit may be small.

In the first part of this chapter, I have reviewed the central ethical question of whether it is ethically justifiable to offer a pregnant woman an intervention that will impose some burdens on her and not provide her with any direct health benefit, in order to improve the chances that a foetus with severe growth problems will have improved growth. On both the one-patient and two-patient models, the conclusion is that such an intervention is permissible but not obligatory. The alternatives (watchful waiting or termination of pregnancy) remain permissible but not obligatory. Where the burden of the intervention to the physical health of the woman is low (as is predicted for the intervention in EVERREST), two principal ethical difficulties can be seen clearly.

First, it is not a priori obvious how much improvement in foetal outcomes may be expected, either in general or in any particular case. This can be put most sharply if one considers the following scenario: a foetus who would, absent intervention, probably miscarry or die within a short time after birth, 'benefits' from the intervention and grows to a 'survivable' birthweight. However, survival with a very low birthweight carries a high probability of significant physical disability. This scenario suggests that at least in some cases, the outcome might be considered 'heroic' from a medical point of view but not actually beneficial to the future child or the mother. This is a controversial view, but it is central to most ethical discussions within neonatal intensive care (McLaughlin et al. 2008; European Critical Care Forum 2010).

Second, though we have assumed that the physical burden of the intervention to the pregnant woman is low, we cannot assume that the *psychological* burden is low. A pregnant woman who faces the loss of her foetus will likely be in distress. She is now offered a chance of saving her pregnancy, but this chance may come with significant risks that the intervention may fail, or confer only modest benefits, and indeed with a chance that her future child will be seriously disabled. This leaves her with a choice that has a considerable burden in itself. Once the choice is made, if she loses the pregnancy nevertheless she may feel "well, at least we tried" – but she may feel some anger and resentment at having gone through this process to no avail. She might even blame the medical team for the outcome. If, on the other hand, her child is born and thrives, she may be profoundly happy and grateful. The most challenging scenario would be where the child survives but has significant problems, and the woman knows that but for her choice to receive the intervention the child would not have survived. How would she feel about this? It is far from clear. The experiences and feelings of parents of children with disabilities vary considerably. But if one's child's disabilities are felt to be somehow caused by one's own choices, then at least sometimes one might feel guilt and grief. On the other hand, it is unlikely that these are the only feelings one would have: only in the most extreme cases do parents of children with disabilities ever think "It would have been better had my child not survived". It should also be noted that if the pregnant woman chooses not to enter the trial, and a bad outcome then occurs, she might feel guilty for not taking an opportunity that might have led to a different and better outcome. She has no way of knowing whether her choice led to the outcome which then occurred. The point of this reflection is not that parents *should* feel one way or another, but rather that this experimental intervention gives pregnant women (and their partners, to some extent) a new, emotionally complex, and morally challenging choice. Any assessment of the benefits and burdens of this intervention must consider these moral and emotional burdens alongside the more narrowly 'clinical' risks and potential benefits (McLaughlin et al. 2008; Mol 2008).

One important practical conclusion of this review of the ethical issues surrounding the intervention itself is that we need a much better idea of how pregnant women in this situation (and their partners) do weigh up and evaluate the issues, and their experience of both the syndrome of foetal growth restriction and of (potentially) making these very difficult choices.[4]

14.2 Is the EVERREST Trial Ethically Acceptable?

If it could be shown of the intervention that it was unethical or unjustifiable in principle, then it is clear that a trial of such an intervention would also be unethical and unjustifiable. The foregoing argument establishes that the EVERREST intervention

[4] Some preliminary empirical work has been done as part of EVERREST, and is in preparation for publication.

is in principle permissible. It also shows that our judgement of whether in practice such an intervention should (or may) be offered depends in a critical way on empirical information about safety, efficacy, expected outcomes and acceptability to patients and society at large. In the ideal case, if the intervention could be shown to improve the outcomes of an identifiable set of pregnancies such that babies born to pregnant women in this set had improved foetal growth and increased birthweights and gestational ages and with improved expected lifespans and less morbidity/disability, then many of the moral concerns touched on in the first part of this chapter would be assuaged. The crucial point would be to define the criteria on which a pregnant woman would be eligible to receive the intervention: does she fall within this identified set of patients whose outcomes will be good or not? These questions are really only answerable through the medium of clinical trials.

The first part of the chapter was necessarily somewhat speculative and reflective. The regulatory framework of clinical trials is more straightforward, in that there is a framework for evaluating the ethics of a trial. The international ethical standard for clinical trials is the World Medical Association's *Declaration of Helsinki* (Ashcroft 2008; World Medical Association 2013). First issued in 1964, it has been revised many times, and the version currently in force was issued in 2008. The *Declaration* is currently under review. Although the *Declaration* has been subject to some controversy in recent years, it remains the only generally accepted international standard, and all countries within the European Union acknowledge its authority as ethical guidance.[5] So far as legal standards governing clinical research in the European Union are concerned, the central legal instruments are the *Clinical Trial Directive* (2001), *Clinical Trial Regulation* (2014), and the *Advanced Therapy Medicinal Products Regulation* (2007) (which is directly relevant to gene transfer). The regulation of medicines and gene transfer products is a complex and technical subject, and here we are concerned only with the ethical aspects of clinical trials (Kimmelman 2010; Jackson 2012; Flear et al. 2013).

Declaration of Helsinki

The *Declaration* covers all medical research, in patients and in healthy volunteers, in general terms. It relies on three central ethical requirements: the responsibility of the physician (as a matter of general medical ethics) for the well-being and safety of his or her patients; the centrality of informed consent; and the need for research to offer a fair and proportionate balance of risk and potential benefit to all participants. It makes no reference to the inclusion (or exclusion) of women – pregnant women – in particular in research or to gene transfer.

---

[5] The guidelines of the Council of International Organizations of Medical Sciences on ethics in biomedical research are well respected, and were at one time much cited, but have not been revised in over ten years and although a revision process was begun in 2012, it has not yet concluded at the time of writing.

In light of the first part of this chapter, it is notable that the *Declaration* provides no guidance on who the patient is, but the natural reading is that the *Declaration* assumes the one-patient model. There has been extensive discussion in the academic literature about what is meant by 'vulnerable' in the *Declaration*, as no specific groups or individuals are identified as vulnerable (World Medical Association 2013, Articles 19–20). While the *Declaration* might be read restrictively to exclude some women (on the grounds of actual or potential pregnancy), note should also be taken of the following statement in the *Declaration*: "Medical progress is based on research that ultimately must include studies involving human subjects" (World Medical Association 2013, Article 5).

Historically, clinical trials frequently did exclude women, either on 'ethical' grounds or to limit legal liability should an intervention prove to be damaging in pregnancy; this position has now, rightly, fallen into disfavour. The problem the *Declaration* leaves us with, however, is that it provides very little guidance that is specific to trials in pregnancy beyond the three rather general principles reproduced above.[6]

Clinical Trials Directive 2001/20/EC and other European Legislation

In the European Union, the governing legislation for clinical trials is the *Clinical Trials Directive*, which has been implemented by all member states in national legislation (in the UK, in the Medicines for Human Use (Clinical Trials) Regulations 2004). This *Directive* in turn adopts the principles of the *International Committee on Harmonisation Good Clinical Practice* guidelines of 1996, an international agreement between the medicines regulators of the United States of America, the European Union, and Japan, and their respective pharmaceutical industry associations. The *Directive* makes prior ethical review and approval by a research ethics committee a legal requirement for all clinical trials of medicinal products, and also harmonises requirements for consent (and authorisation for incapacitated patients and minors) across the Union. It makes mention of the 1996 *Declaration of Helsinki*, and Articles 3 and 4 of the *Directive* specify requirements relating to consent, decision-making for incapacitated subjects and minors, the role of a clinician as supervisor of the care of participants in the trial, and the need for insurance and indemnity in case of research-related injury. But once again, no specific mention is made of the situation where the research participant is a pregnant woman, regardless of whether the research is related to pregnancy.

[6] See in particular a special issue of *IJFAB: The International Journal of the Feminist Association of Bioethics* on vulnerability in bioethics (Rogers et al. 2012), and on the inclusion of women in clinical research in general, Epstein (2007). On the inclusion of pregnant women in research, see Strong (2011).

Table 14.2 *Clinical Trials Regulation* (EU) 536/2014 art.33

Clinical trials on pregnant or breastfeeding women
A clinical trial on pregnant or breastfeeding women may be conducted only where, in addition to the conditions set out in Article 28, the following conditions are met:
(a) the clinical trial has the potential to produce a direct benefit for the pregnant or breastfeeding woman concerned, or her embryo, foetus or child after birth, outweighing the risks and burdens involved; or
(b) if such clinical trial has no direct benefit for the pregnant or breastfeeding woman concerned, or her embryo, foetus or child after birth, it can be conducted only if:
(i) a clinical trial of comparable effectiveness cannot be carried out on women who are not pregnant or breast-feeding;
(ii) the clinical trial contributes to the attainment of results capable of benefitting pregnant or breastfeeding women or other women in relation to reproduction or other embryos, foetuses or children; and
(iii) the clinical trial poses a minimal risk to, and imposes a minimal burden on, the pregnant or breastfeeding woman concerned, her embryo, foetus or child after birth;
(c) where research is undertaken on breastfeeding women, particular care is taken to avoid any adverse impact on the health of the child; and
(d) no incentives or financial inducements are given to the subject except for compensation for expenses and loss of earnings directly related to the participation in the clinical trial.

The *Advanced Therapy Medicinal Products Regulation* (EC) 1394/2007 introduces a legal requirement of traceability of research participants and patients after the conclusion of trials and treatment (art.15), and also explicitly mentions EU and international *human rights* obligations as an interpretative constraint on the Regulation (recital 8). Otherwise, it is silent on ethical requirements, other than to refer back to the *Clinical Trials Directive*.

The *Clinical Trials Regulation* (EU) 536/2014 (European Parliament and The Council of the European Union 2014) will supersede the *Clinical Trials Directive* (European Parliament and The Council of the European Union 2001) when it enters into force in 2016. Article 33 (see Table 14.2) specifically discusses clinical trials in pregnant women and nursing mothers.

The human rights instruments mentioned are the *Charter on Fundamental Rights of the European Union* (2010) and the Council of Europe's *Convention on Human Rights and Biomedicine* (Council of Europe 1997, 2005). The *Charter of Fundamental Rights* (European Union 2010) is chiefly considered with civil, social and economic rights, but refers in Article 3 (The Right to the Integrity of Persons) section 2 to "the free and informed consent of the person concerned, according to the procedures laid down by law" and "the prohibition of eugenic practices, in particular those aiming at the selection of persons." The *Convention on Human Rights and Biomedicine* (Council of Europe) was adopted in 1997 as a complement to the *Convention on Human Rights* (Council of Europe 1950). It is a legal instrument that may be relied upon in human rights litigation before the European Court of Human Rights, but in practice it has been relied upon only rarely. Article 18 (Council of Europe 1997) governs embryo research, but not research on the foetus in utero.

Table 14.3 *Additional Protocol to the Convention on Human Rights and Biomedicine*, concerning Biomedical Research
Article 18 – Research during pregnancy and breastfeeding

1. Research on a pregnant woman which does not have the potential to produce results of direct benefit to her health, or to that of her embryo, foetus or child after birth, may only be undertaken if the following additional conditions are met:

   (i) the research has the aim of contributing to the ultimate attainment of results capable of conferring benefit to other women in relation to reproduction or to other embryos, foetuses or children;

   (ii) research of comparable effectiveness cannot be carried out on women who are not pregnant;

   (iii) the research entails only minimal risk and minimal burden.

2. Where research is undertaken on a breastfeeding woman, particular care shall be taken to avoid any adverse impact on the health of the child.

In 2005, the Council adopted an *Additional Protocol to the Convention on Human Rights and Biomedicine, Concerning Biomedical Research* (Council of Europe 2005). This protocol is broadly similar to the ethical provisions of the *EU Clinical Trial Directive* (European Parliament and The Council of the European Union 2001) and elaborates on the above cited articles in the *Convention on Human Rights in Biomedicine*. For present purposes, Article 18 (Council of Europe 2005; see Table 14.3) is crucial.

This article has been essentially adopted in Art. 33 of the *Clinical Trials Regulation* (European Parliament and The Council of the European Union 2014). The structure of the article is important, because it borrows from the structure of articles in this Protocol and other international instruments enabling research to take place on non-competent participants. This is an interesting construction, because it implicitly treats the foetus as a patient in its own right and thus in this one context at least adopts the two-patient model.

Having examined the international guidance and European legislation governing medical research in general, we are left with relatively little guidance on the necessary ethical conditions for a trial aiming to improve outcomes in pregnancy, or on gene transfer as a potential treatment method in pregnancy.

In the context of gene transfer trials, the debate on *foetal* gene transfer has been relatively quiet, until recently, because regulators worldwide have accepted the proposition that genetic modification of the foetus is (at least in the light of current scientific knowledge) presumptively unsafe and unethical (see Coutelle and Ashcroft 2012). Thus, foetal gene transfer has been considered analogous to germline gene transfer, which is (currently) universally prohibited. Some recent commentators have argued that a good case can be made for some foetal gene transfer and, relatedly, for foetal cell transfer. However, it should be stressed that the current intervention is *not* intended to be foetal gene transfer. The question arising is properly one of the risk that the foetal genome be inadvertently modified, should the gene transfer vector cross the placental barrier into the foetal blood supply in significant amounts.

The general approach taken in the European legislation is the minimal risk approach. Thus, in the research protocol to the *Convention on Human Rights and Biomedicine*, we note that either the research must offer potential direct health benefit to her or to her foetus or child after birth, or it must offer potential health benefit to other women/foetuses in future and be only of minimal risk and burden to each woman and foetus in the trial. I make two observations here. First, it is not clear how we apply the minimal risk standard in research in this context. If the woman has a pregnancy that has a high probability of ending in stillbirth or delivery of a preterm very low birthweight, then it is not clear what 'minimal risk' means. If it means minimal *additional* risk, then this is a very permissive standard, since in comparison with the woman's baseline probability of a poor obstetric outcome, most additional risk *to the foetus* will be small in comparison. Given the prior experience with the type of vector being used in the EVERREST trial, the probability of more than minimal risk *to the mother* can reasonably be assumed to be small. On the other hand, as noted above, the *psychological burden* of the intervention either as a (proven) clinical intervention or in a trial context may be high, although whether it is much higher than the burden of being at a significant risk of stillbirth or delivering a very low birthweight baby is not clear. So a trial might well pass a minimal *risk* threshold but not a minimal *burden* threshold (see Kukla 2016). Moreover, as many commentators have observed in clinical trials generally, the problem of 'therapeutic misconception' will apply in many cases (Lidz et al. 2004). That is, many women may believe (because of emotional stress, the complexity of the issues, or simple hope) that the trial is really treatment and will benefit their foetus, even if it is carefully explained that it is *not* being conducted as treatment but as a clinical experiment. Arguably, however, this intervention is one in which 'the trial is the treatment', since no other effective intervention exists (Ashcroft 2000).

14.3 Conclusion

This review of the ethical guidance available to researchers and research ethics committees (regrading trials in women who are pregnant; of interventions designed to benefit them or their foetuses) is somewhat inconclusive, because the guidance specific to the problem is so scant. The standard advice on informed consent, minimisation of avoidable harm, burden and risk, and taking care to avoid the therapeutic misconception and decision-making under emotional stress are all applicable. But the guidance available on risk assessment is very limited, and not practically helpful. In the end, the issue of psychological burden seems to be the most important – as it is in principle with the intervention itself, as discussed at length in the first part of this chapter.

Both the one- and two-patient models fail to capture well the unique relationship between the pregnant women and the developing foetus. Pregnancy always involves significant physiological connection, with the foetus affecting the women's health and the women's health and behaviour affecting foetal development. Wanted pregnancies are typically characterised by a strong emotional connection to the foetus, and a deep investment in its wellbeing to the point where the interests of the foetus and pregnant women are intertwined.

Given the unique physiological and emotional nature of pregnancy, it is very difficult to undertake risk/potential benefit analysis of the interests of the foetus and the pregnant women. In particular, it is not clear whether research ethics committees can reject protocols solely on the ground of foetal risk. The one-patient model would suggest not, whereas the two-patient model would clearly support rejection of research on the grounds of foetal risk.

Certainly when treatment or research participation is offered, it is the pregnant women who should decide whether to participate. The primary ethical risk in gene transfer trials such as EVERREST is potential exploitation of the pregnant women's vulnerability deriving from her desire to protect the wellbeing of the foetus. The most effective measures to mitigate against this risk relate to informed consent. Researchers must facilitate the woman undertaking her own assessment of the risks and potential benefits not only to the foetus but to her own health. Special attention should be made to address the therapeutic misconception. Support and counselling should be provided to help the woman process the potential psychological burden arising from the moral weight of the decisions that have to be made in the EVERREST trial and their potential long-term impact. Finally, informed consent procedures should protect against potential coercion from the woman's partner or family or her doctors, by allowing her to talk to researchers, doctors, and counsellors on her own.

Acknowledgments The research leading to these results has received funding from the European Union Seventh Framework Programme ([FP7/2007-2013] [FP7/2007-2011]) under grant agreement n° 305823, EVERREST "Does vascular epithelial growth factor safely improve outcome in severe early onset foetal growth restriction?" I'd like to acknowledge the partners on this programme, particularly the lead Principal Investigator Anna David, and the independent ethics advisory board to the programme, in particular the Chair, Jackie Leach Scully, for their comments on the report on which this chapter is based. I would also like to thank Nina Wild for her detailed comments on a draft of this chapter. Most of all I would like to thank Angela Ballantyne for editorial advice and assistance far above and beyond the call of duty.

References

Alberry, M., and P. Soothill. 2007. Management of fetal growth restriction. *Archives of Disease in Childhood. Fetal and Neonatal Edition* 92(1): F62–F67.

ACOG (American College of Obstetricians and Gynecologists). 2013. ACOG practice bulletin no. 134: Fetal growth restriction. *Obstetrics and Gynecology* 121(5): 1122–1133.

Ashcroft, R.E. 2000. Giving medicine a fair trial: Trials should not second guess what patients want. *British Medical Journal* 320(7251): 1686.

Ashcroft, R.E. 2008. Chapter 13: The Declaration of Helsinki. In *The Oxford textbook of clinical research ethics*, eds. E.J. Emanuel, C. Grady, R.A. Crouch, R.K. Lie, F.G. Miller, and D. Wendler, 141–148. Oxford: Oxford University Press.

Chervenak, F.A., and L.B. McCullough. 2011. An ethically justified framework for clinical investigation to benefit pregnant and foetal patients. *American Journal of Bioethics* 11(5): 39–49 (Peer commentaries available in the same edition at 50–68, with authors' response to these commentaries at W3–W7).

Council of Europe. 1950. *European convention for the protection of human rights and fundamental freedoms, as amended by protocols nos. 11 and 14*, ETS 5, 4 November. http://www.echr.coe.int/Documents/Convention_ENG.pdf. Accessed 8 Apr 2016.

Council of Europe. 2005. *Additional protocol to the convention on human rights and biomedicine, concerning biomedical research*, CETS No.195, 25 January. http://www.coe.int/en/web/conventions/full-list/-/conventions/treaty/195. Accessed 7 Apr 2016.

Council of Europe. 1997. *Convention for the protection of human rights and dignity of the human being with regard to the application of biology and medicine: Convention on human rights and biomedicine*, CETS No.164, 04 April. http://www.coe.int/en/web/conventions/full-list/-/conventions/treaty/164. Accessed 7 Apr 2016.

Coutelle, C., and R.E. Ashcroft. 2012. Risks, benefits and ethical, legal and social considerations for translation of prenatal gene transfer to human application. In *Prenatal gene therapy: Concepts, methods, protocols*, eds. C. Coutelle and S. Waddington, 371–387. New York: Humana Press.

Coutts, M.C. 1990. Maternal-fetal conflict. Legal and ethical issues. *Kennedy Institute of Ethics/National Reference Center for Bioethics Literature*. Scope Note 14. 1–13.http://bioethics.georgetown.edu/publications/scopenotes/sn14.pdf. Accessed 11 Apr 2016.

Dickenson, D.L. 2002. *Ethical issues in maternal-fetal medicine*. Cambridge: Cambridge University Press.

Epstein, S. 2007. *Inclusion: The politics of difference in medical research*. Chicago: University of Chicago Press.

European Critical Care Forum. 2010. Meeting minutes for 21st October. In *Outcomes in the eye of the beholder: Making decisions about our tiniest babies*. (Manuscript on file with EVERREST.)

European Parliament and The Council of the European Union. 2001. Directive 2001/20/EC of the European Parliament and of the Council of 4 April 2001 on the approximation of the laws, regulations and administrative provisions of the member states relating to the implementation of good clinical practice in the conduct of clinical trials on medicinal products for human use. *Official Journal of European Community*. L121 | 44 (1 May 2001): 34–44. http://eur-lex.europa.eu/legal-content/EN/TXT/?uri=uriserv:OJ.L_.2001.121.01.0034.01.ENG&toc=OJ:L:2001:121:TOC. Accessed 7 Apr 2016.

European Parliament and The Council of the European Union. 2014. Regulation (EU) No. 536/2014 of the European Parliament and of The Council of 16 April 2014 on clinical trials on medicinal products for human use, and repealing Directive 2001/20/EC. *Official Journal of Europoean Community*. L158 | 57(27 May 2014): 1–76. http://eur-lex.europa.eu/legal-content/EN/TXT/?uri=CELEX%3A32014R0536. Accessed 7 Apr 2016.

European Union. 2010. *Charter of the fundamental rights of the European Union*, 2010/C 83/02. http://eur-lex.europa.eu/LexUriServ/LexUriServ.do?uri=OJ:C:2010:083:0389:0403:en:PDF. Accessed 08 Apr 2016.

Flear, M.L., A.M. Farrell, T.K. Hervey, and T. Murphy (eds.). 2013. *European law and new health technologies*. Oxford: Oxford University Press.

Hervey, T.K., and J.V. McHale. 2004. *Health law and the European Union*. Cambridge: Cambridge University Press.

Jackson, E. 2002. Conception and the irrelevance of the welfare principle. *The Modern Law Review* 65(2): 176–203.

Jackson, E. 2012. *Law and the regulation of medicines*. Oxford: Hart Publishing.

Kimmelman, J. 2010. *Gene transfer and the ethics of first-in-human research: Lost in translation*. Cambridge: Cambridge University Press.

Kukla, R. 2016. Equipoise, uncertainty, and inductive risk in research involving pregnant women. In *Clinical research involving pregnant women*, eds. F. Baylis and A. Ballantyne, 179–196. Cham: Springer.

Lidz, C.W., P.S. Appelbaum, T. Grisso, and M. Renaud. 2004. Therapeutic misconception and the appreciation of risks in clinical trials. *Social Science and Medicine* 58(9): 1689–1697.

MacKenzie, C., and N. Stoljar (eds.). 1999. *Relational autonomy: Feminist perspectives on agency, autonomy and social self*. New York: Oxford University Press.

Mahowald, M.B. 2006. *Bioethics and women: Across the lifespan*. Oxford: Oxford University Press.
Mason, J.K. 2007. *The troubled pregnancy: Legal wrongs and rights in reproduction*. Cambridge: Cambridge University Press.
McLaughlin, J., D. Goodley, E. Clavering, and P. Fisher. 2008. *Families raising disabled children: Enabling care and social justice*. Basingstoke: MacMillan.
Mol, A. 2008. *The logic of care: Health and the problem of patient choice*. London: Routledge.
Munby, J. 2010. Consent to treatment: Patients lacking capacity and children – C: Unborn children. In *Principles of medical law*, 3rd ed, eds. A. Grubb, J. Laing, and J. McHale, 502–508. Oxford: Oxford University Press.
Rogers, W., C. Mackenzie, and S. Dodds, guest (eds.). 2012. Special issue on vulnerability. *International Journal of Feminist Approaches to Bioethics* 5(2): 1–194.
Ross, L.F. 1998. *Children, families and health care decision-making*. Oxford: Oxford University Press.
Savulescu, J. 2001. Procreative beneficence: Why we should select the best children. *Bioethics* 15(5–6): 413–426.
Strong, C. 2011. How should risks and benefits be balanced in research involving pregnant women and fetuses? *IRB* 33(6): 1–5.
Wild, V. 2012. How are pregnant women vulnerable research participants? *International Journal of Feminist Approaches to Bioethics* 5(2): 32–104.
World Medical Association. 2013. *WMA Declaration of Helsinki: Ethical Principles for Medical Research Involving Human Subjects*. http://www.wma.net/en/30publications/10policies/b3/. Accessed 3 Oct 2016.

Chapter 15
Clinical Research Involving Pregnant Women Seeking Abortion Services: United States Perspectives

Lisa H. Harris

Abstract *Many pregnant women have induced abortions. Women seeking abortion are appropriate participants in clinical research requiring embryonic or foetal tissue, research with known risks to a foetus, and research designed to improve induced abortion methods. Critics of clinical research at the time of abortion are concerned that research opportunities will influence women's abortion decisions, help women 'rationalise' their abortion, or exploit their vulnerability. I reject these arguments because they are largely based on stigma and negative stereotypes about women and abortion providers. However, economic and racial justice issues must be considered, since low-income women and women of colour disproportionately experience unintended pregnancy and abortion. I conclude that research regulations, and Institutional Review Board interpretations of them, should reflect knowledge gaps and reproductive justice values, rather than stigma, stereotypes and politics. Concerns about clinical research at the time of abortion must be considered alongside the potential harms of not doing such research.*

When bioethicists and researchers turn their attention to clinical research involving pregnant women, generally the underlying assumption is that pregnant research participants intend to give birth. Many women, however, end their pregnancies for a variety of reasons. Approximately 44 million surgical and medical abortions occur annually around the globe (Sedgh et al. 2012). In the US 21 % of pregnancies end in abortion, and roughly one third of all women have an abortion at some point in their reproductive lives (Jones and Kavanaugh 2011; Jones and Jerman 2014). Therefore, the population of prospective pregnant research participants includes many women who will not continue their pregnancies. While women seeking abortion might participate in clinical research on any topic, they are particularly appropriate research participants for clinical research requiring embryonic or foetal tissue as well as early pregnancy placental or endometrial tissue, research with known risks to a

L.H. Harris, MD, PhD (✉)
Jeffries Professor of Reproductive Health, Associate Professor of Obstetrics and Gynecology, Associate Professor of Women's Studies, University of Michigan, Ann Arbor, MI, USA
e-mail: lhharris@med.umich.edu

F. Baylis, A. Ballantyne (eds.), *Clinical Research Involving Pregnant Women*, Research Ethics Forum 3, DOI 10.1007/978-3-319-26512-4_15

foetus or risk of pregnancy disruption, and clinical studies of abortion techniques. In this chapter I consider research on pregnant women seeking abortion, with a focus on the United States.

In the United States, as in many regions of the world, abortion is contested. Some argue that abortion is ethically unacceptable in some or all circumstances. Others argue that legal and easily accessible abortion is an ethical imperative. These arguments are highly relevant to the ethics of clinical research involving pregnant women seeking abortion. For those who believe that abortion is unethical, abortion-related clinical research is by definition unethical and knowledge gained from such research is tainted. In sharp contrast, for those who believe that abortion is ethically acceptable, and that abortion access is an ethical imperative, abortion-related clinical research is ethically acceptable.

Here, I take as a given that women *will* seek abortion, irrespective of whether they, their communities, their religious mentors or their legislators consider it ethically acceptable. Indeed, there is a long worldwide history of women ending undesired pregnancies in the face of immense self-judgement, shame, social stigma, religious prohibition, and legal restriction. Pregnant women seeking abortion services are entitled to high quality care, which necessitates clinical research. In addition, scientists across many disciplines see enormous potential for knowledge gains through tissue collected or procedures conducted at the time of abortion. For these reasons, it is imperative that we address the ethical dimensions of abortion-related clinical research.

15.1 Abortion Epidemiology

Abortion is common, despite the widespread perception that it is unusual, rare, or even deviant. Because abortion is highly stigmatised, many women who have an abortion do not disclose that history to people in their lives (Kumar et al. 2009). Women's silence about their abortions generates the perception that abortion is uncommon. This widespread (mis)perception generates more stigma. In other words, stigma generates silence, and silence generates stigma in a vicious cycle (Kumar et al. 2009; see also Langston 2016).

Abortion rates follow rates of unintended pregnancy, which in turn are tied to women's access to reliable contraception, reproductive autonomy, and economic well-being. While women of all races, ethnicities and social classes have abortions, in the United States the abortion rate is six-times higher among the poorest women compared to women with more financial resources, and women routinely cite economic disadvantage as a primary reason for having an abortion (Finer et al. 2005; Jones et al. 2010). Women of colour bear a disproportionate burden of economic disadvantage in the United States, and, accordingly, the abortion rate is five-times higher in African-American and two-times higher in Latina women than in white, more affluent women (Jones et al. 2010). Most US abortions (64 %) are sought by women of colour (Jones et al. 2010).

Disparities in abortion rates mean that abortion-related clinical research, or lack thereof, will disproportionately affect low-income women and women of colour. Both *deprivation of gains* in knowledge that research brings, as well as the *risks of research*, will be disproportionately experienced by these groups. Therefore abortion-related clinical research is not just an 'ethics' issue, but is also an issue of socioeconomic, racial, and reproductive justice.

15.2 Pregnancy Intention and Research Eligibility

Recruiting pregnant women for abortion-related clinical research assumes something about which not all bioethicists agree: namely, that a woman's abortion intention can be a legitimate inclusion/exclusion criterion for clinical research. Some believe that abortion intention should not be considered in research recruitment (McCullough et al. 2006). On this view, the level of risk to which a human foetus can be subjected should not depend upon pregnancy continuation, because all human foetuses are owed equal respect (Boonstra 2001).

Strong (2012), Steinbock (1999) and others disagree, pointing out that the very definition of 'risk' for the human foetus is entirely dependent upon a woman's decision to end or continue a pregnancy. If there will be no future person, then there is no person who can be 'at risk'. They go on to argue that a woman's abortion decision is relevant to inclusion in or exclusion from clinical research in the following instances: (i) when the research objective is entirely dependent upon the woman's decision to end or continue her pregnancy; or (ii) when there is more than minimal risk of injury to the developing foetus or the born child that it might become.

15.3 Rationale for Clinical Research Involving Pregnant Women Seeking Abortion Services

There are three important rationales for pursuing clinical research involving pregnant women seeking abortion. First, the research requires embryonic or foetal tissue, or other early pregnancy tissues such as placenta or endometrium. This is the case for foetal tissue therapies aimed at treating a range of chronic debilitating conditions. Second, the research involves known risks to a foetus, or risk of pregnancy disruption. This might include novel in utero interventions, like those intended to treat foetal ailments, and might also include research that does not target the foetus or pregnant uterus specifically, but may nevertheless pose significant risk to a developing foetus or to an ongoing pregnancy. Finally, when disrupting or ending the pregnancy is the *goal* of clinical research, abortion is clearly the ideal setting in which to conduct it. This is the case for clinical research designed to improve the safety and efficacy of abortion services, or to expand the abortion options available to pregnant women. This third goal extends to improving the safety and accessibility of self-induced abortions, abortions performed by providers without formal medical training in low-resource settings, and uterine evacuation before pregnancy is certain (i.e., menstrual regulation). Below, I briefly discuss each of these research objectives, focusing on clinical research designed to improve abortion care.

Scientists long ago recognised the value of research involving foetal tissue. Foetal cells hold therapeutic promise because they divide and grow rapidly, and adapt well to new environments. Indeed, as far back as the 1930s, scientists used tissue from spontaneously aborted foetuses to understand cell biology and early human development. Such research played an important role in the development of polio and rubella vaccines (Boonstra 2001). Clinical research at the time of induced abortion has occurred since the 1973 US Supreme Court decision in Roe v. Wade, which invalidated individual state bans on abortion. In the 1970s, scientists began to explore the role of transplanted foetal tissue obtained from women having abortions for use in treating a number of diseases, including Parkinson's Disease and Type 1 (insulin requiring) diabetes mellitus. Despite several decades of investigation, clinical treatments using foetal tissue lag. American scientists attribute lack of advancement to federal research funding policy, which bans funding for research in which human foetuses are harmed (Beardsley 1992; United States 104th Congress 1996; DHHS 2009). Anti-abortion advocates cite this as evidence that abortion-related research is unimportant and ought to cease. Though foetal tissue therapeutics have not found their way into routine clinical use, researchers continue to find great value in using embryonic, foetal, placental and endometrial tissue from abortions in exploratory research in a range of arenas, including immunology, neurology, urology, and cardiology (Wertz 2002; Thelen et al. 2010; Lepore et al. 2011; Magyar et al. 2015). Most recently, foetal tissue research has provided important information about the pathogenesis of Zika virus infection (Driggers et al 2016; Mlakar et al 2016). In the wake of United States Congressional attacks on the use of foetal tissue in research, both the American Association of Medical Colleges and the American Congress of Obstetricians and Gynecologists and have strongly endorsed its importance (AAMC 2016; ACOG 2016; Swetlitz 2016).

There are other reasons that clinical research in pregnant women seeking abortion services is compelling. When research poses a significant risk of pregnancy disruption or foetal injury, women seeking abortion services may be appropriate participants. For example, early studies of technologies that aimed to benefit foetuses, like foetoscopy (visualisation of the in utero foetus using a small camera), chorionic villus sampling (a method of early prenatal genetic testing), and cordocentesis (sampling foetal blood or treating a foetus through the umbilical cord) were conducted in women seeking abortion (Westin 1954; Ward et al. 1983). Since the pregnancy was destined to end, doctors could gain experience without jeopardising a desired pregnancy. This logic might extend to any clinical research in which there is a known or anticipated risk of pregnancy loss or foetal injury, including research that may not have direct pregnancy or foetal benefits as an aim.

Finally clinical research in pregnant women seeking abortion services is important in order to improve abortion care. Abortion is not a static technique, but rather –

like all medical or surgical procedures – it evolves and changes in light of new evidence. Given that abortion providers, like all healthcare providers, have beneficence-based obligations to provide the best evidence-based care possible, clinical research on abortion services is important (see Baylis and MacQuarrie 2016).

The landscape of both inpatient and outpatient abortion care has changed dramatically over the past 40 years, largely due to rigorous laboratory research and clinical research involving pregnant women seeking abortion services. These changes include first trimester medical abortion with mifepristone and misoprostol, safer, faster and more comfortable techniques for second trimester labour induction, and a shift to outpatient surgical abortion (dilation and evacuation) in the second trimester (Grimes 2008; Winikoff and Sheldon 2012). Well-designed, controlled clinical research studies made these changes possible, resulting in increased options and improvements in care.

Consider, for example, the advent of medical abortion, which transformed first trimester surgical abortion practices. Until 2000, women in the United States only had surgical means of ending first-trimester pregnancies. Rigorous clinical research in Europe and the United States resulted in US Food and Drug Administration (FDA) approval of mifepristone and misoprostol for termination of pregnancy through 49 days gestation (ACOG 2014a). Further research showed that medical abortion could be used safely and effectively through 70 days gestation, using lower medication doses, and requiring fewer physician visits (Winikoff et al. 2012). Based upon this research, in 2016 the FDA changed the labelling of mifepristone to reflect the new evidence (FDA 2016), dramatically increasing the number of women who could choose this method. Currently, approximately 36 % of women who are eligible for either surgical or medical abortion choose medical abortion, indicating that it is the preferred method for a substantial fraction of women seeking abortion (Jones and Jerman 2014). This proportion may increase in light of the recent FDA label change.

Consider next abortion services in the second trimester. A few decades ago, second trimester abortions required hospitalisation and labour induction. Induction was initiated by the injection of highly concentrated saline solutions or prostaglandin compounds into a woman's amniotic cavity (Grimes et al. 1977; Grimes 2008; Bryant et al. 2011). With saline abortion 1.78 % of women suffered serious complications, and the mortality rate was 7 in 100,000 (Grimes 2008). The complication and mortality rate was lower with prostaglandins, but they caused high fever, and significant nausea, vomiting and diarrhoea (Su et al. 2005; Grimes 2008).

Two advances have now transformed second trimester abortion care. First, clinical trials of misoprostol for labour induction, with and without adjunctive mifepristone, have demonstrated faster abortion completion with significantly reduced morbidity compared to older techniques (Su et al. 2005). Next, a surgical method of ending a second trimester pregnancy – dilation and evacuation (D + E) – was shown to be safer and more efficacious than labour induction (Grimes 2008). D + E became the preferred method for second trimester abortion, especially for women who wish to avoid the expense and experience of hospitalisation and the labour process, and for women who do not want to see or hold an intact foetus.

The imperative that abortions be as safe as possible suggests the need for a clinical research agenda aimed at reducing maternal morbidity and mortality from unsafe abortion. Worldwide, illegal abortion is responsible for most unsafe abortion and abortion-related death (Berer 2004). Some 68,000 women die every year from unsafe abortion, accounting for 13 % of global maternal deaths (Haddad and Nour 2009). Where liberalisation of abortion law is unlikely, or where abortion access is shrinking, as in the United States (Cohen 2009; Boonstra and Nash 2014), research to improve the safety and accessibility of self-induced abortion, abortion performed by providers without formal medical training, or uterine evacuation performed before pregnancy is confirmed, is warranted.

15.4 Concerns About Clinical Research Involving Pregnant Women Seeking Abortion Services

When foetal tissue transplantation research was considered by United States federal officials, Secretary of the Department of Health and Human Services (DHHS) Louis Sullivan condemned the research, based largely on his perceptions of how it would affect pregnant women considering abortion. In his words, "permitting the human foetal research at issue will increase the incidence of abortion across the country." He added,

> I am particularly convinced by those who point out that most women arrive at the abortion decision after much soul searching and uncertainty. Providing the additional rationalization of directly advancing the cause of human therapeutics cannot help but tilt some already vulnerable women toward a decision to have an abortion. (Childress 1991)

Since there is no consensus on what the 'right' abortion rate is for a nation, we can dispense with this concern about population-level effects of abortion-related research. However we ought to consider if other concerns have merit, and if *individual* women are impacted in unacceptable ways by clinical research opportunities presented at the time they seek abortion. First, does the opportunity to participate in research impact a woman's abortion decision-making, 'tilting' her to have an abortion in the face of unintended pregnancy? Second, while Sullivan did not address it directly in his comments, other critics at the time raised a related concern that financial incentives for research participation (not just the research opportunity itself) would unduly impact the abortion decision. Third, even if research opportunities or financial incentives do not directly induce a woman to choose abortion, do they allow her to 'rationalise' her abortion decision? Finally, Sullivan's statement raised a fourth concern, that women seeking abortion are 'vulnerable'. Are women seeking abortion indeed in need of special protections? And if so, does that mean they should be left out of research altogether? Though Sullivan appeared most concerned with the vulnerability of women's abortion decision-making to the influence of research opportunities, we ought to also ask if there is something about having an abortion that might interfere with a woman's ability to make a voluntary decision to participate in clinical research. I consider each issue below.

Do Research Opportunities Influence Abortion Decisions?

There are no robust empirical data with which to evaluate this concern. A single study twenty years ago assessed women's opinions on abortion-related clinical research. It found that women, including women who were about to undergo an abortion or had a past abortion, 'overwhelmingly' supported such research (Anderson et al 1994). However, this study did not specifically address the potential impact of a research opportunity on an abortion decision.

Concern about women's decision-making seems predicated upon a belief that a research opportunity could tip a pregnant woman considering her options towards abortion rather than pregnancy continuation. While being able to help advance medical knowledge is indeed compelling, the issues that women consider in the face of unintended pregnancy are also extremely compelling, and connected to a range of personal and significant life issues (Finer et al. 2005). The most commonly cited reasons for having an abortion are: a child would interfere with education, work, or ability to care for dependents (74 %); a woman could not afford a baby now (73 %); and a woman did not want to be a single mother or was having relationship problems (48 %). Nearly 40 % of women had completed desired childbearing at the time of unplanned pregnancy. Other research confirms that economic hardship, partner difficulties and un-readiness for parenting lead women to abortion decisions (Torres and Forrest 1988; Jones et al. 2010). Women rarely report only a single reason. In Finer's study, 89 % reported two reasons and 72 % reported three; the median number of reasons was 4, with some women citing as many as 8. Given the scale and scope of these reasons, it is difficult to imagine that a research opportunity would unduly influence an abortion decision. If it did contribute at all, the best evidence suggests that it would be one of *multiple* factors considered. Women have many other substantive reasons for ending their pregnancies.

Are Abortion Decisions Affected by Financial Incentives for Research Participation?

A second concern is that if researchers pay pregnant women to participate in abortion-related research, as they do for participants in other clinical research, women will choose to end pregnancies that they would have otherwise continued, for financial gain.

This argument is flawed. The majority of US women (60 %) pay for abortion out-of-pocket (Jones et al. 2010). The median cost of an abortion in the first trimester is US$470, and may be as much as US$3,000–$10,000 depending upon the duration of pregnancy, the site of care (hospital or free-standing clinic) and the method used (Jones and Kooistra 2011). For women to have a net financial gain by having an abortion would require that research incentives exceeded the cost of the abortion, or that women were offered a free abortion and a significant enough incentive in addition to undergo a painful procedure. Given the limited funding available for abortion research (discussed below), it is extremely unlikely that researchers could offer financial incentives that large. Moreover, if they did intend to offer financial incentives that exceeded the cost of the abortion, it is unlikely that such incentives would be approved by an Institutional Review Board (IRB). Much more realistically, a research incentive might reduce the cost of an abortion that a woman has already decided to have, but it is hardly conceivable that women would choose to have an abortion solely for financial gain.

Bearing in mind that approximately 20 % of abortions in the US are paid for by state Medicaid funds, some might argue that if research incentives are offered, poor women with Medicaid coverage might choose abortion strictly for financial gain. Besides relying on a demeaning image of poor women as entirely mercenary, this argument neglects some practical issues, namely that there are many possible costs associated with abortion beyond the medical fees. These include childcare for existing children (two-thirds of women have children already at the time of abortion) (Jones et al. 2010), time off of work, transportation costs, overnight accommodations for women in many regions of the US who have no local abortion provider or have a required 1–3 day waiting period between signing abortion consent and having the procedure (Jones et al. 2013). Only a financial incentive for research participation and reimbursement for out-of-pocket expenses could make abortion a potentially profitable choice, even for the poorest group of women.

But even if financial incentives exceeded the costs involved for a woman to have an abortion, the vital point to remember is that financial disadvantage is a primary reason that women choose abortion in the first place. It seems highly unlikely that a research incentive, and not a *baseline condition of poverty*, would motivate termination of an otherwise desired pregnancy. The suggestion that it would motivate termination relies on the far-fetched idea that a woman's financial disadvantage is so extreme that she would have an abortion to earn a modest research stipend, but that otherwise her poverty was not relevant to her decision to end or continue a pregnancy. This argument does not ring true. Rather, the fear that women would have abortions to make money likely reflects broader anxiety about women's reproductive autonomy – anxiety that is particularly pointed with regard to low-income women. Reproductive decisions by poor women often evoke demeaning stereotypes, like that of the "welfare queen" who reportedly has children to increase welfare benefits, despite much evidence to the contrary (Roberts 1997). Perhaps a more relevant consideration regarding financial incentives is whether these are undue inducement for *research participation* (i.e. not for abortion itself), discussed below.

While the practice of asking women to participate in abortion-related research without remuneration may have grown from a fear of exploiting pregnant women, *forbidding payment may actually be what makes research exploitative*, as the researcher stands to gain unfairly, relative to the research participant (Ballantyne 2008; see also Ballantyne and Rogers 2016). A practical, just, and non-exploitative policy would stipulate that women who have decided to end their pregnancy are eligible to participate in abortion-related research and receive a modest financial incentive for doing so.

Finally, this concern about financial inducement must be considered in the broader context of US welfare policies that are likely significantly more determinative of abortion decisions than research opportunities or financial incentives. Welfare family caps – in which welfare benefits do not increase with the birth of subsequent children, and which developed largely in response to the 'welfare queen' stereotype, may be a consideration in low-income women's decisions to end or continue a pregnancy. Family caps have persisted in the US since their introduction in 1992 under welfare reform policy (Smith 2006), apparently without concern among legislators that these caps could push women to have abortions. It is hard to argue that modest research incentives for abortion-related research is more ethically troubling than family cap policies. Legislative or regulatory reluctance to allow modest financial incentives for research participation on the basis of undue inducement seems misplaced and disingenuous.

Does Participation in Abortion-Related Research Help Women 'Rationalise' Abortion?

The concern that participation in abortion-related research allows a woman to 'rationalise' her decision – to tell herself (or others) that her decision was justified, and perhaps to feel good or settled about it – suggests that there is something wrong with feeling that one's abortion was an acceptable choice. Because abortion is highly stigmatised, women who seek abortion are expected to feel bad about it and to internalise the negative attitudes that others hold – that they are irresponsible, anti-maternal, murderers, or a range of other negative labels. Even abortion-rights supporters hesitate to claim that abortion is a "good" choice, preferring language like "least worse option," or "safe, legal and *rare*" (emphasis added) (Weitz 2010). Women and their abortion providers may indeed experience abortion as something good or acceptable – with respect to education, work, caregiving responsibilities, life dreams, or a woman's physical or mental health. However, these potential benefits are not part of mainstream abortion discourse. While many women do internalise negative stereotypes (Norris et al. 2011), and do feel bad about their abortion decision, this ought not be encouraged or required by research policy. Stigma has well-documented negative psychosocial outcomes for women (Major and Gramzow 1999; Kumar et al 2009).

To argue against abortion-related research because a woman might 'rationalise' her abortion decision is, I suggest, to allow stigma to determine what constitutes ethical research practice, and to dictate research policy. This is ironic because stigma's social function is to diminish a person, render him/her less than fully human and unworthy of respect. The field of research ethics developed, among other reasons, to ensure precisely the opposite – that people are treated with appropriate respect. I want to suggest, even further, that as abortion providers and researchers strategise about ways to diminish the burden of stigma experienced by women seeking abortion, they ought to consider whether the opportunity to participate in abortion-related research might actually be a useful stigma reduction intervention.

Are Pregnant Women Considering Abortion 'Vulnerable'?

Given HHS Secretary Sullivan's overriding concern with the impact of research opportunities on abortion decisions, the 'vulnerability' with which he was concerned was likely the vulnerability of a women's abortion decision-making to the undue influence of a research opportunity. However, the inverse question deserves attention: Are women vulnerable as research participants because they are choosing to have an abortion? Does the distress that may be associated with unintended pregnancy undermine a woman's decision-making capacity and ability to offer voluntary consent? Do the many barriers to abortion access or women's experiences of stigma push them to consent to research when they might not otherwise? And in particular, are poor women exploited in abortion-related research, in particular when financial incentives are offered?

There is growing consensus that pregnancy alone does not automatically make a woman 'vulnerable'. However, there may be reasons to see women seeking abortion as different from pregnant women who continue their pregnancies: Women seeking abortion generally have an undesired pregnancy, or a desired pregnancy with a serious complication. Either situation can generate distress. Is it appropriate to approach women for research in this setting? Women seeking abortion also face many logistical and financial barriers to care – does gratitude or relief in finding a caregiver or receiving financial aid for an abortion impact a woman's likelihood to say "yes" to research participation when she otherwise would not? Similarly, does stigma cause women to agree to research participation as a kind of moral redemption for a perceived transgression? Finally, while I argued above that financial incentives for research participation are unlikely to impact an abortion decision, might financial incentives be ethically problematic for a different reason – when they provide the only way for a woman to afford an abortion procedure, and thus push her to participate in research?

Unintended pregnancy is stressful, and women may (but do not always) feel distress as they consider their options. Research conducted at the time of a distressing reproductive event appears to be acceptable to some women, however, (at least those who choose to participate in research), and may even offer benefits. For example, while doctors and researchers worry about approaching potential research participants in times of distress, studies of women experiencing late miscarriage, pregnancy termination for foetal anomaly, or stillbirth show that women appreciate the opportunity to participate in research and thereby contribute to knowledge. They enjoy feelings of altruism; they are grateful to know that other people experience events like their own. They describe that research participation alleviates some of the burden of the distressing life event they are facing (Breeze et al 2011). Even those who experience uncomfortable feelings during research participation say that their distress is outweighed by the benefits of contributing to research. Thus, it appears that the fact of a distressing reproductive life event does not in itself warrant limits on clinical research participation opportunities.

Might barriers to abortion access or abortion stigma compromise voluntary research consent? Women seeking abortion often have tremendous difficulty finding an abortion provider. Eighty-six percent of US obstetrician-gynaecologists do not offer abortion services; therefore most women cannot turn to the doctor who handles their reproductive healthcare needs when they need an abortion (Stulberg et al. 2011). Many must look outside of the county in which they live: 89 % of US counties (in which 38 % of US women live) do not have an abortion provider (Jones and Jerman 2014). When a woman eventually finds an abortion provider, she may fear jeopardising that hard-won access if she declines to participate in research offered.

Similarly, might stigma cause women to consent to research when they otherwise would not, in order to 'make up for' their perceived abortion transgression, or because they feel that they are only deserving of humane abortion care if they consent to research participation? Both phenomena may occur. However neither necessarily means that a patient's vulnerability is exploited. It may in fact be an empowering experience for a woman to be able to 'give back'. Only when women perceive that their care would be threatened in any way if they don't consent to research would it cross the line to become ethically problematic. To minimise the risk of approaching this line, research must be carefully explained, and explanations ought to specifically and explicitly emphasise that a woman deserves the abortion care she receives regardless of whether she consents to research.

Finally, the question of women's vulnerability raises another question about financial research incentives in the setting of abortion. I argued earlier that fears that financial incentives will influence a woman's abortion decision are likely unfounded. However, research incentives raise a separate question about women's vulnerability as research participants. Do modest financial incentives unduly pressure women, especially poor women who may already have difficulty paying for their abortion, to consent to research? And if free or heavily subsidised abortions are offered in exchange for research participation, are women unduly pressured to participate in research, since it might be their only way of accessing abortion?

Empirical data on this question are lacking. However, this issue is not unique to abortion-related research, but would apply equally to all clinical research. Research participants living in all kinds of economic contexts are permitted to participate in clinical research, and are permitted to participate when their motivation is to access healthcare that is otherwise unavailable to them. People living in poverty make difficult decisions about complex trade-offs ('rent or tuition,' 'groceries or medical bills') all the time. They may be quite capable of, and in fact may be *experts* in, making clear decisions about risk and potential benefit in research and many other things. No one has yet declared that having a low income is an exclusion criterion for research participation, and there is no evidence that abortion-related research should be treated differently (see Kaposy 2016). Modest financial incentives (in line with incentives provided for comparable kinds of research in non-pregnant populations) are inducements, but not *undue* inducements, and are ethically appropriate, and in fact, just (Ballantyne 2008).

15.5 US Research Policy And Regulations

Over the past several decades, a range of US organisations and regulatory bodies have considered the issue of clinical research in pregnant women seeking abortion, and engaged with the issues raised here, among others. Here I review three sets of guidelines and regulations – the 1988 report of the US Human Fetal Tissue Transplantation Research Panel (which was never acted upon by federal officials), the National Institutes of Health (NIH) regulations surrounding clinical research in pregnant women seeking abortion, and guidelines of the American Congress of Obstetricians and Gynecologists. I focus on elements of the regulations for which there is no evidence, and which ought to be re-visited.

When US officials first considered the issue of federal funding for foetal tissue transplantation research in the late 1980s, a Federal Human Tissue Transplantation Research Panel concluded in its report to DHHS that such research was acceptable, as long as a number of constraints were in place to ensure that a woman's abortion decision was not unduly influenced by the opportunity to participate in research (Childress 1991; Fletcher 1992). The Panel recommended the following:

- "It is essential…that no fees be paid to the woman to donate."
- "The timing and method of abortion should not be influenced" by the research, and "no abortion should be put off to a later date…nor should any abortion be performed by an alternate method … to supply more useful fetal tissue"
- "informed consent for an abortion should precede informed consent or even the preliminary information for tissue donation"

The Panel's recommendations were never acted upon by the Secretary of the DHHS, for fear of generating controversy over its overall conclusion that abortion-related research was acceptable (Childress 1991).

However, the spirit of the recommendations (and in some cases the identically worded recommendations) is echoed in the US NIH regulations for abortion-related research. These regulations *do* govern contemporary research – both federally-funded research as well as privately-funded research that takes place at institutions that accept federal dollars for other research (DHHS 2011; Harris 2013). First, the regulations state: "no inducements, monetary or otherwise, will be offered to terminate a pregnancy" (DHHS 2009, 45 CFR 46 Subpart B). While it is reasonable to stipulate that a woman should not be paid to have an abortion that she otherwise would not have, inevitably this stipulation becomes linked to the question of whether a pregnant woman can receive any remuneration for research participation. Individual researchers or the IRBs that oversee the research can decide that a financial incentive is (or might appear to be) financial 'inducement' for abortion. As I argued strongly above, the fear that a financial incentive for research participation would cause women to have abortions they otherwise would not have, is unjustified, and might actually be what makes research exploitative. The issue of financial incentive for research participation at the time of abortion ought to be re-visited.

Second, the NIH regulations stipulate: "individuals engaged in the research will have no part in any decisions as to the timing, method, or procedures used to terminate a pregnancy" (DHHS 2009, 45 CFR 46 Subpart B). This regulation has two important implications. First, some abortion-related research might depend upon altering the timing of an abortion procedure – for example, research that seeks to measure foetal uptake of a drug or a short-term outcome of a foetal intervention or therapy. In these cases, it might be necessary to begin the research on the day a woman first presents for an abortion, but to perform the abortion on a different day. While abortion is very safe, its risks (as well as patient expenses) increase with increasing duration of pregnancy. In addition, the abortion process changes as the duration of the pregnancy lengthens. Even small delays in abortion care can dramatically alter a woman's options and experience. For example, medical abortion is available only until 70 days from a woman's last menstrual period. Twenty-four to 48-hour cervical dilation is frequently required after the first trimester.

No formal guidelines for acceptable delays in abortion care exist. I suggest some here, as a starting point. These suggestions presume: (i) that *a woman has determined* that her preference to participate in research (for whatever reasons) outweighs her desire to have her abortion completed immediately; and (ii) that any incremental costs, including transportation or additional childcare, are borne by the researchers.

In my view, it is reasonable for abortion care to be delayed up to 72 hours, if this will not result in a change in abortion procedure. For example, the delay would not require a woman to abandon a plan for a medical abortion, would not turn a single-day procedure into a multiple-day procedure, and would not cause a woman to lose abortion access due to gestational age limits set by her care centre or state law. Further, delays that do involve a change in abortion procedures *may* be acceptable if freely chosen by a woman. Women already have to negotiate many kinds of delays that are *not* of their own choosing. For example, there can be state-mandated waiting periods, (which may be as long as 3 days), and delays related to finding a caregiver, transportation, childcare or money to pay for their abortion. If *un-chosen* delays are routine, even mandated by law in some cases, then *chosen* delays certainly ought to be considered reasonable.

When abortion is delayed after a research intervention is begun, the worry, of course, is that a woman will change her mind about ending her pregnancy and foetal risks of the research intervention (risks that were not relevant when there was to be no future person) suddenly take on new importance. While there are no data on the rate at which women change their mind when there is a delay in accessing abortion for a research intervention, there are relevant and reassuring data from research on medical abortion. In medical abortion, there is a 24–48 hour gap between the time when a woman ingests mifepristone, and the time when she takes a second medication, misoprostol, to complete the abortion. The US company that manufactures mifepristone maintains a registry of all cases in which a woman decides not complete the abortion process after taking mifepristone. Between 2000 and 2012, 0.004 % of women who took mifepristone later chose to continue their pregnancy (Grossman et al. 2015). In other words, it is very unlikely that a pregnant woman will change her mind in the window between starting an intervention to end a pregnancy, and completing the abortion. Fear of this very rare occurrence should not dictate abortion-related research policy.

Beyond the issue of abortion delay, NIH regulations have even more important implications for whether *any* abortion research can be conducted. Clinical research by definition alters something about the timing/method/procedure used to end a pregnancy; that is precisely the point of research interventions – to alter or change something. Thus, even privately-funded abortion-related research could be prohibited when IRBs apply federal regulations using this interpretation. This is not a far-fetched scenario: The IRB of (at least) one major US academic institution has in fact interpreted the regulations this way, ruling out the possibility of doing *any* abortion-related research involving an intervention, and permitting only observational research (Harris 2013; see also Ells and Lyster 2016).

Last, it has been a staple in abortion-related research for abortion consent to take place prior to introduction of and consent for research. On its face this appears to be a reasonable and benign requirement. However I flag here that this requirement puts stock in the idea that research opportunities indeed *do* impact abortion decision-making, a stance for which there is no evidence, and that I have argued against. Even if research did impact decision-making, waiting until abortion consent is signed to introduce research opportunities offers no guarantee that a woman hasn't heard about research through word-of-mouth in her community. In other words, this is a requirement that is not based in evidence, would not necessarily solve the problem it claims to solve (and which may not actually be a problem at all), and neglects the myriad life reasons that women choose to present to an abortion care centre. That said, there is likely no harm in requiring abortion consent to occur first, other than potential paternalism.

Since most US abortions are performed by obstetrician-gynaecologists, the American Congress of Obstetricians and Gynecologist's guidelines on pregnant women and research deserve mention. Its 2007 Ethics Committee Opinion on Research Involving Women is clear on the need for research in women, including research in pregnant women, and rejects the formulation of pregnant women as uniquely vulnerable. However, ACOG's framing of research in pregnant women assumes that a pregnancy will be continued, and does not specifically address the issue of research in pregnant women seeking abortion (ACOG 2007).

ACOG's 2014 *Policy statement on abortion* is clear, however, that clinical research and innovation, including innovation in abortion techniques, should proceed free of unduly burdensome legal, policy or regulatory barriers:

> Medical knowledge and patient care are not static. Innovations in medical practice are critical to the advancement of medicine and the improvement of health. Medical research is the foundation of evidence-based medicine and new research leads to improvements in care. ACOG is opposed to laws and regulations that operate to prevent advancements in medicine. (2014b)

The *Policy* specifically calls out state and federal laws that disrupt evolution of abortion care. By extension, the *Policy* suggests that IRB or research policies that impede knowledge-making from abortion-related research are inappropriate (see also, ACOG 2014a). Finally, the recent ACOG statement on foetal tissue research implies support for research at the time of abortion, since it is the most common way in which foetal tissue would be obtained (ACOG 2016).

15.6 Women of Colour and Abortion-Related Research through a Reproductive Justice Lens

Given disparities in abortion rates, abortion-related research, or lack thereof, will disproportionately affect women of colour. They will disproportionately experience both *deprivation of gains* in knowledge that research brings, as well as the *risks of research*. Therefore, how does one manage this tension? The US history of unethical treatment of women (and men) of colour in research is an important backdrop for considering the ethics of abortion-related research (McCarthy 1994; Reverby 2011). Outside of research settings, women and men of colour have also been subjected to a range of reproductive injustices, including coercive sterilisation under US state laws that permitted it. What unites both research and clinical transgressions, besides their inhumanity, is the ideology that fuelled them – ideology of stratified reproduction, in which the sexual lives, fertility, reproduction, and childbearing wishes of people of colour were disregarded by those with the power to devise and implement research protocols or state policy (Harris and Wolfe 2014).

Therefore, given disparities in abortion rates and stratified experiences of reproduction, I suggest that researchers must ask themselves, at a minimum, if their proposed abortion-related research values or devalues, respects or disrespects the *full range* of fertility and childbearing wishes of those who will be asked to participate. This range may include a desire not to have or to have children now or in the future, as well as a desire to parent in a safe and economically secure environment. This question alone would help move us into a habit of thinking in a reproductive justice-oriented framework.

But even more important than researchers asking themselves this question is that they ask it of the women likely to be most affected by the answer. It is crucial that women and communities likely to be most affected by research policies sit at the table as uncertainties and ambiguities around the meaning of research are worked through. It also means that communities that stand to most gain or lose from abortion-related research have a role in shaping the research questions and methods by which those questions will be answered. This is not an argument against abortion-related research (especially research *on* abortion, where gaps in knowledge most affect communities of colour). Rather, it is a caution that such research must be done carefully, with awareness of the simultaneous over-surveillance and neglect of the fertility and childbearing desires of women and families of colour.

15.7 Concluding Thoughts

We are increasingly moving away from the idea that pregnant women as a class are 'vulnerable' (see Ballantyne and Rogers 2016; Johnson 2016). Instead, we consider them to be 'complex', due to the presence of a foetus and the physiological changes accompanying pregnancy. This new designation means that there are indeed special issues that pregnancy raises for research, but that these issues are not insurmountable. Accordingly we ought to move from a presumption of exclusion in research to one of inclusion (Lyerly et al. 2011).

I extend this line of analysis to pregnant women who will not continue their pregnancies. Because a healthy newborn is not a goal in the setting of abortion, some of the ethical and medical complexities of involving pregnant women in research lessen. However, because abortion is ethically, legally and politically contested, additional complexities arise. Pregnant women seeking abortion are indeed 'complex', but that complexity is largely sociological and political. Many people do not like abortion, work hard to restrict or eliminate access to abortion, and stereotype, and stigmatise women who seek abortion, along with the caregivers who help them.

It is important that we recognise that research protocols and policies governing abortion-related research developed within this contested environment. To the extent possible, research practice should grow from genuine gaps in knowledge, and not from stigma, stereotypes, or the fears that stigma and stereotypes generate. There are compelling reasons to consider abortion-related research as legitimate. And there are harms – to knowledge generation, and to our understanding of pregnant women as trustworthy moral agents – when undue barriers impede such research. As Fletcher pointed out over 20 years ago, in the face of important reasons to conduct abortion-related research, there is an ethical burden on those opposed to such research to demonstrate that there actually is a problem with it (1992). Unproven concerns – especially concerns based in stereotypes and stigma – ought not to trump research gains, and may inadvertently be exploitative. Federal research regulations that limit payment of research incentives to pregnant women participating in abortion-related research, or that restrict clinical research that impacts the timing and method of abortion procedures ought to be revisited.

To be sure, legitimate questions remain, especially about how women experience research in the setting of abortion, and about how to conduct research in a reproductive justice framework – a framework that puts the needs of women of colour and women with few economic resources at the centre of analysis. The answer, though, is not to reflexively deem such research ethically unacceptable, but instead to make a commitment to invite a thoughtful and diverse group of women and men to the table in order to inform research priorities, and to consider the ethical questions such research raises.

Acknowledgements Sincere thanks are owed to Meghan Seewald for assistance with the literature search and references. As well, I would like to thank Stephen Harris for comments on an earlier draft.

References

AAMC (Association of American Medical Colleges). 18 March 2016. Correspondence in support of fetal tissue research. https://www.aamc.org/download/444248/data/statementinsupportoffetaltissueresearch.pdf. Accessed 1 Apr 2016.

ACOG (American College of Obstetricians and Gynecologists). 2007. ACOG committee opinion no. 377: Research involving women. *Obstetrics and Gynecology* 110(3): 731–736 [At the time of writing, ACOG committee opinion no. 377 was in effect. It has since been withdrawn and replaced with American College of Obstetricians and Gynecologists. 2015. ACOG committee opinion no. 646: Ethical considerations for including women as research participants. *Obstetrics and Gynecology* 126: e100–e107].

ACOG (American College of Obstetricians and Gynecologists). 2014a. ACOG practice bulletin no.143: Medical management of first-trimester abortion. *Obstetrics and Gynecology* 123(3): 676–692.

ACOG (American College of Obstetricians and Gynecologists). 2014b. *ACOG College statement of policy: Abortion policy.* http://www.acog.org/-/media/Statements-of-Policy/Public/sop069.pdf?dmc=1&ts=20150622T0902260211. Accessed 1 Apr 2016.

ACOG (American College of Obstetricians and Gynecologists). 2016. *ACOG statement in support of fetal tissue research.* http://www.acog.org/About-ACOG/News-Room/Statements/2016/ACOG-Statement-In-Support-of-Fetal-Tissue-Research. Accessed 1 Apr 2016.

Anderson, F., A. Glasier, J. Ross, and D.T. Baird. 1994. Attitudes of women to fetal tissue research. *Journal of Medical Ethics* 20(1): 36–40.

Ballantyne, A. 2008. Benefits to research subjects in international trials: Do they reduce exploitation or increase undue inducement? *Developing World Bioethics* 8(3): 178–191.

Ballantyne, A., and W. Rogers. 2016. Pregnancy, vulnerability, and the risk of exploitation in clinical research. In *Clinical research involving pregnant women*, eds. F. Baylis and A. Ballantyne, 139–159. Cham: Springer.

Baylis, F., and R. MacQuarrie. 2016. Why physicians and women should want pregnant women included in clinical trials. In *Clinical research involving pregnant women*, eds. F. Baylis and A. Ballantyne, 17–31. Cham: Springer.

Beardsley, T. 1992. Aborting research. *Scientific American* 276(2): 17–18.

Berer, M. 2004. Abortion law, policy and practice in transition. *Reproductive Health Matters* 12(24): 1–8.

Boonstra, H. 2001. Human embryo and fetal research: Medical support and political controversy. *The Guttmacher Report on Public Policy* (4)1: 3–4, 14.

Boonstra, H., and E. Nash. 2014. A surge of state abortion restrictions puts providers – And the women they serve – In the crosshairs. *Guttmacher Policy Review* 17(1): 9–15.

Breeze, A.C., H. Statham, G. Hackett, F. Jessop, and C. Lees. 2011. Attitudes to perinatal postmortem: Parental views about research participation. *Journal of Medical Ethics* 37(6): 364–367.

Bryant, A., D. Grimes, J. Garrett, and G. Stuart. 2011. Second-trimester abortion for fetal anomalies or fetal death: Labor induction compared with dilation and evacuation. *Obstetrics and Gynecology* 117(4): 788–792.

Childress, J. 1991. Deliberations of the human fetal tissue transplantation research panel. In *Biomedical politics*, ed. K.E. Hanna, 215–248. Washington, DC: National Academies Press.

Cohen, S. 2009. Facts and consequences: Legality, incidence and safety of abortion worldwide. *Guttmacher Policy Review* 12(4): 2–6.

DHHS (US Department of Health and Human Services). 2009. *Code of Federal Regulations: Title 45, Part 46, Protection of Human Subjects*. http://www.hhs.gov/ohrp/regulations-and-policy/regulations/45-cfr-46/index.html. Accessed 3 Oct 2016.

DHHS (US Department of Health and Human Services, Office for Human Research Protections). 2011. *Federal wide assurance (FWA) for the protection of human subjects*. http://www.hhs.gov/ohrp/assurances/assurances/filasurt.html. Accessed 1 Apr 2016.

Driggers R.W., C-Y. Ho, E.M. Korhonen, S. Kuivanen, A.J. Jääskeläinen, and T. Smura, et al. 2016. Zika virus infection with prolonged maternal viremia and fetal brain abnormalities. *The New England Journal of Medicine* 374(22): 2142–2151.

Ells, C., and C. Lyster. 2016. Research ethics review of drug trials targeting medical conditions of pregnant women. In *Clinical research involving pregnant women*, eds. F. Baylis and A. Ballantyne, 95–118. Cham: Springer.

FDA (U.S. Food and Drug Administration). 2016. Mifeprex (mifepristone) information. U.S. Department of Health and Human Services, 30 March. http://www.fda.gov/Drugs/DrugSafety/PostmarketDrugSafetyInformationforPatientsandProviders/ucm111323.htm. Accessed 8 Apr 2016.

Finer, L.B., L.F. Frohwirth, L.A. Dauphinee, S. Singh, and A.M. Moore. 2005. Reasons U.S. women have abortions: quantitative and qualitative perspectives. *Perspectives on Sexual and Reproductive Health* 37(3): 110–118.

Fletcher, J.C. 1992. Abortion politics, science, and research ethics: Take down the wall of separation. *The Journal of Contemporary Health Law and Policy* 8: 95–121.

Grimes, D. 2008. The choice of second trimester abortion method: Evolution, evidence and ethics. *Reproductive Health Matters* 16(31): 183–188.

Grimes, D., K. Schulz, W. Cates, and C. Tyler. 1977. Mid-trimester abortion by dilation and evacuation: A safe and practical alternative. *The New England Journal of Medicine* 296(20): 1141–1145.

Grossman, D., K. White, L.H. Harris, M. Reeves, P. Blumenthal, B. Winikoff, and D. Grimes. 2015. Continuing pregnancy after mifepristone and "reversal" of first-trimester medical abortion: A systematic review. *Contraception* 92(3): 206–211.

Haddad, L.B., and N. Nour. 2009. Unsafe abortion: Unnecessary maternal mortality. *Reviews in Obstetrics and Gynecology* 2(2): 122–126.

Harris, L.H. 2013. Abortion politics and the production of knowledge. *Contraception* 88(2): 200–203.

Harris, L.H., and T. Wolfe. 2014. Stratified reproduction and the double edge of history. *Current Opinion in Obsetrics and Gynecology* 26(6): 539–544.

Johnson, L.S.M. 2016. When hypothetical vulnerability becomes actual: Research participation and the autonomy of pregnant women. In *Clinical research involving pregnant women*, eds. F. Baylis and A. Ballantyne, 161–178. Cham: Springer.

Jones, R.K., and J. Jerman. 2014. Abortion incidence and service availability in the United States, 2011. *Perspectives on Sexual and Reproductive Health* 46(1): 3–14.
Jones, R.K., and M.L. Kavanaugh. 2011. Changes in abortion rates between 2000 and 2008 and lifetime incidence of abortion. *Obstetrics and Gynecology* 117(6): 1358–1366.
Jones, R.K., and K. Kooistra. 2011. Abortion incidence and access to services in the United States, 2008. *Perspectives on Sexual and Reproductive Health* 43(1): 41–50.
Jones, R.K., L.B. Finer, and S. Singh. 2010. *Characteristics of U.S. abortion patients, 2008*. New York: Guttmacher Institute.
Jones, R.K., U. Upadhyay, and T. Weitz. 2013. At what cost? Payment for abortion care by U.S. women. *Womens Health Issues* 23(3): e173–e178.
Kaposy, C. 2016. Presumptive inclusion and legitimate exclusion criteria. In *Clinical research involving pregnant women*, eds. F. Baylis and A. Ballantyne, 51–62. Cham: Springer.
Kumar, A., L. Hessini, and E.M.H. Mitchell. 2009. Conceptualising abortion stigma. *Culture, Health and Sexuality* 11(6): 625–639.
Langston, L. 2016. Better safe than sorry: Risk, stigma, and research during pregnancy. In *Clinical research involving pregnant women*, eds. F. Baylis and A. Ballantyne, 33–50. Cham: Springer.
Lepore, A.C., J. O'Donnell, A.S. Kim, T. Williams, A. Tuteja, M.S. Rao, et al. 2011. Human glial-restricted progenitor transplantation into cervical spinal cord of the $SOD1^{G93A}$ mouse model of ALS. *PloS One* 6(10): e25968.
Lyerly, A., M. Little, and R. Faden. 2011. Reframing the framework: Toward fair inclusion of pregnant women as participants in research. *The American Journal of Bioethics* 11(5): 50–52.
Magyar, Z., J. Schönleber, M. Romics, E. Hruby, B. Nagy, B. Sulya, et al. 2015. Expression of VEGF in neonatal urinary obstruction: Does expression of VEGF predict hydronephrosis? *Medical Science Monitor* 21: 1319–1323.
Major, B., and R.H. Gramzow. 1999. Abortion as stigma: Cognitive and emotional implications of concealment. *Journal of Personality and Social Psychology* 77(4): 735–745.
McCarthy, C. 1994. Historical background of clinical trials involving women and minorities. *Academic Medicine* 69(9): 695–698.
McCullough, L., J. Coverdale, and F. Chervenak. 2006. Preventive ethics for including women of childbearing potential in clinical trials. *American Journal of Obstetrics Gynecology* 194(5): 1221–1227.
Mlakar, J., M. Korva, N. Tul, M. Popović, M. Poljšak, J. Mraz, et al. 2016. Zika virus associated with microcephaly. *New England Journal of Medicine* 374: 951–958.
Norris, A., D. Bessett, J. Steinberg, M. Kavanaugh, S. Zordo, and D. Becker. 2011. Abortion stigma: A reconceptualization of constituents, causes, and consequences. *Women's Health Issues* 21(3): 49–54.
Reverby, S. 2011. "Normal exposure" and inoculation syphilis: A PHS "Tuskegee" doctor in Guatemala, 1946–1948. *Journal of Policy History* 23(1): 6–28.
Roberts, D.E. 1997. *Killing the black body: Race, reproduction, and the meaning of liberty*. New York: Random House.
Sedgh, G., S. Singh, I. Shah, E. Ahman, S. Henshaw, and A. Bankole. 2012. Induced abortion: Incidence and trends worldwide from 1995 to 2008. *Lancet* 379(9816): 625–632.
Smith, R. 2006. Family caps in welfare reform: Their coercive effects and damaging consequences. *Harvard Journal of Law and Gender* 29(1): 151–200.
Steinbock, B. 1999. Ethical issues related to the inclusion of pregnant women in clinical trials (II). In *Women and health research: Ethical and legal issues of including women in clinical studies: Volume 2: Workshop and commissioned papers*, eds. A.C. Mastroianni, R. Faden, and D. Federman, 23–28. Washington, DC: The National Academies Press.
Strong, C. 2012. Abortion decisions as inclusion and exclusion criteria in research involving pregnant women and fetuses. *Journal of Medical Ethics* 38(1): 43–47.
Stulberg, D.B., A.M. Dude, I. Dahlquist, and F.A. Curlin. 2011. Abortion provision among practicing obstetrician-gynecologists. *Obstetrics and Gynecology* 118(3): 609–614.

Su, L.L., A. Biswas, M. Choolani, V. Kalaichelvan, and K. Singh. 2005. A prospective, randomized comparison of vaginal misoprostol versus intra-amniotic prostaglandins for midtrimester termination of pregnancy. *American Journal of Obstetrics and Gynecology* 193(4): 1410–1414.

Swetlitz, I. 2016. This scientist works with tissue from aborted fetuses. Congress has come calling. STAT. https://www.statnews.com/2016/03/31/fetal-tissue-congress/. Accessed 11 Apr 2016.

Thelen, T., Y. Hao, A.I. Medeiros, J.L. Curtis, C.H. Serezani, L. Kobzik, et al. 2010. The class A scavenger receptor, macrophage receptor with collagenous structure, is the major phagocytic receptor for Clostridium sordellii expressed by human decidual macrophages. *The Journal of Immunology* 185(7): 4328–4335.

Torres, A., and J.D. Forrest. 1988. Why do women have abortions? *Family Planning Perspectives* 20(4): 169–176.

United States 104th Congress. 1996. *The balanced budget downpayment act, I: Public law 104–99*. January 26: Section 128. http://history.nih.gov/research/downloads/PL104.99.pdf. Accessed 1 Apr 2016.

Ward, R.H., B. Modell, M. Petrou, F. Karagozlu, and E. Douratsos. 1983. Method of sampling chorionic villi in first trimester of pregnancy under guidance of real time ultrasound. *British Medical Journal* 286(6377): 1542–1544.

Weitz, T. 2010. Rethinking the mantra that abortion should be "safe, legal, and rare.". *Journal of Womens History* 22(3): 161–172.

Wertz, D. 2002. Embryo and stem cell research in the USA: A political history. *Trends in Molecular Medicine* 8(3): 143–146.

Westin, B. 1954. Hysteroscopy in early pregnancy. *Lancet* 264(6843): 872.

Winikoff, B., and W. Sheldon. 2012. Use of medicines changing the face of abortion. *International Perspectives on Sexual and Reproductive Health* 38(3): 164–166.

Winikoff, B., I. Dzuba, E. Chong, A. Goldberg, E. Lichtenberg, C. Ball, et al. 2012. Extending outpatient medical abortion services through 70 days of gestational age. *Obstetrics and Gynecology* 120(5): 1070–1076.

Chapter 16
Research on Uterine Transplantation: Ethical Considerations

Ruth M. Farrell and Rebecca Flyckt

Abstract *Unique ethical challenges arise in the context of research involving assisted reproductive technologies where outcomes must be established for both the woman who participates in the research and any children born as a result of that research. Uterine transplantation, the newest experimental procedure to assist women in their family-building efforts, entails a complex combination of fertility procedures and surgeries with the goal of having the uterine recipient achieve a pregnancy and give birth to a healthy child. Some of those procedures, such as in vitro fertilisation and embryo freezing, are now considered established therapeutic interventions for women outside of the context of uterine transplantation. Other procedures, including the transplantation surgery maintenance of a pregnancy in the transplant recipient, and removal of the uterus after pregnancy are more clearly still in a research phase. While there is data on the use of immunosuppressive drugs in pregnancy for individuals with other solid organ transplants, this data is not in the context of uterine transplantation where a distinct set of anatomical and physiological changes are anticipated as a result of pregnancy. Uterine transplantation research raises important ethical challenges for research involving women to advance the science of fertility medicine.*

In this chapter, we examine the unique ethical challenges that arise in the context of research involving reproductive science and assisted reproductive technologies (ARTs) with a particular focus on uterine transplantation. First, we provide a general overview of ethical challenges in the design of research for ARTs. Then, we briefly review the history of *in vitro* fertilisation (IVF) and ARTs, highlighting the

R.M. Farrell, MD, MA (✉)
Departments of Obstetrics and Gynaecology and Bioethics, Cleveland Clinic, Cleveland Clinic Lerner College of Medicine, Cleveland, OH, USA
e-mail: farrelr@ccf.org

R. Flyckt, MD
Departments of Obstetrics and Gynaecology, Section of Reproductive Endocrinology and Infertility, Cleveland Clinic, Fertility Preservation and Cancer Program, Women's Health Institute, Cleveland, OH, USA

F. Baylis, A. Ballantyne (eds.), *Clinical Research Involving Pregnant Women*, Research Ethics Forum 3, DOI 10.1007/978-3-319-26512-4_16

evolution from an absence of formal research to more well-designed trials with ethics oversight. Next, we discuss current research on uterine transplantation, the newest experimental procedure to assist women in their family-building efforts. This experimental intervention – a multidisciplinary collaboration between researchers from different medical specialties – involves a complex combination of fertility procedures and surgeries with the intention of helping the transplant recipient achieve a pregnancy and give birth to a healthy child. Taken together, these interventions raises myriad ethical issues for the donor,[1] the recipient, her partner and her future offspring, but also more broadly for research in reproductive medicine where the aim is to develop new approaches to achieve reproductive goals (Lefkowitz et al. 2013; Johannesson and Enskog 2014; Milliez 2009). In this chapter, we focus narrowly on ethical issues for the transplant recipient and future offspring with particular attention to informed consent while acknowledging the presence of additional ethical considerations for the donor.

As noted above, uterine transplantation involves much more than an innovative surgical intervention with removal of the uterus from a donor and transplantation into a recipient. Indeed, procedures associated with uterine transplantation begin long before surgery is undertaken. If the woman who is to receive a uterine transplant desires a genetic link to the child she hopes will be born as a result of the experimental intervention, she must first undergo controlled ovarian hyperstimulation and egg retrieval followed by IVF and embryo freezing. These steps take place prior to graft placement to ensure that a sufficient number of embryos are available for transfer after the transplanted organ becomes functional. The aim is to avoid possible technical challenges with egg retrieval after uterine transplantation. At this time, there is little data about the number or quality of embryos needed for uterine transplant recipients and there is no agreement on the steps to be taken if the embryo supply is exhausted after multiple failed pregnancy attempts. Only after egg retrieval, IVF, and embryo storage will the experimental transplant surgery be attempted.[2]

Once the transplant is complete and uterine function has been established without evidence of graft rejection, the next step will be to transfer thawed embryos to the recipient's uterus in the hope of establishing a pregnancy. If a pregnancy is established, there will be careful monitoring of the pregnant woman and the foetus. If the pregnancy reaches viability, delivery will be performed by caesarean section at the latest gestational age possible, with the goal of avoiding preterm delivery. While the goal of the transplant procedure will now have been achieved, this will not be the end of the experimental protocol. The transplanted uterus is to be 'ephemeral'. Current protocols call for surgical removal of the uterus following the birth of one or two children. The objective is to eliminate the risks associated with immunosuppression therapy that would be required as long as the allograph remained *in situ* (Johannesson et al. 2015).

[1] At the present time, there is debate about whether a living or deceased donor should be utilised in uterine transplant protocols. Each options presents a different set of risks and benefits, not just to the donor (or donor family in the case of a deceased donor) but also to the recipient.

[2] There is debate about the use of donor oocytes, sperm, or embryos if the recipient or her partner are unable to produce usable gametes, or if pre-transplant IVF procedures are unsuccessful.

The unique aspects of uterine transplantation raise important questions about how new approaches to manage fertility and reproduction are imagined and realised. They also highlight the importance of developing a sound research methodology in which the risks of experimental fertility procedures are carefully identified and balanced against the possible benefits of such procedures, as well as the importance of developing the translational process for bringing experimental fertility procedures to the clinical arena. Finally, uterine transplantation raises new and important questions about the risks that women are willing to accept in an effort to have children through advances in reproductive science and technology. It also touches upon a new paradigm of transplant science in which graft tissues are transplanted to improve the quality of life of the recipient, not as a live-saving procedure.

16.1 Clinical Research and ART

Clinical research plays a central role in the forward motion of medical science and the development of evidence-based medical interventions. By its very nature, clinical research depends upon individuals who are willing to take on personal risks to help answer questions that are of value to both the target research population and the general population (Emanuel et al. 2000). One of the first steps in the design of a scientific study is to define the intended research population and to determine the acceptable levels of risk for this population relative to the research question. This process should include a clear and ethically justifiable method of defining the thresholds for acceptable levels of risk for all involved parties and the protections that need to be in place to ensure the risks are justified, easily identified, and minimised (see Kukla 2016). Answers to the following questions should inform research design: Is a cohort, cross-sectional, case–control or randomised clinical trial most appropriate, ethically and logistically feasible? What variables will be measured as part of the study and outcomes to determine the end of the study? And, what levels of risk are acceptable to achieve those measures (Hulley et al. 2013)?

While sound research design is important in clinical research, it is of paramount importance in research involving reproductive medicine and assisted reproductive technologies. In these areas of research, it is necessary to consider outcome and endpoint metrics for both the women participating in research and any children born as a result of the women's research participation (Lyerly et al. 2011).

To date, much of the debate and discussion on research in reproductive science addresses the involvement of pregnant women in clinical research. Important efforts are underway by ethics scholars to advance the interests of pregnant women in research, both as research participants and as the beneficiaries of new scientific data (Lyerly et al. 2008). Yet, as advances in fertility medicine make evident, it is also important to recognise and address ethical issues that arise as women take part in experimental procedures prior to pregnancy with the goal of becoming pregnant. Consider, for instance, the use of ARTs that involve the manipulation of both the female and male gametes (oocytes and spermatozoa) outside of the body. One of the most common ARTs is IVF, in which eggs are removed from a woman's ovaries, fertilised with sperm in a sterile dish and the resulting embryos are then transferred into a woman's uterus (or frozen for later transfer). Other ARTs include intracytoplasmic sperm injection (ICSI) and preimplantation genetic diagnosis (PGD). The common goal of each of these discrete procedures is to establish a pregnancy to be followed by the birth of a healthy child.

While many of the experimental aspects of ARTs may take place outside of the human body, the impact of any one procedure cannot be assessed until much later. Key risks associated with these procedures may unfold not only at the time of the fertility procedure(s) but also during pregnancy or during the intrapartum period – times during which there are major physiological and anatomical changes to the woman's body. As the goal of ART is pregnancy and the birth of a healthy child, it is also important to consider the risks to resulting offspring. Thus, study design requires deliberate and detailed analysis of how outcomes are determined and measured for both the woman and any potential offspring. Furthermore, there must be careful planning and monitoring of endpoints that may indicate the need to end a trial, thereby ensuring that women are not expected to take on undue risks for the benefit of a future or current pregnancy.

Acknowledging that risks to the offspring could take decades to manifest underscores the need for long-term follow-up as part of the original study design. For example, it is possible that research procedures could not only have a deleterious effect on the health of children born as a result of the procedures, but also on their reproductive capacity. Some men with severe male factor infertility related to genetic causes can successfully conceive using IVF and ICSI. However, their male offspring may not be able to reproduce without ART due to the same genetic condition. Consequently, study design must not only include definitions of thresholds for acceptable levels of risk (including those that are known at the time of the procedure), but also those that may emerge over time.

Historically, healthcare providers and patients have tolerated a degree of uncertainty for many of the currently accepted fertility procedures. The reason for this is that while randomised trials may provide the highest level of scientific evidence (Canadian Task Force 1979), they also present specific practical and ethical challenges, as most randomised controlled trials involve the random assignment to one of two study arms – one group receives the experimental intervention and the other group receives no intervention (i.e., a placebo), or a currently accepted intervention (i.e., standard of care) (see Healy and Mangin 2016). For this reason, the field of reproductive science has most often advanced based on case reports. The challenge with these research methodologies is how best to generate generalizable knowledge. Thus, an important question remains about how to reconcile the uncertainties of research with the promise of what experimental procedures can bring to women using ARTs to build their families.

16.2 Advances in ART

Despite the importance of robust study design in clinical research, key design aspects have not always guided the development of fertility procedures. In the early days of fertility medicine several decades ago, experimental interventions were generally performed outside of formal research protocols and with little or no research ethics oversight. To be sure, much of this early work helped numerous women and men realize their dream of becoming parents, and continues to inform current practice in reproductive medicine. Nonetheless, the importance of well-designed clinical research with appropriate ethics oversight has been recognised and implemented. This trend is particularly important as experimental fertility procedures, such as uterine transplantation, have become increasingly more invasive and present significant risks to women research participants.

The earliest research involving laboratory assisted reproductive technologies occurred in the mid-1930s and involved the fertilisation of oocytes using animal models. American scientist Gregory Pincus pioneered some of these early studies at Harvard University (Pincus and Enzmann 1934). He later faced intense criticism for suggesting the fertilisation technique could be adapted for human use. For many, the origins of human ART trace back to the 1940s when gynaecologist John Rock and laboratory technician Miriam Menken first reported *in vitro* fertilisation of human eggs with sperm (Rock and Menkin 1944). The eggs used for these experiments were mostly collected from ovarian tissue obtained from women who were undergoing surgery for diagnostic or therapeutic purposes – procedures that were not part of a research protocol. During these early years, the Roman Catholic Church strongly opposed research on human conception and in 1949 Pope Pius XII formally denounced the fertilisation of human eggs outside of the body. It has been suggested that Rock abandoned his early work in human IVF due to pressures from the Catholic Church and his university colleagues (Bigger 2012). At the time this research was being done, none of the foundational human subject protections were formally in place. The rise of ethical regulation of research involving humans did not occur until the 1960s. The *Declaration of Helsinki*, which set forth guidelines for biomedical research in humans, appeared in 1964.

The first unequivocal IVF success was reported in 1959 by M. C. Chang in a rabbit model (Chang 1959). By the early 1970s, live births from IVF had been reported in rabbits, hamsters, mice, and guinea pigs (Johnson et al. 2010). According to traditional scientific methodologies, the next step to determine proof of concept for IVF pregnancy in humans would be to conduct extensive research in non-human primate models. Interestingly, this did not occur; research involving IVF procedures in humans followed quickly after small mammalian studies and without the use of a clinical trial paradigm. The lack of primate studies was one reason cited by the UK Medical Research Council for rejecting the application for external funding submitted by IVF pioneers Patrick Steptoe and Robert Edwards (Biggers 2012).[3] This research, for which Edwards was awarded the 2010 Nobel Prize, was responsible for the landmark IVF birth of Louise Brown in July 1978.

Prior to this success, however, there was much early work involving oocyte maturation. This was made possible using slices of ovarian tissue acquired from women undergoing surgery for other indications. It is unclear whether these women provided written informed consent for the use of their reproductive tissues for research purposes. Also, prior to the Browns' success there were unidentified couples who went through unknown numbers of failed IVF attempts. This history is seldom described. In addition, the pre-procedure counselling of the Browns did not include the number of prior attempts with other couples undergoing similar procedures. Although the innovations that led to the birth of Louise Brown were based on decades of scientific work, these occurred without formal research ethics oversight.

Today, it is estimated that over five million children have been born worldwide using IVF (ESHRE 2013). In the United States, approximately 1 % of all births are the result of IVF, and this number is growing with the rising incidence of subfertility and infertility (ESHRE 2013). Although the use of IVF as an infertility treatment has increased dramatically in recent years, the field of ART remains relatively new as the first human birth after IVF was reported less than 40 years ago (Steptoe and Edwards 1978). In retrospect, the birth of Louise Brown and the several million successful births thereafter have confirmed the safety and feasibility of IVF in humans. However, the role of data stemming from procedures developed outside of the purview and protections of clinical trials brings to light some salient ethical challenges inherent in our current understanding of reproductive science that influence the field of ART as reproductive science moves forward. Below we consider these challenges as they apply to the development of twenty-first century reproductive advances, such as uterine transplantation.

16.3 Uterine Transplantation: The Intersection of ART and Biomedical Science

Advances in science and medicine outside of the field of reproductive medicine are paving the way for novel applications of ART. Recently great strides have been made in transplantation medicine, as a result of refined surgical techniques and increasing knowledge about the medical management of transplant recipients in the months and years after the transplantation procedures. These advances paired with cross-disciplinary collaboration have made it possible to not only conceptualise but also apply new surgical and interventional approaches as part of research protocols in reproductive science.

---

[3] In fairness, prior work in monkeys indicated that they were not an optimal research model for IVF. There were difficulties with in vitro fertilisation of primate oocytes as well as lack of responsiveness in some primate species to standard injected fertility medications.

One of the newest experimental ART procedures is uterine transplantation. This procedure was first proposed as a way for women with uterine factor infertility to have children. Women with uterine factor infertility experience infertility secondary to factors associated with the uterus and endometrium. This is a broad category that can include women who, because of congenital or acquired reasons, do not have a functional uterus or have had their uterus removed. A major cause of uterine factor infertility is the Mayer-Rokitansky-Küster-Hauser syndrome (Fritz and Speroff 2011). Women with this condition have a congenital absence of the uterus but normal functioning ovaries, and as such they are ideal candidates for uterine transplantation. Additional candidates include women with other sources of uterine factor infertility, including those who have undergone a hysterectomy for benign and limited malignant conditions. To this point in time, adoption and gestational surrogacy have been invaluable solutions for thousands of women with uterine factor infertility. For some women, however, these options may not be accessible for personal, religious, cultural, or legal reasons (Markens 2007). In such cases, uterine transplantation may be an acceptable alternative strategy for having children. As well, there are other potential benefits to consider. Uterine transplantation would give women with uterine factor infertility the ability to be the primary decision-maker regarding key prenatal decisions that directly impact the health and wellbeing of the pregnancy. This contrasts with other circumstances where prenatal outcomes would be dependent upon the actions and decisions of the birth mother or gestational surrogate. Another potential benefit of research on uterine transplantation is that it could provide greater insight into the causes and management of infertility more broadly.

A series of scientific efforts in animal and human models have led to a body of data that have begun to inform surgical, medical, and ethical aspects of uterine transplantation (Fageeh et al. 2002; Ozkan et al. 2013; Brännström et al. 2014). Many of these studies have been undertaken with research ethics oversight, as well as the publication of procedures and results in peer-reviewed journals, to allow for transparency in experimental methods. The first uterine transplant was performed in Saudi Arabia and, while a pregnancy was established, it did not continue beyond 3 months (Fageeh et al. 2002). A subsequent effort was led by a team from Turkey and, as with the first case, a pregnancy was established but was not viable (Ozkan et al. 2013). At the time of writing, Johannesson and colleagues had documented a total of nine uterine transplants (Johannesson et al. 2015) and at least three others had been reported in the media (for example, Grady 2016).

In this most recent series of uterine transplants, allograph rejection occurred in five women during the first year after transplant, all of whom were successfully treated with steroids or higher doses of immunosuppressive therapy (Johannesson et al. 2015). Two of the transplant recipients experienced a complication that required removal of the uterus (Johannesson et al. 2015). While there has been discussion of possible births following uterine transplantation, only one live birth has been documented to date (Johannesson et al. 2015). However, this birth was not without obstetric complications. The recipient's pregnancy was complicated by preeclampsia and preterm contractions, requiring early delivery by caesarean section at 31 5/7 weeks gestational age (Johannesson et al. 2015). Taken together, these studies have demonstrated the potential of uterine transplantation, however, they raise important ethical and clinical questions (Farrell and Falcone 2015a). There are, for example, important questions about the ethics of pursuing research on non-life-saving transplants (e.g., face and hand). As well, there are important clinical questions about the benefits and harms to children born to recipients of a uterine transplant.

As noted at the outset, uterine transplantation is a complex procedure that encompasses several stages to achieve the intended goal of having the transplant recipient give birth to a child. Use of the term ‘uterine transplantation’, however, has meant that much of the focus is on the surgical aspects of removing the uterus from the donor and placing it in the recipient, with particular attention to ongoing immunosuppression and medical management to restore and continue function. These are not the only research interventions, however. In addition to the surgery and the immunosuppression therapy, there are the fertility procedures required to establish the pregnancy. As a consequence of this, there are multiple potential endpoints across the longitudinal experiment of uterine transplantation, raising specific clinical and ethical challenges of how to ensure that measures are clearly defined at each stage.

What is known from experience with infertility patients is that the chance of pregnancy following IVF is dependent upon several factors. One factor is embryo number. It is not uncommon for multiple embryo transfers to be required to achieve pregnancy and, if the supply of embryos from a single stimulation cycle is insufficient, for the fertility patient to undergo multiple repeated IVF procedures. However, the optimal number of frozen embryos necessary to achieve a successful pregnancy in a uterine transplant recipient is unknown. As the plan set by the women research participants and the researchers is typically for one to two live births before removal of the uterus (Fageeh et al. 2002), it is estimated that ten quality embryos would be an adequate number. Currently, it is unclear what steps would be taken if the supply of embryos were exhausted before a pregnancy was achieved. In part, this is because changes in pelvic anatomy and function following uterine transplantation may increase the risks associated with controlled ovarian stimulation and oocyte retrieval.

Further, the chance of pregnancy following IVF may be a function not only of embryo number but also embryo quality. One measure of embryo quality is based on qualitative metrics (Dokras et al. 1993). During initial embryonic development, some cells may not divide as expected and/or may become fragmented or variable in size. While these morphologic observations are not necessarily associated with an inherent problem in embryonic development, they can be associated with lower pregnancy rates when used in conventional IVF (Erenus et al. 1991). What this means for uterine transplant recipients is unknown.

In addition, women research participants should be aware that such qualitative determinations do not necessarily communicate information about embryonic genetic composition. Studies have shown that chromosomal or genetic factors involved in early embryonic development play a role in continued cell division and successful implantation (Shahine and Lathi 2014). For this reason, preimplantation genetic screening (PGS) is gaining acceptance among fertility patients undergoing IVF as a useful mechanism by means of which to identify embryos that may have a greater likelihood of resulting in a pregnancy (Thornhill et al. 2005). PGS involves removing one or two cells from the preimplantation embryo and then conducting genetic testing of those cells to determine if chromosomal aneuploidies or other genetic variants exist that could impair implantation. These same techniques are utilised for PGD, a procedure by which embryos with or without specific genetic characteristics can be identified and selected for transfer. The information gained from PGD can help fertility patients make value-centred decisions about raising a child with a potentially serious genetic condition prior to pregnancy. While PGS and PGD have not been described in the context of uterine transplantation, it is likely that transplant recipients may have an interest in using these procedures. PGS and PGD should be offered to transplant recipients insofar as they would be offered to any other IVF patients.

Once the transplanted uterus is in place, the endometrial lining of the uterus must be prepared with exogenous hormones to prepare for embryo transfer. Although these medications and their use in fresh or frozen embryo transfer cycles is not new, their use in the context of an experimental uterine transplant protocol is novel. Very little is known about whether there may be subtle differences in uterine receptivity or function in response to hormonal preparation in a transplanted uterus.

As with other experimental fertility procedures, in addition to the known and unknown risks to the graft recipients, there are also the potential risks to the future children. As the goal of uterine transplantation is for recipients to experience gestation and to give birth to healthy children, there is an important longitudinal aspect to this experimental procedure. Clearly more research is needed to explore the various risks at each stage of the process – a process that may span several years to completion.

16.4 Informed Consent for Research Participation

Informed consent for research participation is a necessary first requirement for any individual participating in research. First and foremost, the participant must have a clear understanding of the goals of research and how these goals differ from those of clinical medicine. The primary goal of research is to advance knowledge for the benefit of the general population; it is not to benefit the individual research participant (thought the research may well result in individual benefit). Thus, the informed consent process should include a frank discussion of how clinical research departs from proven therapeutic interventions to avoid therapeutic misconception in which a participant believes that she will receive benefit from the experimental procedure (see Ashcroft 2016). The process should also include a discussion of the potential dual role of physicians and other healthcare team members as caregivers and researchers, and should involve a research participant advocate who is neither a caregiver nor a researcher.

A central component of the informed consent process is the disclosure of information about the risks and potential benefits of research participation. The informed consent process should also include a discussion of alternatives to the proposed intervention, including therapeutic options that are a part of accepted clinical practice as well as other family-building options such as adoption and gestational surrogacy. In addition to disclosure, there must be understanding and voluntariness. Researchers should ensure that participants understand the disclosed information and are free to make an informed decision about whether to contribute to research efforts (see Ballantyne and Rogers 2016).

Given the many uncertainties and potential harms associated with uterine transplantation, it is critical that effective and robust informed consent processes are in place for each stage of the uterine transplant procedure. With the initial and ongoing informed consent discussions, the prospective transplant recipient must understand the various interventions that are required to transplant the uterus, to maintain the functioning of the graft, to establish a pregnancy, to manage the pregnancy, to deliver the baby, and to ensure adequate long-term follow-up of the children. More generally, the transplant recipient must understand that data about the risks and potential benefits of uterine transplantation for her and for her offspring remain uncertain.

As the prospective recipient considers the relevant facts about risks and potential benefits, she should also be encouraged to consider if and how her values and beliefs regarding participation in research may change over time as her role and self-perception changes – first with the establishment of fertility, then through the pregnancy and possibly into motherhood. At all stages of the transplant experiment, the recipient's informed consent for research participation must be obtained.

16.5 Conclusion

The utilisation of ART in the context of uterine transplantation is one arena of reproductive science and technology where uncertainty must be clarified. As well, many questions about the events that unfold after the transplanted uterus resumes function must be answered (Farrell and Falcone 2015b). Answers to these questions must address some of the observations noted among women who become pregnant following ART procedures, such as the possible associations of preeclampsia among donor egg recipients (Klatsky et al. 2010).

For the uterine transplant recipient, the longitudinal experiment of uterine transplantation continues during the antenatal, intrapartum, and postpartum periods. It will be important for recipients to understand the type and range of obstetric complications that may occur during each of these periods because of the altered vasculature of the transplanted uterus. Additionally, the anatomical structure and physiological functioning of the transplanted uterus calls for an understanding about how such obstetric conditions, which have serious short- and long-term consequences for mother and child, may manifest in the transplant pregnancy.

Research will also be needed to examine the plan for the recipient after efforts to childbearing have ended. The ephemeral nature of uterine transplantation raises questions about the sequelae of uterine removal for the recipient, both for her physical and psychosocial well-being. The exact surgical risks associated with uterine removal following transplantation, pregnancy, and caesarean delivery can only be inferred from existing clinical experience. Additionally, the psychological effect of returning the woman to a state of infertility following removal of the transplanted uterus, as well as the impact of this state on her sense of self and relationships, are unknown. The literature about other solid organ recipients who experience organ rejection and failure confirms that there are significant emotional and psychological consequences for these individuals (Ouellette et al. 2009). There may be a very different reaction in the context of uterine transplantation, where ongoing health and function are not dependent upon the uterus but play an important and intimate role in an individual's conception of self and relationships. In addition, because of the profound nature of reproduction, the clinical introduction of this new procedure may have a significant impact not just on the recipient, her child, but also her family. (see Ballantyne and Rogers 2016).

Researchers have an ethical obligation to develop scientific, ethically sound, and responsible studies of uterine transplantation and to use the data from these studies for the evidence-based integration of the procedure into fertility medicine. The field of ART and uterine transplantation should only move forward if women can make informed, voluntary, and value-reflective choices about becoming research participants in their journey to motherhood.

References

Ashcroft, R. 2016. Ethical issues in a trial of maternal gene transfer to improve foetal growth. In *Clinical research involving pregnant women*, eds. F. Baylis and A. Ballantyne, 247–263. Cham: Springer.

Ballantyne, A., and W. Rogers. 2016. Pregnancy, vulnerability, and the risk of exploitation in clinical research. In *Clinical research involving pregnant women*, eds. F. Baylis and A. Ballantyne, 139–159. Cham: Springer.

Biggers, J.D. 2012. IVF and embryo transfer: Historical origin and development. *Reproductive Biomedicine Online* 25: 118–127.

Brännström, M., L. Johannesson, P. Dahm-Kähler, A. Enskog, J. Mölne, N. Kvarnström, et al. 2014. The first clinical uterus transplantation trial: A six-month report. *Fertility and Sterility* 101: 1228–1236.

Canadian Task Force on the Periodic Health Examination. 1979. The periodic health examination. *Canadian Medical Association Journal* 121: 1193–1254.

Chang, M.C. 1959. Fertilization of rabbit ova in vitro. *Nature* 184 (Suppl 7/08 August): 466–467.

Dokras, A., I.L. Sargent, and D.H. Barlow. 1993. Human blastocyst grading: An indicator of developmental potential? *Human Reproduction* 8(12): 2119–2127.

Emanuel, E.J., D. Wendler, and C. Grady. 2000. What makes clinical research ethical? *JAMA* 31(283): 2701–2711.

Erenus, M., C. Zouves, P. Rajamahendran, S. Leung, M. Fluker, and V. Gomel. 1991. The effect of embryo quality on subsequent pregnancy rates after in vitro fertilization. *Fertility and Sterility* 56(4): 707–710.

ESHRE (European Society of Human Reproduction and Embryology). 2013. ART fact sheet. http://www.eshre.eu/Guidelines-and-Legal/ART-fact-sheet.aspx. Accessed 11 Apr 2016.

Fageeh, W., H. Raffa, H. Jabbad, and A. Marzouki. 2002. Transplantation of the human uterus. *International Journal of Gynaecology and Obstetrics* 76(3): 245–251.

Farrell, R.M., and T. Falcone. 2015a. Uterine transplant: New medical and ethical considerations. *Lancet* 385(9968): 581–582.

Farrell, R.M., and T. Falcone. 2015b. Jordan M. Philips keynote speaker – Uterine transplantation. *Journal of Minimally Invasive Gynecology* 22: 311–312.

Fritz, M., and L. Speroff. 2011. *Clinical gynecologic endocrinology and infertility*. Philadelphia: Lippincott Williams and Wilkins.

Grady, D. 2016. First uterus transplant in U.S. bolsters pregnancy hopes of many. *New York Times*. http://www.nytimes.com/2016/02/26/health/uterus-transplant-cleveland-clinic.html. Accessed 4 Apr 2016.

Healy, D., and D. Mangin. 2016. Does my bias look big in this? In *Clinical research involving pregnant women*, eds. F. Baylis and A. Ballantyne, 197–208. Cham: Springer.

Hulley, S., S. Cummings, W. Browner, D.G. Grady, and T.B. Newman. 2013. *Designing clinical research*, 4th ed. Philadelphia: Wolters Kluwer.

Johannesson, L., and A. Enskog. 2014. Experimental uterus transplantation. *Best Practice & Research. Clinical Obstetrics & Gynaecology* 28(8): 1198–2210.

Johannesson, L., N. Kvarnström, J. Mölne, P. Dahm-Kähler, A. Enskog, C. Diaz-Garcia, et al. 2015. Uterus transplantation trial: 1-year outcome. *Fertility and Sterility* 103(1): 199–204.

Johnson, M.H., S.B. Franklin, M. Cottingham, and N. Hopwood. 2010. Why the Medical Research Council refused Robert Edwards and Patrick Steptoe support for research on human conception in 1971. *Human Reproduction* 25(9): 2157–2174.

Klatsky, P.C., S.S. Delaney, A.B. Caughey, N.D. Tran, G.L. Schattman, and Z. Rosenwaks. 2010. The role of embryonic origin in preeclampsia: A comparison of autologous in vitro fertilization and ovum donor pregnancies. *Obstetrics and Gynecology* 116(6): 1387–1392.

Kukla, R. 2016. Equipoise, uncertainty, and inductive risk in research involving pregnant women. In *Clinical research involving pregnant women*, eds. F. Baylis and A. Ballantyne, 179–196. Cham: Springer.

Lefkowitz, A., M. Edwards, and J. Balayla. 2013. Ethical considerations in the era of the uterine transplant: An update of the Montreal criteria for the ethical feasibility of uterine transplantation. *Fertility and Sterility* 100(4): 924–926.

Lyerly, A.D., M.O. Little, and R. Faden. 2008. The second wave: Toward responsible inclusion of pregnant women in research. *International Journal of Feminist Approaches to Bioethics* 1(2): 5–22.

Lyerly, A.D., M.O. Little, and R.R. Faden. 2011. Reframing the framework: Toward fair inclusion of pregnant women as participants in research. *The American Journal of Bioethics* 11(5): 50–52.

Markens, S. 2007. *Surrogate motherhood and the politics of reproduction*. Berkeley: University of California Press.

Milliez, J. 2009. Uterine transplantation. FIGO committee for the ethical aspects of human reproduction and women's health. *International Journal of Gynaecology and Obstetrics* 106(3): 270.

Ouellette, A., M. Achille, and M. Pâquet. 2009. The experience of kidney graft failure: Patients' perspectives. *Qualitative Health Research* 19(8): 1131–1138.

Ozkan, O., M.E. Akar, O. Erdogan, O. Ozkan, and N. Hadimioglu. 2013. Uterus transplantation from a deceased donor. *Fertility and Sterility* 100(6): e41.

Pincus, G., and E.V. Enzmann. 1934. Can mammalian eggs undergo normal development in vitro? *Proceedings of the National Academy of Sciences of the United States of America* 20(2): 121–122.

Rock, J., and M.F. Menkin. 1944. In vitro fertilization and cleavage of human ovarian eggs. *Science* 100(2588): 105–107.

Shahine, L.K., and R.B. Lathi. 2014. Embryo selection with preimplantation chromosomal screening in patients with recurrent pregnancy loss. *Seminars in Reproductive Medicine* 32(2): 93–99.

Steptoe, P.C., and R.G. Edwards. 1978. Birth after the reimplantation of a human embryo. *Lancet* 2(8085): 366.

Thornhill, A.R., C.E. deDie-Smulders, J.P. Geraedts, J.C. Harper, G.L. Harton, S.A. Lavery, et al. 2005. ESHRE PGD consortium: Best practice guidelines for clinical preimplantation genetic diagnosis (PGD) and preimplantation genetic screening (PGS). *Human Reproduction* 20(1): 35–48.

Index

F. Baylis, A. Ballantyne (eds.), *Clinical Research Involving Pregnant Women*, Research Ethics Forum 3, DOI 10.1007/978-3-319-26512-4

The manufacturer's authorised representative in the EU is Springer Nature Customer Service Centre GmbH, Europaplatz 3, 69115 Heidelberg, Germany. If you have any concerns regarding our products, please contact ProductSafety@springernature.com

Printed and bound by CPI Group (UK) Ltd, Croydon, CR0 4YY
30/09/2026
02225234-0011